Self-Learning Control of Finite Markov Chains

CONTROL ENGINEERING
A Series of Reference Books and Textbooks

Editor

NEIL MUNRO, PH.D., D.SC.

Professor
Applied Control Engineering
University of Manchester Institute of Science and Technology
Manchester, United Kingdom

1. Nonlinear Control of Electric Machinery, *Darren M. Dawson, Jun Hu, and Timothy C. Burg*
2. Computational Intelligence in Control Engineering, *Robert E. King*
3. Quantitative Feedback Theory: Fundamentals and Applications, *Constantine H. Houpis and Steven J. Rasmussen*
4. Self-Learning Control of Finite Markov Chains, *A. S. Poznyak, K. Najim, and E. Gómez-Ramírez*

Additional Volumes in Preparation

Robust Control and Filtering for Time-Delay Systems, *Magdi S. Mahmoud*

Classical Feedback Control: With MATLAB, *Boris J. Lurie and Paul J. Enright*

Self-Learning Control of Finite Markov Chains

A. S. Poznyak
Instituto Politécnico Nacional
Mexico City, Mexico

K. Najim
E.N.S.I.G.C.
Process Control Laboratory
Toulouse, France

E. Gómez-Ramírez
La Salle University
Mexico City, Mexico

CRC Press
Taylor & Francis Group
Boca Raton London New York

CRC Press is an imprint of the
Taylor & Francis Group, an **informa** business

First published 2000 by Marcel Dekker, Inc.

Published 2022 by CRC Press
Taylor & Francis Group
6000 Broken Sound Parkway NW, Suite 300
Boca Raton, FL 33487-2742

© 2000 by Taylor & Francis Group, LLC
CRC Press is an imprint of Taylor & Francis Group, an Informa business

No claim to original U.S. Government works

ISBN 13: 978-0-8247-9429-3 (hbk)

This book contains information obtained from authentic and highly regarded sources. Reasonable efforts have been made to publish reliable data and information, but the author and publisher cannot assume responsibility for the validity of all materials or the consequences of their use. The authors and publishers have attempted to trace the copyright holders of all material reproduced in this publication and apologize to copyright holders if permission to publish in this form has not been obtained. If any copyright material has not been acknowledged please write and let us know so we may rectify in any future reprint.

Except as permitted under U.S. Copyright Law, no part of this book may be reprinted, reproduced, transmitted, or utilized in any form by any electronic, mechanical, or other means, now known or hereafter invented, including photocopying, microfilming, and recording, or in any information storage or retrieval system, without written permission from the publishers.

For permission to photocopy or use material electronically from this work, please access www.copyright.com (http://www.copyright.com/) or contact the Copyright Clearance Center, Inc. (CCC), 222 Rosewood Drive, Danvers, MA 01923, 978-750-8400. CCC is a not-for-profit organization that provides licenses and registration for a variety of users. For organizations that have been granted a photocopy license by the CCC, a separate system of payment has been arranged.

Trademark Notice: Product or corporate names may be trademarks or registered trademarks, and are used only for identification and explanation without intent to infringe.

Visit the Taylor & Francis Web site at
http://www.taylorandfrancis.com

and the CRC Press Web site at
http://www.crcpress.com

Library of Congress Cataloging-in-Publication

Pozynak, Alexander S.
    Self-learning control of finite Markov chains / A. S. Pozynak, K. Najim, E. Gomez-Ramirez.
      p. cm. - (Control engineering; 4)
    Includes index.
    ISBN 13: 978-0-8247-9429-3
    1. Markov processes. 2. Stochastic control theory. I. Najim, K. II. Gomez-Ramirez, E. III. Title. IV. Control engineering (Marcel Dekker); 4.
  QA274.7.P69 2000
  519.2'33-dc21   99-048719

To the memory of Professor Ya. Z. Tsypkin

Series Introduction

Many textbooks have been written on control engineering, describing new techniques for controlling systems, or new and better ways of mathematically formulating existing methods to solve the ever-increasing complex problems faced by practicing engineers. However, few of these books fully address the applications aspects of control engineering. It is the intention of this new series to redress this situation.

The series will stress applications issues, and not just the mathematics of control engineering. It will provide texts that present not only both new and well-established techniques, but also detailed examples of the application of these methods to the solution of real-world problems. The authors will be drawn from both the academic world and the relevant applications sectors.

There are already many exciting examples of the application of control techniques in the established fields of electrical, mechanical (including aerospace), and chemical engineering. We have only to look around in today's highly automated society to see the use of advanced robotics techniques in the manufacturing industries; the use of automated control and navigation systems in air and surface transport systems; the increasing use of intelligent control systems in the many artifacts available to the domestic consumer market; and the reliable supply of water, gas, and electrical power to the domestic consumer and to industry. However, there are currently many challenging problems that could benefit from wider exposure to the applicability of control methodologies, and the systematic systems-oriented basis inherent in the application of control techniques.

This new series will present books that draw on expertise from both the academic world and the applications domains, and will be useful not only as academically recommended course texts but also as handbooks for practitioners in many applications domains.

Professors Poznyak, Najim, and Gomez-Ramirez are to be congratulated for another outstanding contribution to the series.

Neil Munro

Preface

The theory of controlled Markov chains originated several years ago in the work of Bellman and other investigators. This theory has seen a tremendous growth in the last decade. In fact, several engineering and theoretical problems can be modelled or rephrased as controlled Markov chains. These problems cover a very wide range of applications in the framework of stochastic systems. The problem with control of Markov chains is establishing a control strategy that achieves some requirements on system performance (control objective). The system performance can be principally captured into two ways:

1. a single cost function which represents any quantity measuring the performance of the system;

2. a cost function in association with one or several constraints.

The use of controlled Markov chains presupposes that the transition probabilities, which describe completely the system dynamics, are previously known. In many applications the information concerning the system under consideration is not complete or not available. As a consequence the transition probabilities are usually unknown or depend on some unknown parameters. In such cases there exists a real need for the development of control techniques which involve adaptability. By collecting and processing the available information, such adaptive techniques should be capable of changing their parameters as time evolves to achieve the desired objective. Broadly speaking, adaptive control techniques can be classified into two categories: indirect and direct approaches. The indirect approach is based on the certainty equivalence principle. In this approach, the unknown parameters are on-line estimated and used in lieu of the true but unknown parameters to update the control accordingly. In the direct approach, the control actions are directly estimated using the available information. In the indirect approach, the control strategy interacts with the estimation of the unknown parameters. The information used for identification purposes is provided by a closed-loop system. As a consequence, the identifiability (consistency of the parameter estimates) cannot be guaranteed, and the certainty equivalence approach may fail to achieve optimal behaviour, even asymptotically.

This book presents a number of new and potentially useful direct adaptive control algorithms and theoretical as well as practical results for both unconstrained and constrained controlled Markov chains. It consists of eight chapters and two appendices, and following an introductory section, it is divided into two parts. The detailed table of contents provides a general idea of the scope of the book.

The first chapter introduces a number of preliminary mathematical concepts which are required for subsequent developments. These concepts are related to the basic description and definitions of finite uncontrolled and controlled Markov chains, the classification of states, and the decomposition of the state space of Markov chains. The coefficient of ergodicity is defined, and an important theorem related to ergodic homogeneous Markov chains is presented. A number of definitions and results pertaining to transition matrices which play a paramount role in the development of Markov chains control strategies are also given. A set of engineering problems which can be modelled as controlled Markov chains are presented in this chapter. A brief survey on stochastic approximation techniques is also given in this chapter. The stochastic approximation techniques constitute the frame of the self-learning control algorithms presented in this book.

The first part of this book is dedicated to the adaptive control of unconstrained Markov chains. It comprises three chapters. The second chapter is dedicated to the development of an adaptive control algorithm for ergodic controlled Markov chains whose transition probabilities are unknown. An adaptive algorithm can be defined as a procedure which forms a new estimate, incorporating new information from the old estimate using a fixed amount of computations and memory. The control algorithm presented in this chapter uses a normalization procedure and is based on the Lagrange multipliers approach. In this control algorithm, the control action is randomly selected. The properties of the design parameters are established. The convergence of this adaptive algorithm is stated, and the convergence rate is estimated.

Chapter 3 describes an algorithm and its properties for solving the adaptive (learning) control problem of unconstrained finite Markov chains stated in chapter 2. The derivation of this learning algorithm is based on a normalization procedure and a regularized penalty function. The algorithms presented respectively in chapter 2 and chapter 3 use a similar normalization procedure which brings the estimated parameter at each instant n into some domain (the unit segment, etc.). They exhibit the same optimal convergence rate.

The primary purpose of chapter 4 is the design of an adaptive scheme for finite controlled and unconstrained Markov chains. This scheme combines the gradient and projection techniques. The notion of partially frozen control strategy (the control action remains unchanged within a given time interval) is introduced. The projection technique, which is commonly used for preserving probability measure, is time-consuming compared to the normalization procedure. This adaptive control algorithm works more slowly than the algorithms presented in chapters 2 and 3.

The results reported in the second part of this book are devoted to the adaptive control of constrained finite Markov chains. A self-learning control algorithm for constrained Markov chains for which transition probabilities are unknown is described and analyzed in chapter 5. A finite set of algebraic constraints is considered. A modified Lagrange function including a regularizing term to guarantee the continuity in the parameters of the corresponding linear programming problem is used for deriving this adaptive algorithm. In this control algorithm the transition probabilities of Markov chain are not estimated. The control policy uses only the observations of the realizations of the loss functions and the constraints. The same problem stated in chapter 5 is solved in chapter 6 on the basis of the penalty function approach. Chapter 7 is dedicated to the control of a class of nonregular Markov chains. The formulation of the adaptive control problem for this class of Markov chains is different from the formulation of the adaptive control problems stated in the previous chapters.

The self-learning algorithms presented in this book are such that at each time n, the control policy is estimated on the basis of learning schemes which are related to stochastic approximation procedures. The learning schemes were originally proposed in an attempt to model animal learning and have since found successful application in the field of adaptive control. The asymptotic properties are derived. They follow from the law of dependent large numbers, martingales theory and Lyapunov function analysis approaches.

It is interesting to note that the area of numerical simulation and computer implementation is becoming increasingly important. The ever present microprocessor is not only allowing new applications but also is generating new areas for theoretical research. Several numerical simulations illustrate the performance and the effectiveness of the adaptive control algorithms developed on the basis of the Lagrange multipliers and the penalty function approaches. These simulations are presented in chapter 8, the last chapter of the book. Two appendices follow. The first appendix is dedicated to stochastic process and to the statements and proofs of theorems and lemmas involved in this book. A set of Matlab^{TM} programs are given in the

PREFACE

second appendix in order to help the reader in the implementation of the above mentioned adaptive control algorithms. This book is filled with more than 150 illustrations, figures and charts to help clarify complex concepts and demonstrate applications.

Professor A. S. Poznyak
Professor K. Najim
Dr. E. Gomez-Ramirez

Notations

Throughout this book we use the following notations.

$\{d_n\}$	control strategy at time n
$\mathbf{e}^M$	unit vector of dimension M
$L_\delta(.,.)$	regularized Lagrange function
$P_{\mu,\delta}(.)$	regularized penalty function
$\mathcal{P}_\varepsilon^\mathbf{C}(\cdot)$	projection operator
$u(i)$	i^{th} control action
U	set of control actions
W_n	Lyapunov function
$x(i)$	i^{th} state
X	state space
Φ_n^0	loss function
Φ_n^m	constraints ($m = 1, ..., M$)
$(\Omega, \mathcal{F}, P)$	probability space
π_{ij}^l	probability of transition from state $x(i)$ to state $x(j)$ under the control action $u(l)$
Π_n	transition matrix of components $(\pi_{ij})_n$
$\sim$	proportional

Contents

Series Introduction v
Preface vii

1 Controlled Markov Chains   1
  1.1 Introduction 1
  1.2 Random sequences 1
      1.2.1 Random variables 2
      1.2.2 Markov sequences and chains 5
  1.3 Finite Markov chains 6
      1.3.1 State space decomposition 6
      1.3.2 Transition matrix 8
  1.4 Coefficient of ergodicity 12
  1.5 Controlled finite Markov chains 17
      1.5.1 Definition of controlled chains 18
      1.5.2 Randomized control strategies 19
      1.5.3 Transition probabilities 20
      1.5.4 Behaviour of random trajectories 22
      1.5.5 Classification of controlled chains 24
  1.6 Examples of Markov models 26
  1.7 Stochastic approximation techniques 31
  1.8 Numerical simulations 32
  1.9 Conclusions 40
  1.10 References 40

I Unconstrained Markov Chains

2 Lagrange Multipliers Approach   47
  2.1 Introduction 47
  2.2 System description 48
  2.3 Problem formulation 51

2.4	Adaptive learning algorithm	52
2.5	Convergence analysis	57
2.6	Conclusions	65
2.7	References	65

3 Penalty Function Approach — 69
3.1	Introduction	69
3.2	Adaptive learning algorithm	69
3.3	Convergence analysis	76
3.4	Conclusions	85
3.5	References	85

4 Projection Gradient Method — 87
4.1	Introduction	87
4.2	Control algorithm	87
4.3	Estimation of the transition matrix	91
4.4	Convergence analysis	98
4.5	Rate of adaptation and its optimization	107
4.6	On the cost of uncertainty	111
4.7	Conclusions	112
4.8	References	113

II Constrained Markov Chains

5 Lagrange Multipliers Approach — 117
5.1	Introduction	117
5.2	System description	118
5.3	Problem formulation	121
5.4	Adaptive learning algorithm	122
5.5	Convergence analysis	129
5.6	Conclusions	137
5.7	References	138

6 Penalty Function Approach — 141
6.1	Introduction	141

6.2	System description and problem formulation	142
6.3	Adaptive learning algorithm	144
6.4	Convergence analysis	154
6.5	Conclusions	163
6.6	References	163

7 Nonregular Markov Chains — 167

7.1	Introduction	167
7.2	Ergodic Markov chains	167
7.3	General type Markov chains	182
7.4	Conclusions	186
7.5	References	186

8 Practical Aspects — 189

8.1	Introduction	189
8.2	Description of controlled Markov chain	190
	8.2.1 Equivalent Linear Programming Problem	190
8.3	The unconstrained case (example 1)	192
	8.3.1 Lagrange multipliers approach	193
	8.3.2 Penalty function approach	202
8.4	The constrained case (example 1)	210
	8.4.1 Lagrange multipliers approach	210
	8.4.2 Penalty function approach	219
8.5	The unconstrained case (example 2)	228
	8.5.1 Lagrange multipliers approach	228
	8.5.2 Penalty function approach	237
8.6	The constrained case (example 2)	245
	8.6.1 Lagrange multipliers approach	245
	8.6.2 Penalty function approach	254
8.7	Conclusions	263

Appendix A	265
Appendix B	281
Index	297

Chapter 1

Controlled Markov Chains

1.1 Introduction

The first purpose of this chapter is to introduce a number of preliminary mathematical concepts which are required for subsequent developments. We start with some definitions concerning random variables, expectation and conditional mathematical expectation. The basic description and definitions of finite uncontrolled and controlled Markov chains will be given. The classification of states and the decomposition of the state space of Markov chains are described in details. Homogeneous and non-homogeneous controlled chains are considered. A number of definitions and results pertaining to transition matrices which play a paramount role in the development of Markov chains control strategies are also developed.

The significance of any definition, of course, resides in its consequences and applications, and so we turn to such questions in the next chapters which are dedicated to adaptive control of finite Markov chains [1-3].

The second part of this chapter presents various practical and theoretical problems which can be modelled or related to finite Markov chains. A brief survey on stochastic approximation techniques is given in the third part of this chapter. In fact the control algorithms presented in this book are closely connected with these optimization (estimation) techniques. Finally we present some numerical simulations dealing with Markov chains.

1.2 Random sequences

In this section we recall some definitions related to random processes theory which are important and useful in the study of stochastic systems. These fundamental mathematical background will be used throughout this book.

1.2.1 Random variables

Let $\Omega = \{\omega\}$ be a set of elementary events ω which represents the occurrence or nonoccurrence of a phenomenon.

Definition 1 *The system $\mathcal{F}$ of subsets of Ω is said to be **the σ-algebra associated with** Ω, if the following properties are fulfilled:*

1. $\Omega \in \mathcal{F}$;

2. *for any sets $A_n \in \mathcal{F}$ $(n = 1, 2, ...)$*
$$\bigcup_{n=1}^{\infty} A_n \in \mathcal{F}, \bigcap_{n=1}^{\infty} A_n \in \mathcal{F};$$

3. *for any set $A \in \mathcal{F}$*
$$\overline{A} := \{\omega \in \Omega \mid \omega \notin A\} \in \mathcal{F}.$$

Consider, as an example, the case when Ω is a subset X of the real axis R^1, i.e.,
$$\Omega = X \subseteq R^1$$
and define the set $A := (a, b)$ as the semi-open interval $[a, b) \in R^1$. Then the σ - algebra $\mathcal{B}(X)$ constructed from all possible intervals (a, b) of the real axis R^1 is called the **Borel σ - algebra** generated by all intervals belonging to the subset X.

It is possible to demonstrate that this Borel σ - algebra coincides with the σ - algebra generated by the class of all open intervals $(a, b) \in R^1$ (see Halmos [4]).

Definition 2 *The pair $(\Omega, \mathcal{F})$ represents the **measurable space**.*

Definition 3 *The function $P = P(A)$ of sets $A \in \mathcal{F}$ is called **probability measure** on $(\Omega, \mathcal{F})$ if it satisfies the following conditions:*

1. *for any $A \in \mathcal{F}$*
$$P(A) \in [0, 1];$$

2. *for any sequence $\{A_n\}$ of sets $A_n \in \mathcal{F}$ $(n = 1, 2, ...)$ such that*
$$A_n \bigcap_{n \neq m} A_m = \emptyset$$

we have
$$P\left(\bigcup_{n=1}^{\infty} A_n\right) = \sum_{n=1}^{\infty} P(A_n).$$

1.2. RANDOM SEQUENCES

Often, the number $P(A)$ is called **the probability** of the event A. From a practical point of view, probability is concerned with the occurrence of events.

Example 1 *Let $X = [a_-, a^+]$, then*

$$P(A = [a,b] \in X) = \frac{b-a}{a^+ - a_-} \quad (uniform\ measure)$$

Example 2 *Let $X = [0, \infty)$, then*

$$P(A = [a,b] \in X) = \frac{1}{\sqrt{\pi}} \int_a^b e^{-\frac{x^2}{2}} dx \quad (Gaussian\ measure)$$

Definition 4 *The triple $(\Omega, \mathcal{F}, P)$ is said to be the **probability space**.*

Random variables will be defined in the following.

Definition 5 *A real function $\xi = \xi(\omega)$, $\omega \in \Omega$ is called **random variable** defined on the probability space $(\Omega, \mathcal{F}, P)$, if it is $\mathcal{F}$ - measurable, i.e., for any $x \in (-\infty, \infty)$*

$$\{\omega \mid \xi(\omega) \leq x\} \in \mathcal{F}.$$

We say that two random variables $\xi_1(\omega)$ and $\xi_2(\omega)$ are equal with probability one (or, almost surely) if

$$P\{\omega \mid \xi_1(\omega) = \xi_2(\omega)\} = 1.$$

This fact can be expressed mathematically as follows

$$\xi_1(\omega) \stackrel{a.s.}{=} \xi_2(\omega).$$

Definition 6 *Let $\xi_1, \xi_2, ..., \xi_n$ be random variables defined on $(\Omega, \mathcal{F}, P)$. The minimal σ - algebra $\mathcal{F}_n$ which for any $x = (x_1, ..., x_n)^T \in R^n$ contains the events*

$$\{\omega \mid \xi_1(\omega) \leq x_1, ..., \xi_n(\omega) \leq x_n\}$$

*is said to be **the σ - algebra associated (or, generated by) to the random variables** $\xi_1, \xi_2, ..., \xi_n$. It is denoted by*

$$\mathcal{F}_n = \sigma(\xi_1, \xi_2, ..., \xi_n).$$

In the subsequent discussion two important operators, *expectation* and *conditional Mathematical Expectation*, are of profound importance.

Definition 7 *The Lebesgue integral (see [5])*

$$E\{\xi\} := \int_{\omega \in \Omega} \xi(\omega) P\{d\omega\}$$

*is said to be the **mathematical expectation** of a random variable $\xi(\omega)$ given on $(\Omega, \mathcal{F}, P)$.*

Usually, there exists dependence (relationship) between random variables. Therefore, the next definition deals with the definition of conditional mathematical expectation.

Definition 8 *The random variable $E\{\xi \mid \mathcal{F}_0\}$ is called the **conditional mathematical expectation** of the random variable $\xi(\omega)$ given on $(\Omega, \mathcal{F}, P)$ with respect to the σ - algebra $\mathcal{F}_0 \subseteq \mathcal{F}$ if*

1. *it is $\mathcal{F}_0$ - measurable, i.e.,*

$$\{\omega \mid E\{\xi \mid \mathcal{F}_0\} \leq x\} \in \mathcal{F}_0 \; \forall x \in R^1$$

2. *for any set $A \in \mathcal{F}_0$*

$$\int_{\omega \in A} E\{\xi \mid \mathcal{F}_0\} P\{d\omega\} = \int_{\omega \in A} \xi(\omega) P\{d\omega\}$$

(here the equality must be understood in the Lebesgue sense).

The basic properties of the operator $E\{\xi \mid \mathcal{F}_0\}$ will be presented in the following.

Let $\xi = \xi(\omega)$ and $\theta = \theta(\omega)$ be two random variables given on $(\Omega, \mathcal{F}, P)$, θ an $\mathcal{F}_0$ - measurable ($\mathcal{F}_0 \subseteq \mathcal{F}$) then (see [5])

1.
$$E\{\theta \mid \mathcal{F}_0\} \stackrel{a.s.}{=} \theta;$$

2.
$$E\{\theta\xi \mid \mathcal{F}_0\} \stackrel{a.s.}{=} \theta E\{\xi \mid \mathcal{F}_0\};$$

3.
$$E\{E\{\xi \mid \mathcal{F}_1\} \mid \mathcal{F}_0\} \stackrel{a.s.}{=} E\{\xi \mid \mathcal{F}_0\} \text{ if } \mathcal{F}_0 \subseteq \mathcal{F}_1 \subseteq \mathcal{F}.$$

1.2. RANDOM SEQUENCES

Notice that if ξ selected to be equal to the characteristic function of the event $A \in \mathcal{F}$, i.e.,

$$\xi(\omega) = \chi(\omega, A) := \begin{cases} 1 & \text{if the event } A \text{ has been realized} \\ 0 & \text{if not} \end{cases}$$

from the last definition we can define **the conditional probability** of this event under fixed $\mathcal{F}_0$ as follows:

$$P\{A \mid \mathcal{F}_0\} := E\{\chi(\omega, A) \mid \mathcal{F}_0\}.$$

Having considered random variables and some of their properties, we are now ready for our next topic, the description of Markov sequences and chains.

1.2.2 Markov sequences and chains

Definition 9 *Any sequence $\{x_n\}$ of random variables $x_n = x_n(\omega)$ ($n = 1, 2, ...$) given on $(\Omega, \mathcal{F}, P)$ and taking value in a set X is said to be a **Markov sequence** if for any set $A \in B(X)$ and for any time n the following property (**Markov property**) holds:*

$$P\{x_{n+1} \in A \mid \sigma(x_n) \wedge \mathcal{F}_{n-1}\} \stackrel{a.s.}{=} P\{x_{n+1} \in A \mid \sigma(x_n)\}$$

where $\sigma(x_n)$ is the σ - algebra generated by x_n, $\mathcal{F}_{n-1} = \sigma(x_1, ..., x_{n-1})$ and $(\sigma(x_n) \wedge \mathcal{F}_{n-1})$ is the σ - algebra constructed from all events belonging to $\sigma(x_n)$ and $\mathcal{F}_{n-1}$.

By simple words, this property means that any distribution in the future depends only on the value x_n realized at time n and is independent on the past values $x_1, ..., x_{n-1}$. In other words, the Markov property means that the present state of the system determines the probability for one step into the future.

This Markov property represents a probabilistic analogy of the familiar property of usual dynamic systems described by the recursive relation

$$x_{n+1} = T(n; x_n, x_{n-1}, ..., x_1)$$

when

$$T(n; x_n, x_{n-1}, ..., x_1) = T(n; x_n).$$

This last identity means that the present state x_n of the system contains all relevant information concerning the future state x_{n+1}. In other words, any other information given concerning the past of this system up to time n is superfluous as far as future development is concerned.

Having defined a Markov sequence we can introduce the following concept.

Definition 10 *If the set X, defining any possible values of the random variables x_n, is countable then the Markov sequence $\{x_n\}$ is called a **Markov chain**. If, in addition, this set contains only finite number K of elements ("atoms"), i.e.,*
$$X = \{x(1), ..., x(K)\}$$
*then this Markov sequence is said to be a **finite Markov chain**.*

Hereafter we will deal only with finite Markov chains and we shall be concerned with different problems related to the development of adaptive control strategies for these systems.

1.3 Finite Markov chains

We start with a general description of finite state space and will present its decomposition which will be intensively used in our future studies.

1.3.1 State space decomposition

In this subsection we shall consider a classification of the states of a given finite Markov chain.

Definition 11 *Let $X = \{x(1), ..., x(K)\}$ be a finite set of states. A state $x(i) \in X$ is said to be*

1. *a **non - return state** if there exists a transition from this state to another one $x(j) \in X$ but there is no way to return back to $x(i)$;*

2. *an **accessible (reachable) state** from a state $x(j) \in X$ if there exist a finite number n such that the probability for the random state x_n of a given finite Markov chain to be in the state $x(i) \in X$ starting from the state $x_1 = x(j) \in X$ is more than zero, i.e.,*

$$P\{x_n = x(i) \mid x_1 = x(j)\} \stackrel{a.s.}{>} 0.$$

We will denote this fact as follows

$$x(j) \Rightarrow x(i).$$

Otherwise we say that the considered state is inaccessible from the state $x(j)$.

1.3. FINITE MARKOV CHAINS

It is clear that if a state $x(i)$ is reachable from $x(j)$ ($x(j) \Rightarrow x(i)$) and, in turn, a state $x(k)$ is reachable from $x(i)$ ($x(i) \Rightarrow x(k)$) then evidently the state $x(k)$ is reachable from $x(j)$ ($x(j) \Rightarrow x(k)$).

Definition 12 *Two states $x(j)$ and $x(i)$ are said to be a **communicating** states if each of them is accessible from the other one. We will denote this fact by*
$$x(j) \Leftrightarrow x(i).$$

It is evident that from the facts
$$x(j) \Leftrightarrow x(i) \text{ and } x(i) \Leftrightarrow x(k)$$
it follows
$$x(j) \Leftrightarrow x(k).$$

Communicating states share various properties [6]. Communication ($\Leftrightarrow$) is clearly an equivalence relationship since it is reflexive, symmetric, and transitive.

Definition 13 *A state $x(i)$ is called **recurrent** if, when starting there, it will be visited infinitely often with probability one; otherwise the state is said to be **transient**.*

Definition 14 *A state $x(i)$ is said to be an **absorbing** state if the probability to remain in state $x(i)$ is positive, and the probability to move from any state $x(j)$, $j \neq i$, to the state $x(i)$ is equal to zero.*

Definition 15 *The class $X(i)$ is said to be the jth **communicating class** of states if it includes all communicating states of a given finite Markov, i.e., it includes all states such that*
$$x(i) \Leftrightarrow x(j) \Leftrightarrow \cdots \Leftrightarrow x(m) \Leftrightarrow x(k).$$

Based on this definition we can conclude that the set X of states of a finite Markov chain can be presented as the union of a finite number L ($L \leq K$) of disjoint communicating classes $X(i)$ plus the class $X(0)$ of non-return states, i.e.,
$$X = X(0) \cup X(1) \cup \cdots \cup X(L), \tag{1.1}$$
$$X(i) \underset{i \neq j}{\cap} X(j) = \emptyset. \tag{1.2}$$

The relations (1.1) and (1.2) represent the **state space decomposition** of the state space X for a finite Markov chain. Figure 1.1 illustrates this fact.

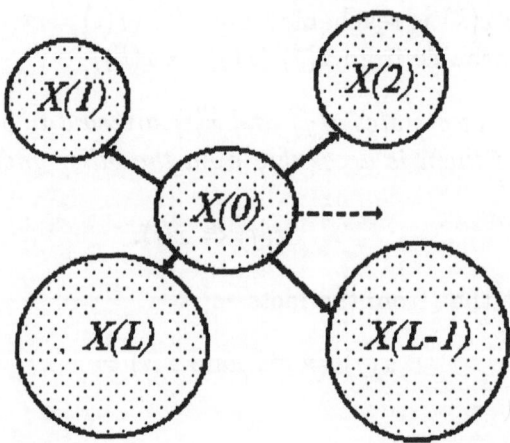

Figure 1.1: State space decomposition.

1.3.2 Transition matrix

Here we will present the general structure of the matrix describing at each time n, the transition probabilities from one state of a given finite Markov chain to another one. Several important definitions will be introduced based on this structure presentation.

Definition 16 *A matrix $\Pi_n \in R^{K \times K}$ is said to be the **transition matrix** at time n of a given Markov chain with finite number K of states if it has the form*

$$\Pi_n = \begin{bmatrix} (\pi_{11})_n & (\pi_{12})_n & \cdot & \cdot & (\pi_{1K})_n \\ (\pi_{21})_n & (\pi_{22})_n & \cdot & \cdot & (\pi_{2K})_n \\ \cdot & & & & \cdot \\ & & \cdot & & \\ (\pi_{K1})_n & (\pi_{K2})_n & \cdot & \cdot & (\pi_{KK})_n \end{bmatrix} \quad (1.3)$$

where each element $(\pi_{ij})_n$ represents the probability (one-step transition probability) for this finite Markov chain to go from the state $x_n = x(i)$ to the next state $x_{n+1} = x(j)$, i.e.,

$$(\pi_{ij})_n := P\{x_{n+1} = x(j) \mid x_n = x(i)\} \ (i,j = 1,...,K). \quad (1.4)$$

Because each element $(\pi_{ij})_n$ (1.4) of the transition matrix Π_n (1.3) is a probability of the corresponding event, we conclude that

$$(\pi_{ij})_n \in [0,1], \ \sum_{j=1}^{K} (\pi_{ij})_n = 1 \ (i = 1,...,K). \quad (1.5)$$

1.3. FINITE MARKOV CHAINS

The k-step transition probability from one state to another one, corresponds to the probability of transition from the considered state at the ith epoch (instant) to the other considered state at the $(i+k)^{th}$ epoch. Notice that the fundamental relationships connecting the transition probabilities are the Chapman-Kolmogorov equations.

The distribution of a given process is completely determined by the transition probabilities (transition matrix), and the initial distribution.

Definition 17 *Any matrix $\Pi_n \in R^{K \times K}$ (1.3) with elements $(\pi_{ij})_n$ (1.4) satisfying the condition (1.5) is said to be a **stochastic matrix**.*

So, any transition matrix of a finite Markov chains is a stochastic matrix. It is obvious that by inspection of condition (1.5) that a stochastic matrix exhibits the following properties:

1. the norm of a stochastic matrix is equal to one;

2. the modulus of the eigenvectors of a stochastic matrix are less or equal to one;

3. any stochastic matrix has 1 as an eigenvalue;

4. if λ is an eigenvalue of modulus equal to 1, and of multiplicity order equal to k, then the vector space generated by the eigenvectors associated with this eigenvalue (λ) is of dimension k.

This completes our discussion of stochastic matrices.

States have been classified according to their connectivity to other states. This classification leads to the following Markov chains classification.

Definition 18 *A finite Markov chain is said to be*

1. *a **homogeneous** (stationary or time homogeneous) chain if its associated transition matrix is stationary, i.e., $\Pi_n = \Pi$;*

2. *a **non-homogeneous** chain if its associated transition matrix Π_n is nonstationary.*

Let us consider a homogeneous finite Markov chain with its corresponding transition matrix Π. Taking into account the decomposition (1.1) and (1.2) which remains invariable with time for any homogeneous finite Markov chain we may conclude that the following **structure presentation** (canonical form) for the corresponding transition matrix holds:

$$\Pi = \begin{bmatrix} \Pi^1 & 0 & \cdot & \cdot & 0 & 0 \\ 0 & \Pi^2 & \cdot & \cdot & 0 & 0 \\ \cdot & \cdot & \cdot & \cdot & \cdot & \cdot \\ 0 & 0 & \cdot & \cdot & \Pi^L & 0 \\ \Pi^{01} & \Pi^{02} & \cdot & \cdot & \Pi^{0(L-1)} & \Pi^{0L} \end{bmatrix} \quad (1.6)$$

where

- Π^l $(l = 1, ..., L)$ is a transition matrix corresponding to the lth group of communicating states (each state from this group can be reached from any other state belonging to the same group by a finite number of transitions) or, in other words, describing the transition probabilities within the communicating class $X(l)$ of states;

- Π^{0l} $(l = 1, ..., L)$ is a transition matrix describing the transition probabilities from a group of nonessential (non - return, transient) states (that to be started and never return back) to the lth group $X(l)$ of communicating states.

Definition 19 *For a homogeneous chain each lth group $X(l)$ $(l = 1, ..., L)$ of communicating states is also said to be l^{th} **ergodic subclass** of states. The index L corresponds to the number of ergodic subclasses.*

Definition 20 *It turns out that any transition matrix Π^l $(l = 1, ..., L)$ corresponding to l^{th} ergodic subclass can be represented in the following irreducible form [7]:*

$$\Pi^l = \begin{bmatrix} 0 & \Pi^l_{12} & 0 & \cdot & 0 & 0 \\ 0 & \cdot & \Pi^l_{23} & 0 \cdot & 0 & 0 \\ \cdot & \cdot & \cdot & \cdot & \cdot & \cdot \\ 0 & 0 & \cdot & \cdot & \Pi^l_{r_l-1,r_l} & 0 \\ \Pi^l_{r_l,1} & 0 & \cdot & \cdot & 0 & 0 \end{bmatrix}. \quad (1.7)$$

Definition 21 *The index r_l is said to be the **periodicity (period) index** of the lth ergodic subclass.*

The structure (1.7) reflects the fact that within each lth ergodic subclass $X(l)$ $(l = 1, ..., L)$ of states there exist transitions from a subgroup of states to another one corresponding to the deterministic cyclic scheme (see figure 1.2).

1.3. FINITE MARKOV CHAINS

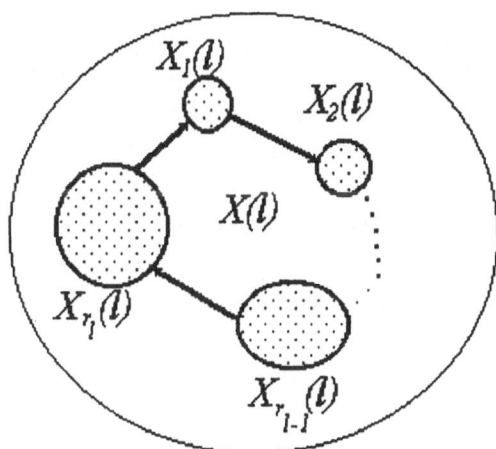

Figure 1.2: Space set decomposition for an ergodic subclass containing several cyclic subclasses.

Definition 22 *If for a given l^{th} ergodic subclass $X(l)$ ($l = 1, ..., L$) of states the corresponding periodicity index r_l is equal to one, i.e.*

$$r_l = 1$$

then the corresponding transition matrix Π^l is said to be simple "primitive."

Definition 23 *If an homogeneous finite Markov chain has only one ergodic subclass and has no group of non-return states, i.e.,*

$$L = 1, \ X(0) = \emptyset$$

*it is said to be an **ergodic homogeneous** finite Markov chain.*

Remark 1 *For any homogeneous finite Markov chain there exists a time n_0 such that the probabilities of transition from any initial states $x_1 = x(i)$ to the state $x_{n_0} = x(j)$ are strictly positive, i.e.,*

$$(\tilde{\pi}_{ij})_{n_0} > 0 \ (i, j = 1, ..., K)$$

where

$$(\tilde{\pi}_{ij})_{n_0} := P\{x_{n_0} = x(j) \mid x_1 = x(i)\} = (\Pi^{n_0})_{ij}.$$

We complete this subsection with a definition concerning aperiodic Markov chains.

Definition 24 *An ergodic homogeneous finite Markov chain is said to be **aperiodic** or **regular** if the corresponding transition matrix is simple "primitive," i.e.,*

$$L = 1, \ r_1 = 1 \ X(0) = \emptyset.$$

In other words: i) an ergodic subclass (set of states) is a collection $X(l)$ of recurrent states with the probability that, when starting in one of the states in $X(l)$, all states will be visited with probability one; ii) a Markov chain is ergodic if it has only one subclass, and that subclass is ergodic; iii) a Markov chain is regular if it has only one closed subclass and that subclass is ergodic. In addition, any other subclass is transient.

We have mainly based the classification of states in a given Markov chain on the basis of the transition matrix.

The coefficient of ergodicity which plays an important role in the study of Markov chains is introduced in the next section.

1.4 Coefficient of ergodicity

In this section we discuss the conditions which guarantee the convergence of the state distribution vectors to their stationary distribution.

According to the previous definitions, we conclude that for any time n and for any finite Markov chain with transition matrix

$$\Pi = [\pi_{ij}]_{i,j=1,...,K}$$

containing K states, the following basic relation holds:

$$p_{n+1} = \Pi^T p_n \quad (n = 1, 2, ...)$$

where the state distribution vector p_n is defined by

$$p_n^T = (p_n(1), ..., p_n(K)), \ p_n(i) = P\{x_n = x(i)\} \ (i = 1, ..., K).$$

Definition 25 *The state distribution vector*

$$(p^*)^T = (p^*(1), ..., p^*(K))$$

*is called the **stationary distribution** of a homogeneous Markov chain with a given transition matrix*

$$\Pi = [\pi_{ij}]_{i,j=1,...,K}$$

if it satisfies the following algebraic relations

$$p^*(j) = \sum_{i=1}^{K} \pi_{ij} p^*(i) \ (i = 1, ..., K). \tag{1.8}$$

The next definition concerns a fundamental tool in the study of Markov chains, namely the coefficient of ergodicity.

1.4. COEFFICIENT OF ERGODICITY

Definition 26 *For an homogeneous finite Markov chain, the parameter $k_{erg}(n_0)$ defined by*

$$k_{erg}(n_0) := 1 - \frac{1}{2} \max_{i,j=1,\ldots,K} \sum_{m=1}^{K} \left|(\tilde{\pi}_{im})_{n_0} - (\tilde{\pi}_{jm})_{n_0}\right| < 1$$

is said to be the coefficient of ergodicity of this Markov chain at time n_0, where

$$(\tilde{\pi}_{im})_{n_0} = P\{x_{n_0} = x(m) \mid x_1 = x(i)\} = (\Pi^{n_0})_{im}$$

is the probability to evolve from the initial state $x_1 = x(i)$ to the state $x_{n_0} = x(m)$ after n_0 steps.

Remark 2 *The coefficient of ergodicity $k_{erg}(n_0)$ can be calculated as (see [8-9]) given by*

$$k_{erg}(n_0) \geq \min_{i,j} \sum_{m=1}^{K} \min\left\{(\tilde{\pi}_{im})_{n_0}, (\tilde{\pi}_{jm})_{n_0}\right\}.$$

Its lower estimate is given by:

$$k_{erg}(n_0) = \max_{j=1,\ldots,K} \min_{i=1,\ldots,K} (\tilde{\pi}_{ij})_{n_0}.$$

If all the elements $(\tilde{\pi}_{ij})_{n_0}$ of the transition matrix Π^{n_0} are positive, then the coefficient of ergodicity $k_{erg}(n_0)$ is also positive. The converse is not true. In fact, there exist ergodic Markov chains with elements $(\tilde{\pi}_{ij})_{n_0}$ equal to zero, but with positive coefficient of ergodicity $k_{erg}(n_0)$ (see, for example, Rozanov [10]).

The next theorem concerns the properties of homogeneous Markov chains having at some time n_0 a strict positive coefficient of ergodicity $k_{erg}(n_0) > 0$.

Theorem 1 *(Rozanov [10]) For a given ergodic homogeneous Markov chain, if there exists a time n_0 such that*

$$k_{erg}(n_0) > 0$$

then

1.
$$\lim_{n \to \infty} p_n(i) := p^*(i) \quad (i = 1, \ldots, K)$$

exist, where the vector p^* describes a stationary distribution with positive components;

2. for any initial state distribution p_1,

$$\sup_{p_1} |p_n(i) - p^*(i)| \leq C \exp\{-D \cdot n\} \qquad (1.9)$$

where

$$C = \frac{1}{1 - k_{erg}(n_0)} \text{ and } D = \frac{1}{n_0} \ln C. \qquad (1.10)$$

Proof. Let us consider the following sequences

$$r_n(j) := \min_i (\tilde{\pi}_{ij})_n \text{ and } R_n(j) := \max_i (\tilde{\pi}_{ij})_n \quad (j = 1, ..., K).$$

Taking into account that

$$(\tilde{\pi}_{ik})_1 = \pi_{ik},$$

it follows

$$r_{n+1}(j) := \min_i (\tilde{\pi}_{ij})_{n+1} = \min_i \sum_k (\tilde{\pi}_{ik})_1 (\tilde{\pi}_{kj})_n$$

$$\geq \min_i \sum_k (\tilde{\pi}_{ik})_1 \min_k (\tilde{\pi}_{kj})_n = \min_i \sum_k \pi_{ik} r_n(j) = r_n(j)$$

and

$$R_{n+1}(j) = \max_i (\tilde{\pi}_{ij})_{n+1} = \max_i \sum_k (\tilde{\pi}_{ik})_1 (\tilde{\pi}_{kj})_n$$

$$\leq \max_i \sum_k (\tilde{\pi}_{ik})_1 \max_k (\tilde{\pi}_{kj})_n = \max_i \sum_k \pi_{ik} R_n(j) = R_n(j).$$

So, the sequences $\{r_n(j)\}$ and $\{R_n(j)\}$ are respectively monotonically increasing and decreasing, i.e.,

$$r_1(j) \leq r_2(j) \leq ... \leq r_n(j) \leq ...$$

and

$$R_1(j) \geq R_2(j) \geq ... \geq R_n(j) \geq ...$$

Then, for any states $x(\alpha)$ and $x(\beta)$ we have

$$\sum_k (\tilde{\pi}_{\alpha k})_{n_0} - \sum_k (\tilde{\pi}_{\beta k})_{n_0}$$

$$= \sum_{k=1}^{+} \left[(\tilde{\pi}_{\alpha k})_{n_0} - (\tilde{\pi}_{\beta k})_{n_0} \right] + \sum_{k=1}^{-} \left[(\tilde{\pi}_{\alpha k})_{n_0} - (\tilde{\pi}_{\beta k})_{n_0} \right] = 0$$

1.4. COEFFICIENT OF ERGODICITY

where $\sum_{k=1}^{+}$ and $\sum_{k=1}^{-}$ represent respectively the summation with respect to the terms $\left[(\tilde{\pi}_{\alpha k})_{n_0} - (\tilde{\pi}_{\beta k})_{n_0}\right]$ which are positive and negative or equal to zero. Evidently,

$$\max_{\alpha,\beta} \sum_{k=1}^{+} \left[(\tilde{\pi}_{\alpha k})_{n_0} - (\tilde{\pi}_{\beta k})_{n_0}\right] = 1 - k_{erg}(n_0).$$

Based on the previous relations, we derive

$$R_{n_0}(j) - r_{n_0}(j) = \max_{\alpha} (\tilde{\pi}_{\alpha j})_{n_0} - \min_{\beta} (\tilde{\pi}_{\beta j})_{n_0}$$

$$= \max_{\alpha,\beta} \sum_{k=1}^{+} \left[(\tilde{\pi}_{\alpha j})_{n_0} - (\tilde{\pi}_{\beta j})_{n_0}\right]$$

$$\leq \max_{\alpha,\beta} \sum_{k=1}^{+} \left[(\tilde{\pi}_{\alpha k})_{n_0} - (\tilde{\pi}_{\beta k})_{n_0}\right] = 1 - k_{erg}(n_0).$$

We also have

$$R_{n_0+n}(j) - r_{n_0+n}(j) = \max_{\alpha,\beta} \sum_{k=1}^{+} \left[(\tilde{\pi}_{\alpha j})_{n_0+n} - (\tilde{\pi}_{\beta j})_{n_0+n}\right]$$

$$= \max_{\alpha,\beta} \sum_{k=1} \left[(\tilde{\pi}_{\alpha k})_{n_0} - (\tilde{\pi}_{\beta k})_{n_0}\right] (\tilde{\pi}_{kj})_n$$

$$\leq \max_{\alpha,\beta} \left\{\sum_{k=1}^{+} \left[(\tilde{\pi}_{\alpha k})_{n_0} - (\tilde{\pi}_{\beta k})_{n_0}\right] R_n(j) \right.$$

$$\left. + \sum_{k=1}^{+-} \left[(\tilde{\pi}_{\alpha k})_{n_0} - (\tilde{\pi}_{\beta k})_{n_0}\right] r_n(j)\right\}$$

$$= \max_{\alpha,\beta} \sum_{k=1}^{+} \left[(\tilde{\pi}_{\alpha k})_{n_0} - (\tilde{\pi}_{\beta k})_{n_0}\right] [R_n(j) - r_n(j)]$$

$$= [1 - k_{erg}(n_0)] [R_n(j) - r_n(j)].$$

From this recursive inequality, we derive

$$R_{Nn_0}(j) - r_{Nn_0}(j) \leq [1 - k_{erg}(n_0)]^N, \quad N = 1, 2, \ldots \quad (1.11)$$

Here we used the estimate

$$0 \leq R_{n_0}(j) - r_{n_0}(j) = \max_{\alpha} (\tilde{\pi}_{\alpha j})_{n_0} - \min_{\beta} (\tilde{\pi}_{\beta j})_{n_0} \leq 1.$$

From the last estimation (1.11), it follows that the sequences $\{r_n(j)\}$ and $\{R_n(j)\}$ have the same limit, i.e.,

$$p^*(j) := \lim_{n \to \infty} r_n(j) = \lim_{n \to \infty} R_n(j).$$

We also have
$$\left|(\tilde{\pi}_{ij})_n - p^*(j)\right| \le R_n(j) - r_n(j) \le [1 - k_{erg}(n_0)]^{\frac{n}{n_0}-1}.$$

So, for any initial probability distribution $p_1(.)$, we get
$$|p_n(j) - p^*(j)| = \left|\sum_i p_1(i)(\tilde{\pi}_{ij})_n - p^*(j)\right|$$
$$\le \sum_i p_1(i)\left|(\tilde{\pi}_{ij})_n - p^*(j)\right|$$
$$\le \sum_i p_1(i)[R_n(j) - r_n(j)] = R_n(j) - r_n(j) \le [1 - k_{erg}(n_0)]^{\frac{n}{n_0}-1}.$$

All the estimates obtained above can be rewritten in the form (1.9) where the parameters C and D are given by (1.10).

To finish this proof, let us now show that the limit vector p^* satisfies the system of algebraic equation (1.8). For any m
$$\sum_{j \le m} p^*(j) = \lim_{n \to \infty} \sum_{j \le m} p_n(j) \le 1.$$

It follows
$$\sum_{j=1}^{K} p^*(j) < \infty.$$

Then, from the following inequality
$$p_{n+1}(j) \ge \sum_{i=1}^{K} p_n(i)\pi_{ij}$$

we derive
$$p^*(j) \le \sum_{i=1}^{K} p^*(i)\pi_{ij}.$$

Summing up these inequalities, we obtain
$$\sum_{j=1}^{K} p^*(j) \ge \sum_{j=1}^{K}\sum_{i=1}^{K} p^*(i)\pi_{ij} = \sum_{i=1}^{K} p^*(i) \sum_{j=1}^{K} \pi_{ij} = \sum_{i=1}^{K} p^*(i).$$

It is clear that the last expression is a strict equality; hence, all the previous inequalities are strict equalities. So relation (1.8) is proved.

Now, let us show that
$$p^*(i) > 0 \text{ and } \sum_{i=1}^{K} p^*(i) = 1.$$

1.5. CONTROLLED FINITE MARKOV CHAINS

Based on the Toeplitz lemma (see lemma 8 of Appendix A), we deduce

$$p^*(j) = \lim_{n \to \infty} (\tilde{\pi}_{ij})_n = \lim_{n \to \infty} \frac{1}{n} \sum_{k=1}^{n} (\tilde{\pi}_{ij})_{n_0 \cdot k} =$$

$$= \lim_{n \to \infty} \frac{1}{n} \sum_{k=1}^{n} \left(\Pi^{n_0 \cdot k} \right)_{ij} > 0.$$

Let us consider the following initial probability distribution

$$p_1(i) = p^*(i) / \sum_{j=1}^{K} p^*(j) \quad (i = 1, ..., K).$$

These initial probabilities satisfy the following relation

$$p_1(j) = \sum_{i=1}^{K} \pi_{ij} p_1^*(i).$$

But for the stationary initial distribution, we have

$$p_n(j) = p_1(j) = \lim_{n \to \infty} p_n(j),$$

from which, we conclude that

$$\sum_{i=1}^{K} p^*(i) = 1.$$

The theorem is proved. ∎

In the next section we shall be concerned with the so-called controlled finite Markov chains which represent the basic model investigated in this book.

1.5 Controlled finite Markov chains

We start by discussing the properties of controlled finite Markov chains. This discussion will be followed by the consideration and classification of control strategies (or policies).

1.5.1 Definition of controlled chains

In general, the behaviour of a controlled Markov chain is similar to the behaviour of a controlled dynamic system and can be described as follows. At each time n the system is observed to be in one state x_n. Whenever the system is in the state x_n one decision u_n (control action) is chosen according to some rule to achieve the desired control objective. In other words, the decision is selected to guarantee that the resulting state process performs satisfactorily. Then, at the next time $n+1$ the system goes to the state x_{n+1}. In the case when the state and action sets are finite, and the transition from one state to another is random according to a fixed distribution, we deal with *Controlled Finite Markov Chains*.

Definition 27 *A controlled homogeneous finite Markov chain is a dynamic system*
   described by the triplet $\{X, U, \Pi\}$ *where:*

1. X *denotes the set* $\{x(1), x(2), ..., x(K)\}$ *of states of the Markov chain;*

2. U *denotes the set* $\{u(1), u(2), ..., u(N)\}$ *of possible control actions;*

3. $\Pi = \left[\pi_{ij}^l\right]$ *the transition probabilities. The element* π_{ij}^l *($i = 1, ..., K; j = 1, ..., K$ and $l = 1, ..., N$) represents at time n ($n = 1, 2, ...$), the probability of transition from state $x(i)$ to state $x(j)$ under the action $u(l)$:*

$$\pi_{ij}^l = P\{x_{n+1} = x(j) \mid x_n = x(i), u_n = u(l)\}. \tag{1.12}$$

We assume that all the random sequences are defined on the probability space $(\Omega, \mathcal{F}, P)$.

Definition 28 *We say that a controlled homogeneous finite Markov chain is a **communicating** chain, if for any two states $x(i)$ and $x(j)$ of this chain there exists a deterministic causal strategy*

$$\{u_n\} \ (u_n = u_n(x_1, u_1; ...; x_{n-1}, u_{n-1}; x_n))$$

such that for some n the conditional probability corresponding to the transition from $x(i)$ to $x(j)$ would be positive, i.e.,

$$P\{x_n = x(j) \mid x_1 = x(i) \wedge \sigma(x_1, u_1; ...; x_{n-1}, u_{n-1})\} \stackrel{a.s.}{>} 0$$

We shall now be concerned with control strategies.

1.5.2 Randomized control strategies

Basically, to introduce control actions in a system means to couple the system to its environment. Under the notion of environment, we will consider the external conditions and influences [11]. This comment and the previous statements and mnemonics will be reinforced as the reader proceeds through the book. The definition of a randomized control policy is given in the following.

Definition 29 *A sequence of random stochastic matrices $\{d_n\}$ is said to be a **randomized control strategy** if*

1. *it is causal (independent on the future), i.e., $d_n = \left[d_n^{il}\right]_{i=1,...,K;l=1,...,N}$ is $\mathcal{F}_{n-1}$-measurable, where*

$$\mathcal{F}_{n-1} := \sigma\left(x_1, u_1, ; ....; x_{n-1}, u_{n-1}\right) \qquad (1.13)$$

   is the σ-algebra generated by the random variables $(x_1, u_1; ....; x_{n-1}, u_{n-1})$;

2. *the random variables $(u_1, ..., u_{n-1})$ represent the realizations of the applied control actions, taking values on the finite set $U = \{u(1), ..., u(N)\}$, which satisfy the following property:*

$$d_n^{il} = \Pr\{u_n = u(l) \mid x_n = x(i) \wedge \mathcal{F}_{n-1}\}. \qquad (1.14)$$

Different classes of control policies will be defined in the following.

Definition 30 *Let us denote by*
*(i) Σ the class of all **randomized strategies**, i.e.,*

$$\Sigma = \{\{d_n\}\}; \qquad (1.15)$$

*(ii) Σ_s the class of all **randomized stationary strategies**, i.e.,*

$$\Sigma_s = \{\{d_n\} : d_n = d\}$$

*(iii) Σ_+ the class of all randomized and **non-singular** (non-degenerated) stationary strategies, i.e.,*

$$\Sigma_+ = \left\{\{d_n\} : d_n^{il} = d^{il} > 0 \ (i = 1, ..., K; l = 1, ..., N)\right\}. \qquad (1.16)$$

It is clear that

$$\Sigma_+ \subseteq \Sigma_s \subseteq \Sigma.$$

Control criteria

We have presented a classification of the control policies. Notice that each control action incurs a stream of random costs. In the framework of controlled Markov chains, the behaviours of interest are in some sense similar to the behaviours associated with quadratic control systems. They can be classified into two main categories: finite (short-run, short-term) and infinite horizon (long-run, long-term) control problems [3]. The main criteria used by several authors are:

1. Total cost;

2. Discounted cost (devalued cost) in the discounted cost, the future reward is discounted per unit time by a discount factor which can be compared to the forgetting factor used in least squares method to reduce the influence of old data;

3. Normalized discounted cost;

4. Average cost;

5. Sample path average cost.

1.5.3 Transition probabilities

According to (1.12) and (1.14), for any fixed strategy $\{d_n\} \in \Sigma$, the conditional transition probability matrix $\Pi(d_n)$ can be defined as follows

$$\Pi(d_n) = \left[\pi_n^{ij}(d_n)\right]_{i=1,\ldots,K;j=1,\ldots,K}$$

where

$$\pi_n^{ij} := P\{x_{n+1} = x(j) \mid x_n = x(i) \wedge \mathcal{F}_{n-1}\} \stackrel{a.s.}{=} \sum_{l=1}^{N} \pi_{ij}^l d_n^{il}. \qquad (1.17)$$

It is well known that for any fixed randomized stationary strategy $d \subseteq \Sigma_+$ the controlled Markov chain becomes an uncontrolled Markov chain with the **transition matrix** given by (1.17) which in general, has the following structure analogous to (1.6):

$$\Pi(d) = \begin{bmatrix} \Pi^1(d) & 0 & \cdot & \cdot & \cdot & 0 & 0 \\ 0 & \Pi^2(d) & \cdot & \cdot & \cdot & 0 & 0 \\ \cdot & \cdot & \cdot & \cdot & \cdot & \cdot & \cdot \\ 0 & \cdot 0 & \cdot & \cdot & \cdot & \Pi^L(d) & 0 \\ \Pi^{01}(d) & \Pi^{02}(d) & \cdot & \cdot & \cdot & \Pi^{0(L-1)}(d) & \Pi^{0L}(d) \end{bmatrix} \qquad (1.18)$$

where

1.5. CONTROLLED FINITE MARKOV CHAINS

- $\Pi^l(d)$ $(l = 1, ..., L)$ is a transition matrix corresponding to the lth group of communicating states $X(l)$;

- $\Pi^{0l}(d)$ $(l = 1, ..., L)$ is a transition matrix describing the transition probabilities from a group of nonessential states $X(0)$ to $X(l)$.

Analogously, each functional matrix $\Pi^l(d)$ $(l = 1, ..., L)$ can be expressed as follows:

$$\Pi^l(d) = \begin{bmatrix} 0 & \Pi^l_{12}(d) & 0 & \cdot & \cdot & 0 & 0 \\ 0 & \cdot & \Pi^l_{23} & 0 & \cdot & 0 & 0 \\ \cdot & \cdot & \cdot & \cdot & \cdot & \cdot & \cdot \\ 0 & 0 & \cdot & \cdot & \cdot & \Pi^l_{r_l-1,r_l}(d) & 0 \\ \Pi^l_{r_l,1}(d) & 0 & \cdot & \cdot & \cdot & 0 & 0 \end{bmatrix}. \quad (1.19)$$

It is clear that for any randomized strategy $\{d_n\}$ the corresponding transition matrix $\Pi(d)$ (1.18) changes its properties from time to time. It can correspond, for example, to ergodic homogeneous finite Markov chain, then to a chain with two ergodic subclasses, to a chain with five ergodic subclasses and so on.

The next lemma proved by V. Sragovitch [12] (see also [13]) clarifies the notion of a communicating homogeneous controlled chain and states the conditions when a given chain is a communicating chain.

Lemma 1 *A controlled homogeneous finite Markov chain is a communicating chain if and only if there exists a non-degenerated stationary strategy $\{d\} \in \Sigma_+$ such that the corresponding transition matrix $\Pi(d)$ (1.18) would be **irreducible** (corresponds to a single cyclic subclass $(L = 1)$), i.e., the matrix $\Pi(d)$ for this fixed d can not by renumbering the states, be presented in the form*

$$\Pi(d) = \begin{bmatrix} Q & R \\ S & T \end{bmatrix}$$

where Q and T are quadratic matrices, and R and T non-quadratic matrices satisfying the condition that at least one of them is equal to zero.

Proof. 1) *Necessity.* Assume that the given chain is a communicating chain, i.e., for any states $x_1 = x(i_1)$ and $x_n = x(i_n)$ there exists some intermediate states $x_2 = x(i_1), ... x_{n-1} = x(i_{n-1})$ and the corresponding control actions $u_1 = u(l_1), ... u_{n-1} = u(l_{n-1})$ such that

$$\pi^{l_1}_{i_1,i_2} > 0, ..., \pi^{l_{n-1}}_{i_{n-1},i_n} > 0.$$

In view of this fact and because of the linearity of $\Pi(d)$ (1.18) with respect to d it follows that for any non-degenerated randomized stationary strategy $\{d\} \in \Sigma_+$ the probability of such transition from $x_1 = x(i_1)$ to $x_n = x(i_n)$ would be positive. Indeed, according to the Markov property and in view of the Bayes' rule [14] we have

$$P\{x_n = x(i_n) \mid x_1 = x(i_1)\} = \prod_{t=2}^{n} P\{x_t = x(i_t) \mid x_{t-1} = x(i_{t-1})\} = \quad (1.20)$$

$$= \prod_{t=2}^{n} (\Pi(d))_{i_{t-1}, i_t} = \prod_{t=2}^{n} \sum_{l=1}^{N} \pi_{i_{t-1}, i_t}^{l} d^{i_{t-1}, l} \geq \prod_{t=2}^{n} \pi_{i_{t-1}, i_t}^{l_{t-1}} d^{i_{t-1}, l_{t-1}} > 0.$$

Taking into account that this chain is finite we derive that any pair of states is communicating one. So, this chain is a communicating chain.

2) *Sufficiency.* Assume now that there exists a strategy $\{d\} \in \Sigma_+$ such that the corresponding transition matrix is irreducible. But it means that there exist a states $x_1 = x(i_1)$, $x_2 = x(i_1)$, ..., $x_{n-1} = x(i_{n-1})$ such that all the corresponding transitions $\pi_{i_{t-1}, i_t}^{l_{t-1}}$ are positive and, as a result, using the previous formula (1.20) we state that

$$P\{x_n = x(i_n) \mid x_1 = x(i_1)\} > 0$$

which corresponds to the definition of a communicating chain.

The lemma is proved. ■

This is a striking result. Based on this lemma we can conclude that the structure $\Pi(d_n)$ (1.18) of any controlled Markov chain under any random strategy $\{d_n\} \in \Sigma$ remains unchanged, i.e., for any nonstationary strategy the elements of the diagonal subblocks of $\Pi(d_n)$ can change (ergodic subclasses and the class of non-return states can appear and disappear) but the distribution of zero blocks remains unchanged.

Therefore, to define the structure of any transition matrix it is sufficient to define it only, for example, within a simple class of non-degenerated stationary random strategies $\{d\} \in \Sigma_+$.

Now we shall be concerned with the behaviour of the random trajectories associated with the states of a Markov chain.

1.5.4 Behaviour of random trajectories

The previous lemma gives a chance to forecast the behaviours of the random sequence $\{x_n\}$ within the set X of states.

Denote by $X^+(l)$ $(l, ..., L)$ the lth **ergodic subset** (or the **communicating component**) of states corresponding to the transition matrix $\Pi(d)$

1.5. CONTROLLED FINITE MARKOV CHAINS

for a non-singular (non-degenerated) stationary strategy $\{d\} \in \Sigma_+$. The corresponding subclass of non-return states will be denoted by $X^+(0)$. It is evident that

$$X = X^+(0) \cup X^+(1) \cup \cdots \cup X^+(L), \qquad (1.21)$$

$$X^+(i) \underset{i \neq j}{\cap} X^+(j) = \emptyset. \qquad (1.22)$$

We now have the following lemma.

Lemma 2 *For any controlled homogeneous finite Markov chain with any distribution of initial state and for any nonstationary randomized strategy $\{d_n\} \in \Sigma$, a set Ω of elementary events ω can be decomposed into subsets according to*

$$\Omega = \Omega^+(0) \cup \Omega^+(1) \cup \cdots \cup \Omega^+(L)$$

$$\Omega^+(i) \underset{i \neq j}{\cap} \Omega^+(j) = \emptyset$$

such that for almost all $\omega \in \Omega^+(l)$ there exists a finite random time $n_l = n_l(\omega)$ starting from which the random trajectory $\{x_n\}$ stays forever within the set $X^+(l)$, i.e.,

$$x_n = x_n(\omega) \in X^+(l) \ (l = 1, ..., L).$$

Proof. The proof is reported from [13]. Let $\Omega^+(0)$ be the set of Ω such that for any elementary events $\omega \in \Omega^+(0)$ the corresponding trajectory stays within $X^+(0)$ all times $n = 1, 2, ...$.

Let us consider the set

$$\overline{\Omega} = \Omega \setminus \Omega^+(0).$$

Then, evidently there exists a finite time $n_l(\omega)$ such that starting from $X^+(0)$ the trajectory $x_{n_l(\omega)}$ evolves to the component $X^+(l)$ $(l = 1, ..., L)$. Let us prove that it never leaves it. We have

$$P\{x_{n_l+1} \notin X^+(l) \mid x_{n_l} \in X^+(l)\}$$

$$= E\{\chi(x_{n_l+1} \notin X^+(l) \wedge x_{n_l} \in X^+(l))\}$$

$$= E\{\chi(x_{n_l+1} \notin X^+(l))\chi(x_{n_l} \in X^+(l))\}$$

$$= E\{E\{\chi(x_{n_l+1} \notin X^+(l))\chi(x_{n_l} \in X^+(l)) \mid \sigma(x_t, d_t : t = 1, ...n_l)\}\}$$

$$= E\{\chi(x_{n_l} \in X^+(l))E\{\chi(x_{n_l+1} \notin X^+(l)) \mid \sigma(x_t, d_t : t = 1, ...n_l)\}\}.$$

But

$$\chi\left(x_{n_l} \in X^+\left(l\right)\right) E\left\{\chi\left(x_{n_l+1} \notin X^+\left(l\right)\right) \mid \sigma\left(x_t, d_t : t = 1, ...n_l\right)\right\}$$

$$= \chi\left(x_{n_l} \in X^+\left(l\right)\right) \sum_{x(j) \notin X^+(l)} \sum_{x(i) \in X^+(l)} \sum_{l=1}^{N} \pi_{ij}^l d_{n_l}^{il} = 0$$

because (see (1.20))

$$\sum_{l=1}^{N} \pi_{ij}^l d_{n_l}^{il} = 0 \ \forall x(i) \in X^+(l), x(j) \notin X^+(l)$$

for any strategies $\{d_n\} \in \Sigma$. So,

$$P\left\{x_{n_l+1} \notin X^+(l) \mid x_{n_l} \in X^+(l)\right\} \stackrel{a.s.}{=} 0.$$

The lemma is proved. ■

This lemma results in a host of interesting results.

The main aim of the next subsection is to give a classification of controlled Markov chains.

1.5.5 Classification of controlled chains

Based on the previous lemma we may introduce the following definition:

Definition 31 *If*

- *there exists a stationary non-degenerated strategy $\{d\} \in \Sigma_+$ such that the corresponding transition matrix $\Pi(d)$ (1.6) has the structure corresponding only to the single communicating component $X^+(1)$ ($L = 1$) without non - return states*

$$L = 1, \ X^+(0) = \emptyset,$$

*the controlled Markov chain is said to be **ergodic** or a **communicating** chain;*

- *in addition, the periodicity index is also equal to one*

$$L = 1, \ r_1 = 1, \ X^+(0) = \emptyset,$$

*the ergodic controlled Markov chain is said to be **aperiodic** or **regular**.*

1.5. CONTROLLED FINITE MARKOV CHAINS

In view of this definition and the properties of controlled Markov chains described above, we will define the following basic structures:

- controlled finite Markov chains of **general type** (see figure 1.3)

$$L \geq 2,\ X^+(0) \neq \emptyset;$$

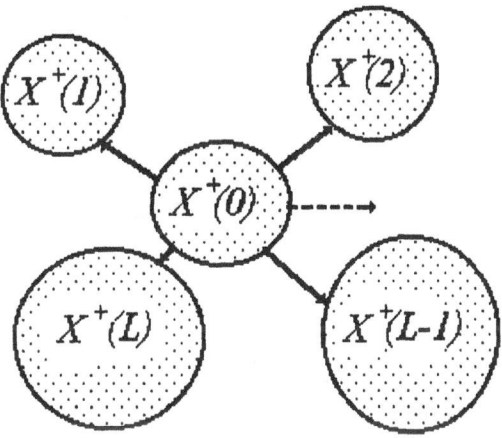

Figure 1.3: Controlled homogeneous finite Markov chains of general type.

- **ergodic (or communicating)** homogeneous finite Markov chains (see figure 1.4)

$$L = 1,\ r_1 \geq 2,\ X^+(0) \neq \emptyset;$$

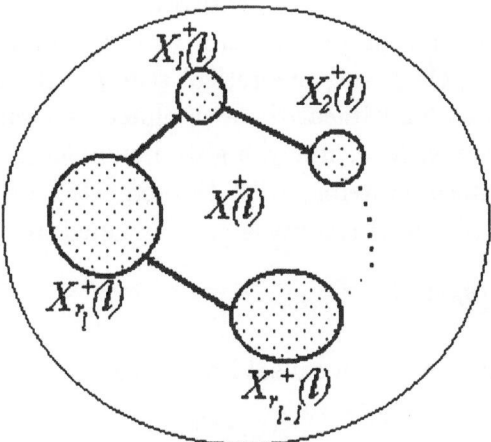

Figure 1.4: Controlled ergodic (communicating) homogeneous Markov chains.

- **aperiodic** or **regular** controlled finite Markov chains (see figure 1.5)

$$L = 1,\ r_1 = 1,\ X^+(0) \neq \emptyset.$$

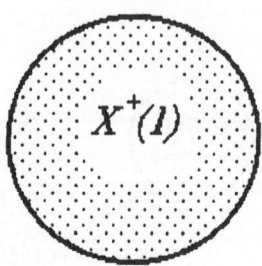

Figure 1.5: Aperiodic controlled finite Markov chain.

The attention given to these structures is due to their very interesting intrinsic properties.

Various systems which can be described or related to finite Markov chains are presented in the next section.

1.6 Examples of Markov models

Markov chains with finite-states and finite-decisions (actions) have been used as a control model of stochastic systems in various applications (pattern recognition, speech recognition, networks of queues, telecommunications, biology and medicine, process control, learning systems, resource allocation, communication, etc.) and theoretical studies. It has been argued that in a sense a Markov chain linearizes a nonlinear system. By the use of probabilistic state transitions, many highly nonlinear systems can be accurately modelled as linear systems, with the expected dividends in mathematical convenience [15]. Some examples are described in what follows.

Example 3 *Black-box models.*

It is well known that a causal linear stable time-invariant discrete system is described by

$$y_n = \sum_{k=0}^{\infty} h_k u_{n-k}$$

where $\{y_n\}$ and $\{u_n\}$ are the output and input signal sequences, respectively.

1.6. EXAMPLES OF MARKOV MODELS

The sequence of Markov parameters $\{h_k\}$ represents the impulse response of the system.

The ARMAX models and the non-linear time series models (under some conditions) are Markov chains or can be rephrased as Markov chains [16-17].

Another problem concerns the systems modelled by parametric models of regression type, that alternate between different dynamic modes. For example: 1) A supersonic aircraft has very different dynamics for different velocities; 2) A manufacturing process where the raw material can be of some different typical qualities. The dynamics of such systems can be captured by letting the parameter vector θ belong to a finite set

$$\theta_n \in [\theta_1, \theta_2, ..., \theta_N].$$

Let us assume that there exists a stochastic variable $\xi_n \in [1, 2, ..., N]$ which controls the variation of the parameter vector θ_n

$$\theta_n = \theta_i \text{ if } \xi_t = i, \ i = 1, ..., N.$$

A nice way to describe how frequent the different θ_i's are and the chances (probability) of transition from one θ_i to another is to model ξ_n as an N-state Markov chain [18-19].

Example 4 *Process Control and optimization*

1. Chemical reactor.

A study concerning the control of a discrete-time Markov process has been considered in [20]. The computation of an optimal feedback controller has been done for a heat treatment process involving an endothermic reaction for temperatures below $800°C$, and an exothermic reaction for higher temperatures. The following dynamic model has been considered for control purposes [20]

$$x_{n+1} = x_n \left\{1.005 + 0.015 \tanh\left[0.1\left(x_n - 803.446\right)\right]\right\} +$$
$$+ 0.00333 u_n + \zeta_n^1 + \zeta_n^2 + \zeta_n^3 + \zeta_n^4$$

where x_n and u_n represent respectively the temperature and the heat input at time n. ζ_n^i ($i = 1, ..., 4$) are independent samples from normal zero-mean distributions with the following respective standard deviations:

$$\sigma_1 = 0.0002 x_n \left|x_n - 800\right|, \ \sigma_2 = 0.005 x_n y \left[1 + |x_n - 800|^{\frac{1}{2}}\right]^{-1}$$

$$\sigma_3 = 0.0005 \left|u_n\right| \text{ and } \sigma_4 = 1, \ y = 1 \text{ for } x_n > 800, y = 0 \text{ otherwise.}$$

The desired operating temperature is a point of unstable equilibrium. It has been stated [20] that the control of this thermal process is in some sense analogous to the problem of maintaining an inverted pendulum in an upright position. The temperature range of interest was divided into nine quantized intervals.

2. Two-layer control structure.

In this control structure, the lower and high layers contain the local controller and the supervisor, respectively. The parameters of the controller can be modified by the supervisor when change occurs in the plant dynamics or in the environment. Under some conditions, this control problem has been formulated and solved by Forestier and Varaiya [21] in the framework of Markov chains.

3. Simulated annealing method.

Simulated annealing method is suitable for the optimization of large scale systems and multimodal functions, and is based on the principles of thermodynamics involving the way liquids freeze and crystallize [22]. It has been shown that simulated annealing generates a non-homogeneous Markov chain [23]

Example 5 *Learning automata.*

There exist several connections between stochastic learning automata [11, 24] and finite controlled Markov chains. Tsetlin [25] has shown that the behaviour of a variable-structure learning automata operating in a random media can be described by a non-homogeneous Markov chain.

The gore game [25] is a symmetric game played by N identical learning automata. Each automaton consists of two actions and m states. It has been shown [25] that the behaviour of this group of automata is described by a homogeneous Markov chain with $(2m)^N$ states.

A study concerning the problem of controlling Markov chains using decentralized learning automata have been carried by Wheeler and Narendra [26]. In this study, a learning automaton is associated with each state of the Markov chain and acts only when the chain is in that state. The probability distribution related to the considered automaton occurs only after the chain returns to that state and is based on some performance index.

Example 6 *Networks management (telephone traffic routing).*

The problem related with routing in telecommunication networks is a representative example of problems associated with networks management. A telephone network is a circuit switched network [26]. A message (information) is transmitted from node to node till it reaches its destination [27]. In [28], it has been also shown that the problem of determining a routing and flow control policy of networks of queues can sometimes be formulated as a Markov decision process.

1.6. EXAMPLES OF MARKOV MODELS

Example 7 *Inventory system. In an inventory system (replacement parts, etc.). The stock level and the amount ordered at time t ($t = 1, 2, ...$) represent the state and the control respectively.*

Example 8 *Statistical alignment (synchronization) of narrow polar antenna diagrams in communication systems.*

The radio antennas commonly used in communication systems when working within the microwaves frequency band may have very narrow polar diagrams with a width about $1 - 2^0$. Let us consider two space stations A and B equipped with receiver-transmitter devices and approximately oriented to each other (see figure 1.6)

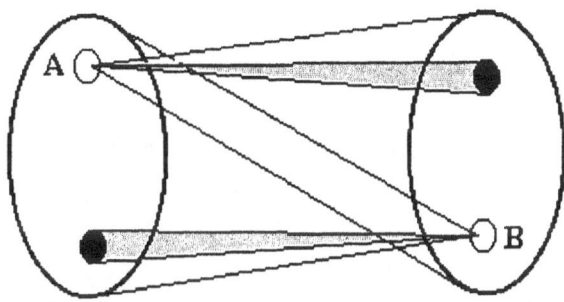

Figure 1.6: Two space stations with their corresponding polar diagrams.

The polar diagrams of each station can move within their associated scanning zone and, the transmitters are continuously emitting. The communication procedure is realized as follows: when one station, for example station A detects the signal transmitted by station B (and as a consequence detects its direction) it stops the scanning process and starts to receive the information to be transmitted in this direction. During this transmission period, station B continues its random scanning until it detects the position of station A. At this time, the alignment process stops and these stations are considered as synchronized.

This process has been modelled by a Markov chain consisting of four states $\{x(1), x(2), x(3), x(4)\}$ *[29]*. These states are associated with the following behaviours:

State $x(1)$ corresponds to the situation when both diagrams coincide exactly in their directions; the stations are synchronized and the transmission of information can start. This is an absorbing state;

State $x(2)$ corresponds to the situation when the polar diagrams are oriented randomly in the space and the synchronization is not realized; This is a non-return state;

State $x(3)$ concerns the situation when station A finds the signal transmitted by station B, stops the scanning process and starts to receive and to transmit the desired information. At the same time station B continues its random scanning. This is also a non-return state;

State $x(4)$ corresponds to the situation when station B finds the signal transmitted by station A, stops the scanning process and starts to receive and transmit the desired information. At the same time, station A continues its random scanning. This is also a non-return state.

The block diagram of this Markov chain is represented in figure 1.7.

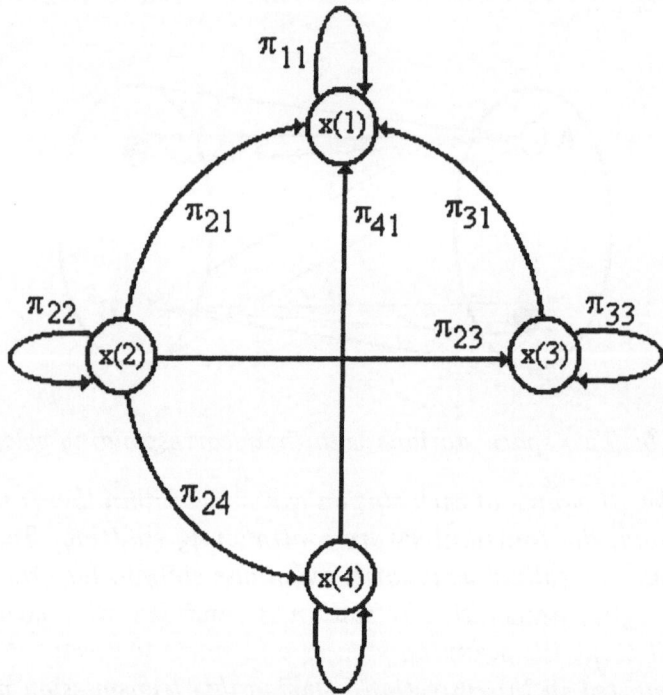

Figure 1.7: The state block diagram of the communication system.

The transition matrix Π is

$$\Pi = \begin{bmatrix} 1 & 0 & 0 & 0 \\ \pi_{21} & \pi_{22} & \pi_{23} & \pi_{24} \\ \pi_{31} & 0 & \pi_{33} & 0 \\ \pi_{41} & 0 & 0 & \pi_{44} \end{bmatrix}$$

The average time of the setting up of the communication (alignment or synchronization time) can be estimated by

$$T_{comm} = \frac{\pi_{41}\pi_{31} + \pi_{23}\pi_{41} + \pi_{24}\pi_{31}}{(\pi_{21} + \pi_{23} + \pi_{24})\pi_{31}\pi_{41}}$$

Notice that the transition matrix can be adapted to minimize the synchronization time T_{comm}.

A brief survey on stochastic approximation techniques is given in the next section.

1.7 Stochastic approximation techniques

Stochastic approximation (SA) techniques are well known recursive procedures for solving many engineering problems (finding roots of equations, optimization of multimodal functions, neural and neuro-fuzzy systems synthesis, stochastic control theory, etc.) in the presence of noisy measurements.

Let us consider the following estimation problem [30]: Determine the value of the vector parameter c which minimizes the following function:

$$f(c) = \int_x Q(x,c) P(x) dx = E_x \{Q(x,c)\} \qquad (1.23)$$

where $Q(x,c)$ is a random functional not explicitly known, x is a sequence of stationary random vectors, and $P(x)$ represents the probability density function which is assumed to be unknown. The optimal value c^* of the vector parameter c which minimizes (1.23) is the solution of the following equation (necessary condition of optimality):

$$\nabla_c f(c) = E_x \{\nabla_c Q(x,c)\}$$

where $\nabla_c f(c)$ represents the gradient of the functional $f(c)$ with respect to the vector parameter c.

If the function $Q(x,c)$ and the probability density function $P(x)$ are assumed to be unknown, it follows that the gradient $\nabla_c f(c)$ can not be calculated. The optimal value c^* of the vector parameter c can be estimated using the realizations of the function $Q(x,c)$ as follows:

$$c_n = c_{n-1} - \gamma_n \nabla_c Q(x,c). \qquad (1.24)$$

This is the stochastic approximation technique [30].

Stochastic approximation techniques are inspired by the gradient method in deterministic optimization. The first studies concerning stochastic approximation techniques were done by Robbins and Monro [31] and Kiefer and Wolfowitz [32] and were related to the solution, and the optimization of regression problems. These studies have been extended to the multivariable case by Blum [33]. Several techniques have been proposed by Kesten [34] and Tsypkin [30] to accelerate the behaviour of stochastic approximation algorithms. Tsypkin [30] has shown that several problems related to pattern recognition, control, identification, filtering, etc. can be treated in an unified manner as learning problems by using stochastic approximation techniques. These techniques belong to the class of random search techniques [16, 35-48].

One of several advantages of random search techniques is that these techniques do not require the detailed knowledge of the functional relationship between the parameters being optimized and the objective function being minimized that is required in gradient based techniques. The other advantage is their general applicability, i.e., there are almost no conditions concerning the function to be optimized (continuity, etc.), and the constraints. For examples, Najim et al. [44], have developed an algorithm for the synthesis of a constrained long-range predictive controller based on neural networks. The design of an algorithm for training under constraints, distributed logic processors using stochastic approximation techniques has been done by Najim and Ikonen [49].

The methods used for obtaining with probability one convergence results as well as useful estimates of the convergence rate are the powerful martingale based method, the theory of large deviations, and the ordinary differential equation technique (ODE). Stochastic processes such as martingales arise naturally whenever one needs to consider mathematical expectation with respect to increasing information patterns (conditional expectation). The theory of large deviations has been developed in connection with the averaging principle by Freidlin [50]. For example, this theory has been used to get a better picture of the asymptotic properties of a class of projected algorithms [51]. The ODE technique is based on the connection between the asymptotic behaviour of a recursively defined sequence (recursive algorithm) and the stability behaviour of a corresponding differential equation, i.e., heuristically, if the correction factor is assimilated to a sampling period Δt, equation (1.24) leads to an ordinary differential equation. The ODE contains information about convergence of the algorithm as well as about convergence rates and behaviour of the recursive algorithm.

Now, to illustrate the behaviour of finite controlled Markov chains, some numerical simulations are presented in the next section.

1.8 Numerical simulations

In this section we are concerned with some simulation results dealing with finite controlled Markov chains. Let us consider a Markov chain containing 5 states and 6 control actions. The associated transition probability matrices

1.8. NUMERICAL SIMULATIONS

π_{ij}^l are

$$\pi_{ij}^1 = \begin{bmatrix} 0 & 0.5 & 0 & 0.5 & 0 \\ 0.25 & 0 & 0.25 & 0.25 & 0.25 \\ 0 & 1 & 0 & 0 & 0 \\ 0.7 & 0.1 & 0.1 & 0 & 0.1 \\ 0 & 0 & 1.0 & 0 & 0 \end{bmatrix}, \pi_{ij}^2 = \begin{bmatrix} 0 & 0.2 & 0.2 & 0.2 & 0.4 \\ 0 & 0 & 0.3 & 0.3 & 0.4 \\ 0.1 & 0.2 & 0 & 0.2 & 0.5 \\ 0.1 & 0.2 & 0.2 & 0 & 0.5 \\ 0.2 & 0.2 & 0.2 & 0.4 & 0 \end{bmatrix},$$

$$\pi_{ij}^3 = \begin{bmatrix} 0 & 0.3 & 0.3 & 0.3 & 0.1 \\ 0.3 & 0 & 0.3 & 0.3 & 0.1 \\ 0.3 & 0.3 & 0 & 0.3 & 0.1 \\ 0.3 & 0.3 & 0.3 & 0 & 0.1 \\ 0.3 & 0.3 & 0.3 & 0.1 & 0 \end{bmatrix}, \pi_{ij}^4 = \begin{bmatrix} 0 & 0.1 & 0.1 & 0.1 & 0.7 \\ 0.1 & 0 & 0.2 & 0.2 & 0.5 \\ 0.2 & 0.2 & 0 & 0.3 & 0.3 \\ 0.3 & 0.3 & 0.3 & 0 & 0.1 \\ 0.3 & 0.3 & 0.3 & 0.1 & 0 \end{bmatrix},$$

$$\pi_{ij}^5 = \begin{bmatrix} 0 & 0.5 & 0.5 & 0 & 0 \\ 0.5 & 0 & 0.1 & 0.2 & 0.2 \\ 0.1 & 0 & 0 & 0.7 & 0.2 \\ 0.6 & 0 & 0.2 & 0 & 0.2 \\ 0.7 & 0 & 0.3 & 0 & 0 \end{bmatrix}, \pi_{ij}^6 = \begin{bmatrix} 0 & 0.25 & 0.25 & 0.25 & 0.25 \\ 0.25 & 0 & 0.25 & 0.25 & 0.25 \\ 0.25 & 0.25 & 0 & 0.25 & 0.25 \\ 0.25 & 0.25 & 0.25 & 0 & 0.25 \\ 0.25 & 0.25 & 0.25 & 0.25 & 0 \end{bmatrix},$$

Let us consider a uniform initial distribution

$$p_1(i) = \frac{1}{5}$$

The following results illustrate the behaviour of the considered controlled Markov chain for different fixed stationary strategies $d = \begin{bmatrix} d^{il} \end{bmatrix}$ ($i = 1, ..., 5; l = 1, ..., 6$).

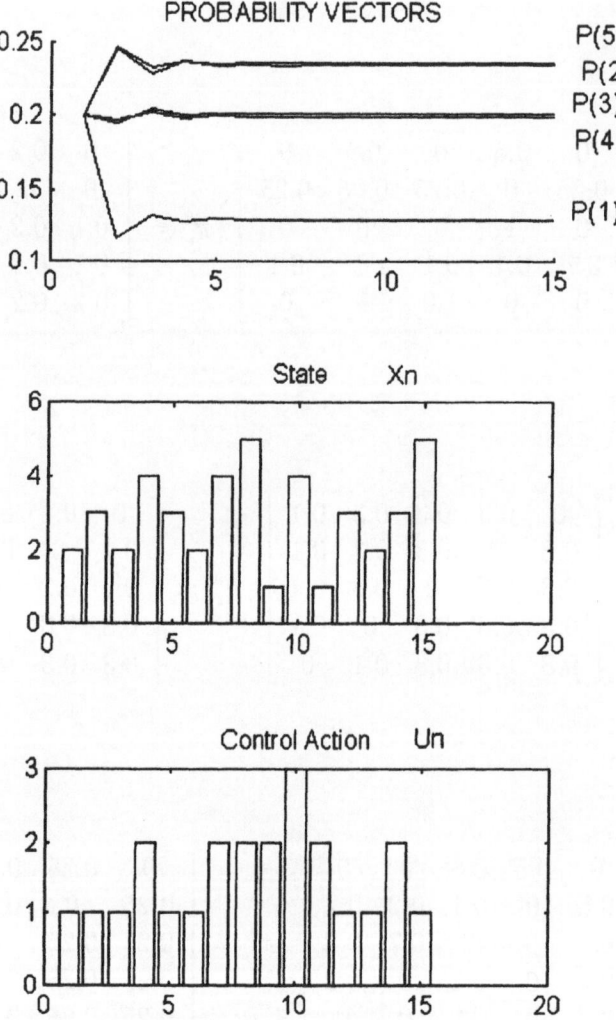

Figure 1.8: Evolution of the probability vector, state and control actions.

Figure 1.8 represents the evolution of the probability vector, the state and the control action, for the following stationary strategy d

$$d = \begin{bmatrix} 0.0368 & 0.1702 & 0.0489 & 0.3267 & 0.2741 & 0.1433 \\ 0.2292 & 0.1028 & 0.0754 & 0.2214 & 0.0543 & 0.3168 \\ 0.2330 & 0.0369 & 0.0274 & 0.0125 & 0.0424 & 0.6478 \\ 0.0780 & 0.2824 & 0.2859 & 0.0969 & 0.1229 & 0.1339 \\ 0.0670 & 0.2217 & 0.2299 & 0.0145 & 0.2545 & 0.2123 \\ 0.0044 & 0.2943 & 0.0785 & 0.2382 & 0.0614 & 0.3232 \end{bmatrix}$$

1.8. NUMERICAL SIMULATIONS

For the following control strategy

$$d = \begin{bmatrix} 0.1244 & 0.1772 & 0.2458 & 0.2066 & 0.2460 & 0 \\ 0.4424 & 0.1128 & 0.1357 & 0.2145 & 0.0946 & 0 \\ 0.2250 & 0.4235 & 0.600 & 0.01031 & 0.1214 & 0.0670 \\ 0.4408 & 0.4557 & 0.1035 & 0 & 0 & 0 \\ 0.0379 & 0.2387 & 0.1150 & 0.4328 & 0.1756 & 0 \\ 0.3365 & 0.4166 & 0.2468 & 0 & 0 & 0 \end{bmatrix},$$

the behaviour of the probability vector, the states and the control actions are depicted in figure 1.9.

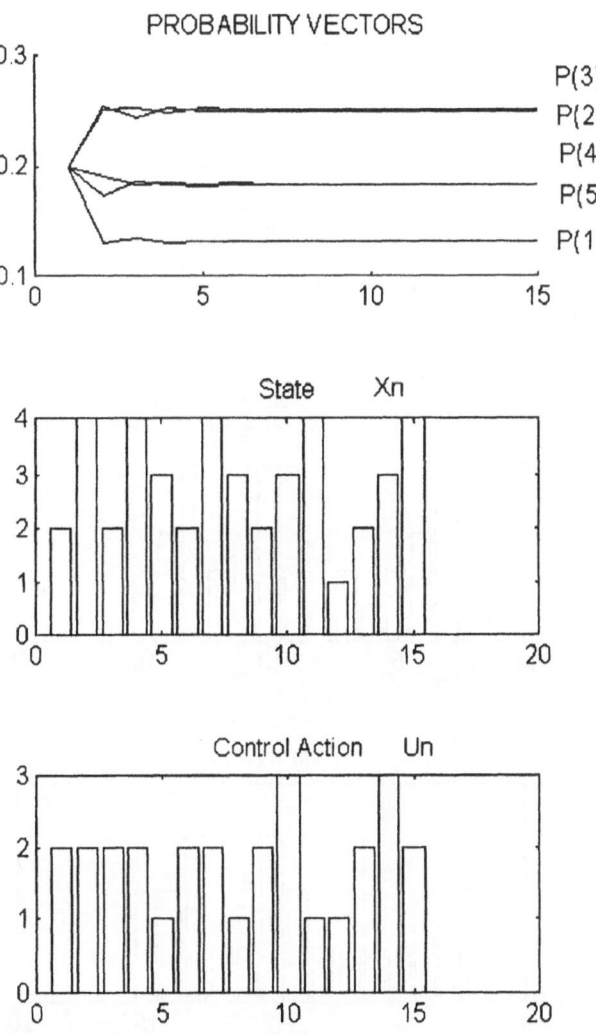

Figure 1.9: Evolution of the probability vector, the states and the control actions.

For the considered Markov chain, the following control strategy

$$d = \begin{bmatrix} 0.1889 & 0.2865 & 0.3536 & 0.1711 & 0 & 0 \\ 0.2116 & 0.4582 & 0.0723 & 0.0197 & 0.2382 & 0 \\ 0.1286 & 0.1094 & 0.4188 & 0.1737 & 0.1696 & 0 \\ 0.0510 & 0.4336 & 0.2733 & 0.1522 & 0.0899 & 0 \\ 0.1952 & 0.1600 & 0.1521 & 0.2738 & 0.2189 & 0 \\ 0.1950 & 0.1678 & 0.4672 & 0.1700 & 0 & 0 \end{bmatrix},$$

has been also implemented. Figure 1.10 shows the evolution of the components of the probability vector as well as the states and the control actions.

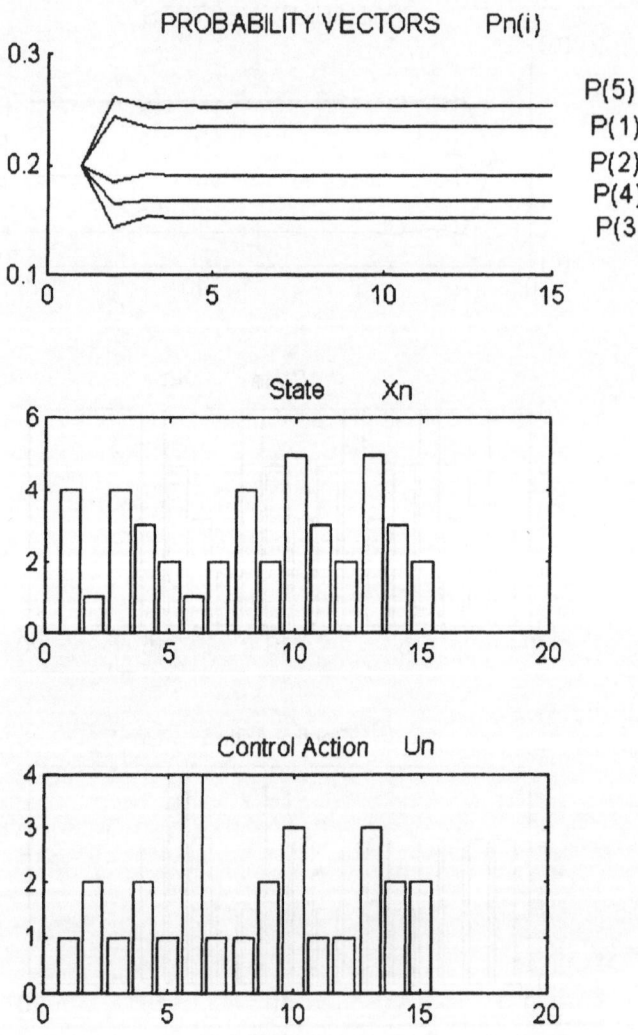

Figure 1.10: Evolution of the probability vector, the states and the control actions.

1.8. NUMERICAL SIMULATIONS

The following figure (figure 1.11)

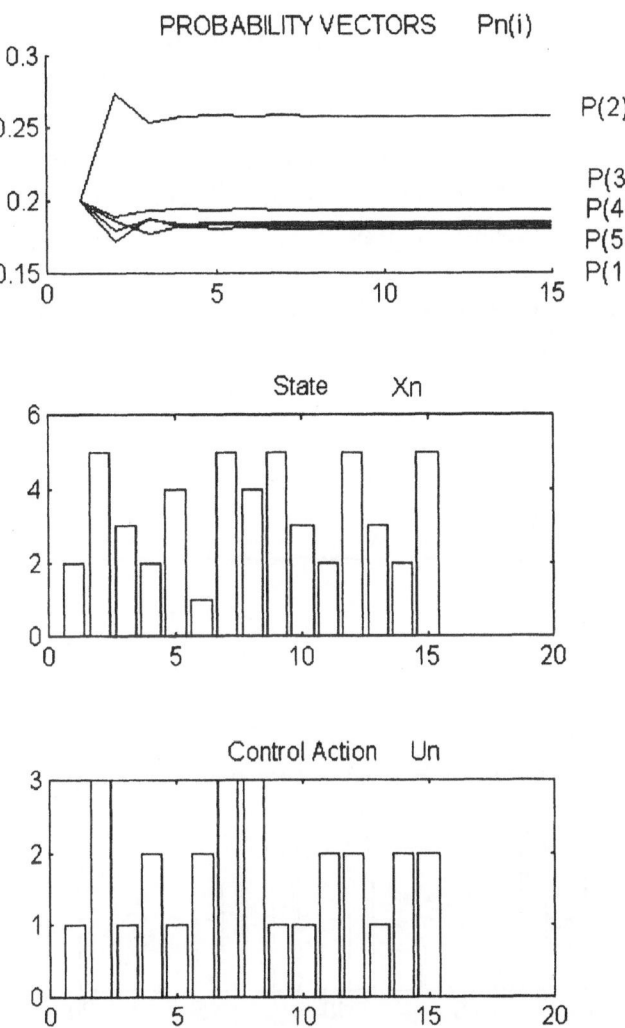

Figure 1.11: Evolution of the probability vector, the states and the control actions.

is related to the following matrix d

$$d = \begin{bmatrix} 0.4457 & 0.2887 & 0.2656 & 0 & 0 & 0 \\ 0.2677 & 0.1423 & 0.1801 & 0.0889 & 0.1826 & 0.1383 \\ 0.2714 & 0.1320 & 0.2486 & 0.2236 & 01245 & 0 \\ 0.2951 & 0.3959 & 0.3090 & 0 & 0 & 0 \\ 0.0477 & 0.1061 & 0.3572 & 0.0082 & 0.0962 & 0.3846 \\ 0.0081 & 0.1276 & 0.2415 & 0.4455 & 0.1773 & 0 \end{bmatrix}$$

38   CHAPTER 1. CONTROLLED MARKOV CHAINS

In the following simulations, the following control strategy

$$d = \begin{bmatrix} 0.1220 & 0.7436 & 0.1344 & 0 & 0 & 0 \\ 0.8640 & 0.1360 & 0 & 0 & 0 & 0 \\ 0.9485 & 0.515 & 0 & 0 & 0 & 0 \\ 0.1160 & 0.5131 & 0.3709 & 0 & 0 & 0 \\ 0.0662 & 0.0492 & 0.6089 & 0.2757 & 0 & 0 \\ 0.3635 & 0.6049 & 0.0315 & 0 & 0 & 0 \end{bmatrix},$$

has been implemented. Figure 1.12 corresponds to this control strategy.

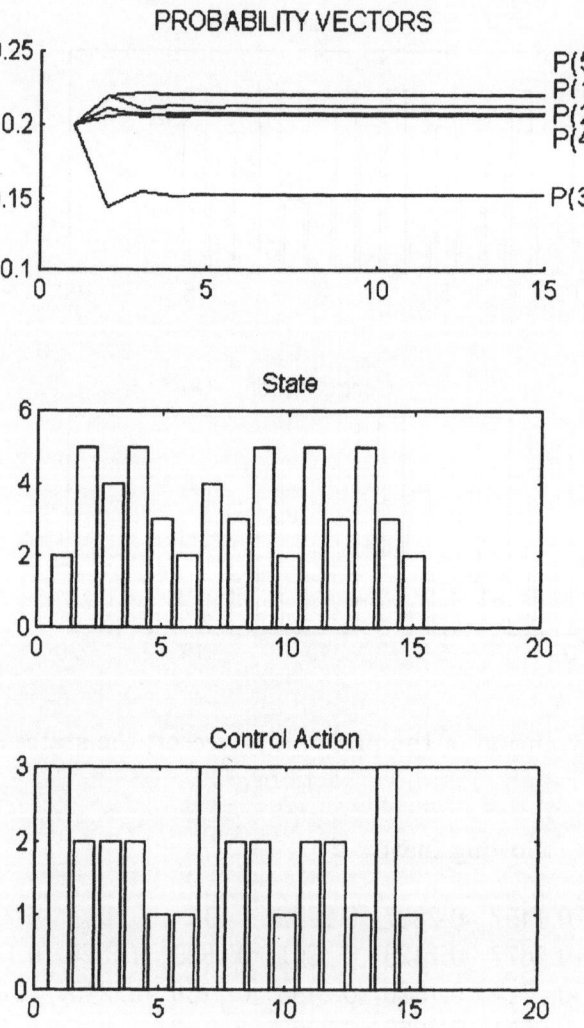

Figure 1.12: Evolution of the probability vector, the states and the control actions.

1.8. NUMERICAL SIMULATIONS

Finally, for the control strategy

$$d = \begin{bmatrix} 0.6507 & 0.2692 & 0.0801 & 0 & 0 & 0 \\ 0.4663 & 0.5337 & 0 & 0 & 0 & 0 \\ 0.9993 & 0.0007 & 0 & 0 & 0 & 0 \\ 0.3706 & 0.0691 & 0.5603 & 0 & 0 & 0 \\ 0.2171 & 0.3339 & 0.1175 & 0.1761 & 0.1554 & 0 \\ 0.1868 & 0.8132 & 0 & 0 & 0 & 0 \end{bmatrix},$$

figure 1.13 shows the evolution of the components of the probability vector as well as the states and the control actions.

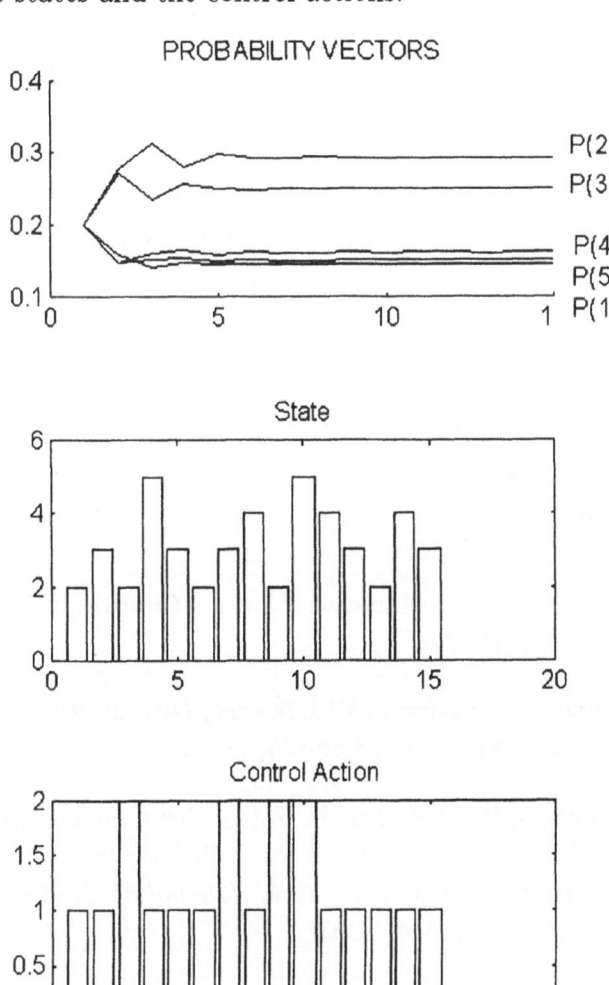

Figure 1.13: Evolution of the probability vector, the states and the control actions.

We can observe that different stationary control strategies d lead to different stationary (final) state distributions

$$p^*(i) := \lim_{n \to \infty} p_n(i), \ i = 1, ..., 5$$

which can practically cover all the unit segment $[0, 1]$. This fact is due to the continuous dependence of the state distribution on the stationary strategy d.

1.9 Conclusions

This chapter has surveyed some of the basic definitions and concepts related to controlled finite Markov chains and stochastic approximation techniques. We shall frequently call upon the results of this chapter in the next chapters. A brief survey on stochastic approximation techniques has also been given. These techniques represent the frame of the self-learning (adaptive) control algorithms developed in this book. Several adaptive control algorithms for both unconstrained and constrained Markov chains will presented and analyzed in the remainder of this book. An adaptive algorithm (recursive) can be defined as a procedure which forms a new estimate, incorporating new information (realizations), from the old estimate using a fixed amount of computations and memory.

1.10 References

1. O. Hernandez-Lerma, *Adaptive Markov Control Processes*, Springer-Verlag, London, 1989.

2. O. Hernandez-Lerma and J. B. Lasserre, *Discrete-time Markov Control Processes*, Springer-Verlag, London, 1996.

3. A. Arapostathis, V. S. Borkar, E. Fernandez-Gaucherand, M. K. Ghosh and S. I. Marcus, Discrete-time controlled Markov processes with average cost criterion: a survey, *SIAM Journal of Control and Optimization*, vol. 31, pp. 282-344, 1993.

4. P. R. Halmos, *Measure Theory*, D. Van Nostrand Co., Princeton, N. J., 1950.

5. R. B. Ash, *Real Analysis and Probability*, Academic Press, New York, 1972.

1.10. REFERENCES

6. J. Bather, Optimal decision procedures for finite Markov chains, Part II: Communicating systems, *Advances in Applied Probability*, vol. 5, pp. 521-540, 1973.

7. J. G. Kemeny, and J. L. Snell, *Finite Markov Chains*, Springer-Verlag, Berlin, 1976.

8. E. Seneta, *Nonnegative Matrices and Markov Chains*, Springer-Verlag, Berlin, 1981.

9. D. J. Hartfiel and E. Seneta, On the theory of Markov set-chains, *Adv. Appl. Prob.* vol. 26, pp. 947-964, 1994.

10. Yu. A. Rozanov, *Random Processes*, (in Russian) Nauka, Moscow, 1973.

11. K. Najim and A. S. Poznyak, *Learning Automata Theory and Applications*, Pergamon Press, London, 1994.

12. V. G. Sragovitch, *Adaptive Control*, (in Russian) Nauka, Moscow, 1981.

13. A. V. Nazin and A. S. Poznyak, *Adaptive Choice of Variants*, (in Russian) Nauka, Moscow, 1986.

14. A. N. Shiryaev, *Probability*, Springer-Verlag, New York, 1984.

15. J. Slansky, Learning systems for automatic control, *IEEE Trans. Automatic Control*, vol. 11, pp. 6-19, 1966.

16. M. Duflo, *Random Iterative Models*, Springer-Verlag, Berlin, 1997.

17. D. Tjøstheim, Non-linear time series and Markov chains, *Adv. Appl. Prob.*, vol. 22, pp. 587-611, 1990.

18. G. Lindgren, Markov regime models for mixed distributions and switching regressions, *Scand. J. Statistics*, vol. 5, pp. 81-91, 1978.

19. M. Millnert, *Identification and control of systems subject to abrupt changes*, Dissertation n^0. 82, Department of Electrical Engineering, Linköping University,

20. J. S. Riordon, An adaptive automaton controller for discrete-time Markov processes, *Automatica*, vol. 5, pp. 721-730, 1969.

21. J. P. Forestier and P. Varaiya, Multilayer control of large Markov chains, *IEEE Trans. Automatic Control*, vol. 23, pp. 298-305, 1978.

22. F. Romeo and A. Sangiovanni-Vincentelli, A theoretical framework for simulated annealing, *Algorithmica*, vol. 6, pp. 302-345, 1991.

23. N. Wojciech, Tails events of simulated annealing Markov chains, *J. Appl. Prob.*, vol. 32, pp. 867-876, 1995.

24. A. S. Poznyak and K. Najim, *Learning Automata and Stochastic Optimization*, Springer-Verlag, Berlin, 1997.

25. M. L. Tsetlin, *Automaton Theory and Modeling of Biological Systems*, Academic Press, New York, 1973.

26. R. M. Jr. Wheeler and K. S. Narendra, Decentralized learning in finite Markov chains, *IEEE Trans. Automatic Control*, vol. 31, pp. 519-526, 1986.

27. K. S. Narendra and M. A. L. Thathachar, *Learning Automata An Introduction*, Prentice-Hall, Englewood Cliffs, N.J., 1989.

28. S. Jr. Stidham and R. Weber, A survey of Markov decision models for control of networks of queues, *Queueing Systems*, vol. 13, pp. 291-314, 1993.

29. V. A. Kazakov, *Introduction to Markov Processes and some Radiotechnique Problems*, (in Russian) Sovetskoye Radio, Moscow, 1973.

30. Ya. Z. Tsypkin, *Foundations of the Theory of Learning Systems*, Academic Press, New York, 1973.

31. H. Robbins and S. Monro, A stochastic approximation method, *Ann. Math. Statistics*, vol. 22, n^0. 1, pp. 400-407, 1951.

32. J. Kiefer and J. Wolfowitz, Stochastic estimation of the maximum of a regression, *Ann. Math. Stat.* vol. 23, pp. 462-466, 1952.

33. J. A. Blum, Multidimensional stochastic approximation method, *Ann. Math. Statistics*, vol. 25, n^0. 1, pp. 737-744, 1954.

34. H. Kesten, Accelerated stochastic approximation, *Ann. Math. Statistics*, vol. 29, pp. 41-59, 1958.

35. H. J. Kushner and E. Sanvicente, Stochastic approximation for constrained systems with observation noise on the system and constraints, *Automatica*, vol. 11, pp. 375-380, 1975.

1.10. REFERENCES

36. J. B. Hiriart-Urruty, Algorithms for penalization type and dual type for the solution of stochastic optimization problems with stochastic constraints, In. J. R. Barra et al. (Ed.), *Recent Developments in Statistics*. pp. 183-219. North-Holland, Amsterdam, 1977.

37. H. J. Kushner and D. S. Clark, *Stochastic Approximation Methods for Constrained and Unconstrained Systems*, Springer-Verlag, Berlin, 1978.

38. Ya. Z. Tsypkin, *Adaptive and Learning in Automatic Systems*, Academic Press, New York, 1971.

39. J. C. Spall, Multivariate stochastic approximation using a simultaneous perturbation gradient approximation, *IEEE Trans. Auto. Control*, vol. 37, pp. 332-341, 1992.

40. L. Ljung, G. Pflug and H. Walk, *Stochastic Approximation and Optimization of Random Systems*, Springer-Verlag, Berlin, 1992.

41. K. Najim and A. S. Poznyak, Neural networks synthesis based on stochastic approximation algorithm, *Int. J. of Systems Science*, vol. 25, pp. 1219-1222, 1994.

42. A. S. Poznyak, K. Najim and M. Chtourou, Use of recursive stochastic algorithm for neural networks synthesis, *Appl. Math. Modelling*, vol. 17, pp. 444-448, 1993.

43. J. C. Spall and J. A. Cristion, Non-linear adaptive control using neural networks: estimation with a smoothed form of simultaneous perturbation gradient approximation, *Statistica Sinica*, vol. 4, pp.1-27, 1994.

44. K. Najim, A. Rusnak, A. Meszaros and M. Fikar, Constrained Long-Range Predictive Control Based on Artificial Neural Networks, *Int. J. of Systems Science*, vol. 28, n^0. 12, pp. 1211-1226, 1997.

45. H. Walk, Stochastic iteration for a constrained optimization problem, *Commun. Statist.-Sequential Analysis*, vol. 2, pp. 369-385, 1983-84.

46. A. Benveniste, M. Metivier and P. Priouret, 1990, *Stochastic Approximations and Adaptive Algorithms*, Springer-Verlag, Berlin, 1990.

47. G. Pflug, Stepsize rules, stopping times and their implementation in stochastic quasigradient algorithms. In Y. and R. Wets (Ed.), *Numerical Techniques for Stochastic Optimization*, Springer-Verlag, Berlin, pp. 137-160, 1988.

48. C. C. Y. Dorea, Stopping rules for a random optimization method, *SIAM J. Control and Optimization*, vol. 28, pp. 841-850, 1990.

49. K. Najim and E. Ikonen, Distributed logic processor trained under constraints using stochastic approximation techniques, *IEEE Trans. On Systems Man and Cybernetics*.

50. M. I. Freidlin, The averaging principle and theorems on large deviations, *Russian Math. Surveys*, vol. 33, pp. 117-176, 1978.

51. P. Dupuis and H. J. Kushner, Asymptotic behavior of constrained stochastic approximations via the theory of large deviations, Probability Theory and Related Fields, vol. 75, pp. 223-244, 1987.

Part I

Unconstrained Markov Chains

Chapter 2

Lagrange Multipliers Approach

2.1 Introduction

Markov chains have been widely studied [1-6]. Many engineering problems can be modelled as finite controlled Markov chains whose transition probabilities depends on the control action. The control actions are generated to achieve some desired goal (control objective) such the maximization of the expected average reward or the minimization of a loss function. The control problem related to Markov chains with known transition probabilities has been extensively studied by several authors [1-2, 7-8] and solved on the basis of dynamic programming and linear programming. Many studies have been devoted to the control of Markov chains whose transition probabilities depend upon a constant and unknown parameter taking values in a finite set [9-20] or a time-varying parameter with a certain period [21]. In these studies, the self-tuning approach (certainty equivalence) has been considered (the unknown parameters are estimated and the control strategy is designed as if the estimated parameters were the true system parameters [22]). The maximum likelihood estimation procedure has been used by several authors. In [15] the problem of adaptive control of Markov chains is treated as a kind of multiarmed bandit problem. The certainty equivalence control with forcing [23] approach has been used in [15] and [16] to derive adaptive control strategies for finite Markov chains. In this control approach, at certain *a priori* specified instants, the system is forced (forcing or experimenting phase) by using other control actions in order to escape false identification traps. The forcing phase is similar to the introduction of extra perturbations in adaptive systems, to obtain good excitation (persistent excitation which is a uniform identifiability condition).

In [24] and [25] the problem of adaptive control of Markov chains is addressed by viewing it as a multi-armed bandit problem.

Controlling a Markov chain may be reduced to the design of a control policy which achieves some optimality of the control strategy. In this study the optimality is associated with the minimization of a loss function which is assumed to be bounded. This chapter presents a novel adaptive learning control algorithm for ergodic controlled Markov chains whose transition probabilities are unknown. In view of the fact that the requirements of a given control system can always be represented as an optimization problem, the adaptive learning control algorithm developed in the sequel is based on the Lagrange multipliers approach [26]. Lagrange multipliers are prominent in optimality conditions and play an important role in methods involving duality and decomposition. In this control algorithm the transition probabilities of the Markov chain are not estimated. The control policy is adjusted using the Bush-Mosteller reinforcement scheme [26-27] as a stochastic approximation procedure [17]. The Bush-Mosteller reinforcement scheme [28] is commonly used in the design of stochastic learning automata to solve many engineering problems. It should be noted that our approach here differs significantly from the previous one (see references therein), in that we do not assume that the transition probabilities depend upon an unknown parameter.

The system to be controlled is described in the next section.

2.2 System description

The design of an adaptive learning control algorithm for controlled Markov chains will be based on the minimization of a loss function. Let us first introduce some definitions concerning the loss function. The loss sequence (control objective) associated with a controlled Markov chain will be assumed to be bounded and is defined as follows:

Definition 1 *The sequence $\{\eta_n\}$ is said to be **a loss sequence** if:*
(i)
$$\sup_n |\eta_n| \stackrel{a.s.}{\leq} \sigma < \infty; \tag{2.1}$$

(ii) for any $n = 1, 2, ...$
$$E\{\eta_n \mid x_n = x(i), u_n = u(l)\} \stackrel{a.s.}{=} v_{il}. \tag{2.2}$$

We assume that the loss sequence $\{\eta_n\}$ associated with an ergodic controlled Markov chain, satisfies also the following Markov property:

$$E\{\eta_n \mid x_n = x(i), u_n = u(l) \wedge \mathcal{F}_{n-1}\} \tag{2.3}$$

2.2. SYSTEM DESCRIPTION

$$\stackrel{a.s.}{=} E\{\eta_n \mid x_n = x(i), u_n = u(l)\} \stackrel{a.s.}{=} v_{il}.$$

It means that the average values of loss sequence ("*external environment*" properties) does not depend on any process history and are determined only by the selected control action in the corresponding state $(x_n = x(i), u_n = u(l))$. In some sense, this fact reflects the property of stationarity of "*the external environment.*"

Definition 2 *On the dynamic trajectories $\{x_n\}$ and $\{u_n\}$ of a given controlled Markov chain we define a sequence $\{\Phi_n\}$ of **loss functions** Φ_n by*

$$\Phi_n := \frac{1}{n}\sum_{t=1}^{n}\eta_t. \qquad (2.4)$$

Generally, for any nonstationary random control strategy $\{d_n\} \in \Sigma$ the sequence $\{\Phi_n\}$ (2.4) may have no limits (in any probability sense). Nevertheless, there exist a lower and an upper bound for their partial limit points. These points belong to an interval $[\Phi_*, \Phi^*]$ which is given by the following lemma.

Lemma 1 *For any ergodic controlled Markov chain with any distribution of initial states*

$$\Phi_* := \min_{d \in \mathbf{D}} V(d) \stackrel{a.s.}{\leq} \varliminf_{n \to \infty} \Phi_n \leq \varlimsup_{n \to \infty} \Phi_n \stackrel{a.s.}{\leq} \max_{d \in \mathbf{D}} V(d) := \Phi^* \qquad (2.5)$$

where

$$V(d) = \sum_{i=1}^{K}\sum_{l=1}^{N} v_{il} d^{il} p_i(d) \qquad (2.6)$$

the vector $p^T(d) = (p_1(d), ..., p_K(d))$ satisfies the following linear algebraic equation

$$p(d) = \Pi^T(d) p(d) \qquad (2.7)$$

and the set $\mathbf{D}$ of stochastic matrices d is defined by

$$d \in \mathbf{D} := \left\{ d \mid d^{il} \geq 0, \sum_{l=1}^{N} d^{il} = 1 \ (i=1,...,K; l=1,...,N) \right\}. \qquad (2.8)$$

Proof. Let us rewrite Φ_n as follows:

$$\Phi_n = \sum_{i=1}^{K}\sum_{l=1}^{N}\left(\frac{1}{n}\sum_{t=1}^{n}\eta_t\right)\chi(x_n = x(i), u_n = u(l)) = \sum_{i=1}^{K}\sum_{l=1}^{N}\widehat{v}_n^{il}\widehat{\theta}_n^{il}$$

where
$$\widehat{v}_n^{il} := \frac{\sum_{t=1}^{n} \eta_t \chi\left(x_t = x(i), u_t = u(l)\right)}{1 + \sum_{t=1}^{n} \chi\left(x_t = x(i), u_t = u(l)\right)},$$

$$\widehat{\theta}_n^{il} := \frac{1}{n}\left[1 + \sum_{t=1}^{n} \chi\left(x_t = x(i), u_t = u(l)\right)\right].$$

If for a random realization $\omega \in \Omega$ we have
$$\sum_{t=1}^{\infty} \chi\left(x_t = x(i), u_t = u(l)\right) < \infty$$
then $\widehat{\theta}_n^{il}(\omega) \to 0$.

Consider now the realizations $\omega \in \Omega$ for which
$$\sum_{t=1}^{\infty} \chi\left(x_t = x(i), u_t = u(l)\right) = \infty.$$

Taking into account the properties of the controlled Markov chain and its associated loss sequence, and applying the law of large numbers for dependent sequences [29] we conclude that for all indexes

$$\left(\widehat{v}_n^{il} - v_{il}\right) \xrightarrow[n \to \infty]{a.s.} 0,$$

$$\left(\frac{\sum_{t=1}^{n} \chi\left(x_t = x(i), x_{t+1} = x(j)\right)}{1 + \sum_{t=1}^{n} \chi\left(x_t = x(i)\right)} - \pi_{ij}\left(\widehat{d}_n^{il}\right)\right) \xrightarrow[n \to \infty]{} 0$$

where
$$\widehat{d}_n^{il} := \frac{\sum_{t=1}^{n} \chi\left(x_t = x(i), u_t = u(l)\right)}{1 + \sum_{t=1}^{n} \chi\left(x_t = x(i)\right)}.$$

Hence, we obtain
$$\widehat{\theta}_n^{il} = \frac{1}{n}\left[1 + \frac{\sum_{t=1}^{n} \chi\left(x_t = x(i), u_t = u(l)\right)}{1 + \sum_{t=1}^{n} \chi\left(x_t = x(i)\right)}\left(1 + \sum_{t=1}^{n} \chi\left(x_t = x(i)\right)\right)\right]$$

$$= \widehat{d}_n^{il} \frac{1}{n} \sum_{t=1}^{n} \chi\left(x_t = x(i)\right) + O\left(\frac{1}{n}\right) = \widehat{d}_n^{il} p_i\left(\widehat{d}_n^{il}\right) + O\left(\frac{1}{n}\right)$$

where the components $p_i(\cdot)$ satisfies (2.7).

From the previous equalities we conclude that

$$\Phi_n - \sum_{i=1}^{K}\sum_{l=1}^{N} v_{il}\widehat{d}_n^{il} p_i\left(\widehat{d}_n^{il}\right) = \Phi_n - V\left(\widehat{d}_n^{il}\right) \overset{a.s.}{\underset{n\to\infty}{\to}} 0.$$

which directly leads to (2.5). ∎

The adaptive control problem of finite Markov chains will be formulated in the next section.

2.3 Problem formulation

Based on lemma 1, the problem related to the minimization of the asymptotic realization of the loss function within the class Σ of random strategies

$$\overline{\lim_{n\to\infty}}\, \Phi_n \to \inf_{\{d_n\}\in\Sigma}\ (a.s.) \qquad (2.9)$$

can be solved in the class Σ_s of stationary strategies, and the minimal value of the asymptotic realization of the loss function is equal to Φ_* (2.5), i.e.,

$$\inf_{\{d_n\}\in\Sigma}\overline{\lim_{n\to\infty}}\,\Phi_n \overset{a.s.}{=} \inf_{\{d\}\in\Sigma_s}\overline{\lim_{n\to\infty}}\,\Phi_n \overset{a.s.}{=} \Phi_* := \min_{d\in\mathbf{D}} V(d). \qquad (2.10)$$

So, the following adaptive control problem will be solved:

Based on the available observations

$$(x_1, u_1, \eta_1; \ldots; x_{n-1}, u_{n-1}, \eta_{n-1}; x_n)$$

construct a sequence $\{d_n\}$ of random matrix

$$d_n = d_n(x_1, u_1, \eta_1; \ldots; x_{n-1}, u_{n-1}, \eta_{n-1}; x_n)$$

such that the sequence $\{\Phi_n\}$ of loss functions reaches its minimal value Φ_* (2.10).

The next section deals with the adaptive learning control algorithm.

2.4 Adaptive learning algorithm

The increase in the power and the performance of computer systems has acted mainly as a catalyst for the development of adaptive control systems. The adaptive ideas have matured to a point where several implementations are now commercially available for different purposes (process control, signal processing, pattern recognition, etc.). An adaptive algorithm (recursive) can be defined as a procedure which forms a new estimate, incorporating new information (realizations), from the old estimate using a fixed amount of computations and memory. To develop an adaptive learning controller for Markov chains, let us consider the following nonlinear programming problem

$$V(d) \to \min_{d \in \mathbf{D}} \qquad (2.11)$$

where the function $V(d)$ and the set $\mathbf{D}$ are defined by (2.6) and (2.8).

The standard transformation [1]

$$c^{il} = d^{il} p_i(d) \qquad (2.12)$$

transforms this problem into the following linear programming problem

$$\tilde{V}(c) := \sum_{i=1}^{K} \sum_{l=1}^{N} v_{il} c^{il} \to \min_{c \in \mathbf{C}} \qquad (2.13)$$

where the set $\mathbf{C}$ is given by

$$\mathbf{C} = \left\{ c \mid c = \left[c^{il}\right],\ c^{il} \geq 0,\ \sum_{i=1}^{K}\sum_{l=1}^{N} c^{il} = 1, \right. \qquad (2.14)$$

$$\left. \sum_{l=1}^{N} c^{jl} = \sum_{i=1}^{K}\sum_{l=1}^{N} \pi_{ij}^{l} c^{il}\ (i,j = 1, ..., K; l = 1, ..., N) \right\}.$$

In view of (2.12) we conclude that

$$\sum_{l=1}^{N} c^{il} = p_i(d)\ (i = 1, ..., K).$$

Notice that for ergodic controlled Markov chains there exists a unique final (stationary) distribution $p_i(d)$ $(i = 1, ..., K)$ (irreducibility of its associated transition matrix), and for aperiodic controlled Markov chains it is a non-singular one [30]:

$$\sum_{l=1}^{N} c^{il} \geq \min_{d \in \mathbf{D}} p_i(d) := c_- > 0 \qquad (2.15)$$

2.4. ADAPTIVE LEARNING ALGORITHM

and hence, in this case we can define the elements d^{il} of the matrix d as follows

$$d^{il} = c^{il} \left(\sum_{l=1}^{N} c^{il}\right)^{-1} \quad (i = 1, ..., K; l = 1, ..., N). \quad (2.16)$$

As a consequence, the solution $c = \left[c^{il}\right]$ of the problem (2.13) would be unique.

The Lagrange multipliers approach [26] will be used to solve the optimization problem (2.13), (2.14) in which the values v_{il} and π_{ij}^l are not a priori known, and the available information at time n corresponds to x_n, u_n, η_n.

Let us introduce the vectors $\mathbf{c}$ and $\boldsymbol{\lambda}$

$$\mathbf{c}^T := \left(c^{11}, ..., c^{1N}; ...; c^{K1}, ..., c^{KN}\right), \quad \boldsymbol{\lambda}^T := (\lambda_1, ..., \lambda_K)$$

and the following regularized Lagrange function

$$L_\delta(\mathbf{c}, \boldsymbol{\lambda}) := \sum_{i=1}^{K} \sum_{l=1}^{N} v_{il} c^{il} - \sum_{j=1}^{K} \lambda_j \left[\sum_{l=1}^{N} c^{jl} - \sum_{i=1}^{K} \sum_{l=1}^{N} \pi_{ij}^l c^{il}\right] \quad (2.17)$$

$$+ \frac{\delta}{2} \left(\sum_{i=1}^{K} \sum_{l=1}^{N} \left(c^{il}\right)^2 - \sum_{j=1}^{K} \lambda_j^2\right), \quad \delta > 0$$

which is given on the set $S_0^{KN} \times R^K$, where the simplex S_0^{KN} is defined as follows:

$$S_\varepsilon^{KN} := \left\{\mathbf{c} \mid c^{il} \geq \varepsilon \geq 0, \ \sum_{i=1}^{K} \sum_{l=1}^{N} c^{il} = 1 \ (i = 1, ..., K; l = 1, ..., N)\right\}. \quad (2.18)$$

The saddle point of this regularized Lagrange function will be denoted by

$$\left(\mathbf{c}_\delta^*, \boldsymbol{\lambda}_\delta^*\right) := \arg \min_{\mathbf{c} \in S_0^{KN}} \max_{\boldsymbol{\lambda} \in R^K} L_\delta(\mathbf{c}, \boldsymbol{\lambda}). \quad (2.19)$$

Due to the strict convexity of the function $L_\delta(\mathbf{c}, \boldsymbol{\lambda})$ $(\delta > 0)$, this saddle point is unique and possesses the Lipshitz property with respect to parameter δ [26]:

$$\left\|\mathbf{c}_{\delta_1}^* - \mathbf{c}_{\delta_2}^*\right\| + \left\|\boldsymbol{\lambda}_{\delta_1}^* - \boldsymbol{\lambda}_{\delta_2}^*\right\| \leq Const \, |\delta_1 - \delta_2|. \quad (2.20)$$

It has been shown in [26] that if $\delta \to 0$ the saddle point $\left(\mathbf{c}_\delta^*, \boldsymbol{\lambda}_\delta^*\right)$ converges to the solution of the optimization problem (2.13) which has the minimal norm (in the aperiodic case this point is unique):

$$\mathbf{c}_\delta^* \underset{\delta \to 0}{\to} \mathbf{c}^{**} := \arg \min_{c^*, \lambda^*} \left(\|\mathbf{c}^*\|^2 + \|\boldsymbol{\lambda}^*\|^2\right) \quad (2.21)$$

(the minimization is done over all saddle points of the nonregularized Lagrange functions).

Based on the results which we have developed thus far, we are now in a position to present an algorithm for the adaptive control of unconstrained Markov chains.

To find the saddle point $\left(c_\delta^*, \lambda_\delta^*\right)$ (2.19) of the function $L_\delta(c, \lambda)$ (2.17) when the parameters $\left(v_{il}, \pi_{ij}^l\right)$ are unknown, we will use the stochastic approximation technique [17] which will permit us to define a recursive procedure

$$c_{n+1} = c_{n+1}\left(x_n, u_n, \eta_n, x_{n+1}, c_n\right)$$

generating the sequence $\{c_n\}$ which converges in some probability sense to the solution c^{**} of the initial problem. This procedure performs the following steps:

Step 1 (normalization procedure): use the available information

$$x_n = x(\alpha), u_n = u(\beta), \eta_n, x_{n+1} = x(\gamma), \eta_n, c_n\left(c_n^{il} > 0\right), \lambda_n$$

to construct the following function

$$\xi_n := \eta_n - \lambda_n^\alpha + \lambda_n^\gamma + \delta_n c_n^{\alpha\beta} \tag{2.22}$$

and normalize it using the following affine transformation

$$\zeta_n := \frac{a_n \xi_n + b_n}{c_n^{\alpha\beta}} \tag{2.23}$$

where the numerical sequences $\{a_n\}, \{b_n\}$ are given by

$$a_n := \left(2\frac{(\sigma + 2\lambda_n^+)}{\varepsilon_n} + \frac{N \cdot K}{N \cdot K - 1}\delta_n\right)^{-1}, \quad b_n := a_n\left(\sigma + 2\lambda_n^+\right). \tag{2.24}$$

The positive sequences $\{\varepsilon_n\}, \{\delta_n\}$ and $\{\lambda_n^+\}$ will be specified below.

Step 2 (learning procedure): calculate the elements c_{n+1}^{il} using the following recursive algorithm

$$c_{n+1}^{il} = \begin{cases} c_n^{\alpha\beta} + \gamma_n^c(1 - c_n^{\alpha\beta} - \zeta_n) & i = \alpha \wedge l = \beta \\ c_n^{il} - \gamma_n^c\left(c_n^{il} - \dfrac{\zeta_n}{N \cdot K - 1}\right) & i \neq \alpha \vee l \neq \beta \end{cases}, \tag{2.25}$$

$$\lambda_{n+1}^j = \left[\lambda_n^j + \gamma_n^\lambda \psi_n^j\right]_{-\lambda_{n+1}^+}^{\lambda_{n+1}^+}, \tag{2.26}$$

$$\psi_n^j = \chi\left(x(j) = x_{n+1}\right) - \sum_{l=1}^N c_n^{jl} - \delta_n \lambda_n^j, \tag{2.27}$$

2.4. ADAPTIVE LEARNING ALGORITHM

where the operators $[y]_{-\lambda_{n+1}^+}^{\lambda_{n+1}^+}$ and $\chi(x(j) = x_{n+1})$ are defined as follows:

$$[y]_{-\lambda_{n+1}^+}^{\lambda_{n+1}^+} = \begin{cases} y & \text{if } y \in \left[-\lambda_{n+1}^+, \lambda_{n+1}^+\right] \\ \lambda_{n+1}^+ & \text{if } y > \lambda_{n+1}^+ \\ -\lambda_{n+1}^+ & \text{if } y < -\lambda_{n+1}^+ \end{cases}$$

and

$$\chi(x(j) = x_{n+1}) = \begin{cases} 1 & \text{if } x(j) = x_{n+1} \\ 0 & \text{otherwise} \end{cases}.$$

The deterministic sequences $\{\gamma_n^c\}$ and $\{\gamma_n^\lambda\}$ will be specified below.

Step 3 (new action selection): construct the stochastic matrix

$$d_{n+1}^{il} = c_{n+1}^{il} \left(\sum_{k=1}^{N} c_{n+1}^{ik}\right)^{-1} \quad (i = 1, ..., K, l = 1, ..., N) \tag{2.28}$$

and according to

$$\Pr\{u_{n+1} = u(l) \mid x_{n+1} = x(\gamma) \wedge \mathcal{F}_n\} = d_{n+1}^{\gamma l}$$

generate randomly a new discrete random variable u_{n+1} as in learning stochastic automata implementation [26-27], and get the new observation (realization) η_{n+1} which corresponds to the transition to state x_{n+1}.

Step 4 return to Step 1.

The next lemma shows that the normalized function ζ_n belongs to the unit segment $(0, 1)$ and that

$$\mathbf{c}_{n+1} \in S_{\varepsilon_{n+1}}^{KN} \text{ if } \mathbf{c}_n \in S_{\varepsilon_n}^{KN}.$$

Lemma 2 *If*

1. *in the procedure (2.23)*

$$c_n^{\alpha\beta} \geq \varepsilon_n, \; \delta_n \downarrow 0 \tag{2.29}$$

then

$$\zeta_n \in \left[\zeta_n^-, \zeta_n^+\right] \subset [0, 1] \tag{2.30}$$

where

$$\zeta_n^- = a_n \delta_n, \; \zeta_n^+ = 1 - \frac{a_n \delta_n}{N \cdot K - 1}; \tag{2.31}$$

2. *in the recurrence (2.25)*

$$\zeta_n \in \left[\zeta_n^-, \zeta_n^+\right] \subset [0, 1], \; \gamma_n^c \in [0, 1], \; \mathbf{c}_n \in S_{\varepsilon_n}^{KN}$$

then

$$\mathbf{c}_{n+1} \in S_{\varepsilon_{n+1}}^{KN}. \tag{2.32}$$

Proof. To prove (2.30) it is enough to notice that

$$\zeta_n = \frac{a_n \left(\eta_n + \lambda_n^\alpha - \lambda_n^\gamma\right) + b_n}{c_n^{\alpha\beta}} + a_n \delta_n$$

$$\geq \frac{-a_n \left(\sigma + 2\lambda_n^+\right) + b_n}{c_n^{\alpha\beta}} + a_n \delta_n = a_n \delta_n = \zeta_n^- > 0$$

and

$$\zeta_n \leq \frac{a_n \left(\sigma + 2\lambda_n^+\right) + b_n}{\varepsilon_n} + a_n \delta_n$$

$$= a_n \left(\frac{2}{\varepsilon_n} \left(\sigma + 2\lambda_n^+\right) + \delta_n\right) = \zeta_n^+.$$

Notice that the procedure (2.25) corresponds to the Bush-Mosteller reinforcement scheme [26-27], and simple algebraic calculations demonstrate that (2.30) holds and $\mathbf{c}_{n+1} \in S_0^{KN}$. Indeed, from (2.25) it follows

$$c_{n+1}^{il} \geq (1 - \gamma_n^c) \, c_n^{il} + \gamma_n^c \min\left\{1 - \zeta_n^+; \frac{\zeta_n^-}{N \cdot K - 1}\right\}$$

$$\geq (1 - \gamma_n^c) \, c_n^{il} + \gamma_n^c \min\left\{1 - \zeta_n^+; \frac{\zeta_n^-}{N \cdot K - 1}\right\}.$$

To fulfill the following condition

$$\zeta_n^+ = 1 - \frac{\zeta_n^-}{N \cdot K - 1} < 1. \qquad (2.33)$$

a_n has to be selected as in (2.24).

Taking into account relation (2.33), the last inequality can be rewritten as follows

$$c_{n+1}^{il} \geq (1 - \gamma_n^c) \, c_n^{il} + \gamma_n^c \frac{\zeta_n^-}{N \cdot K - 1} \geq \min\left\{c_n^{il}, \frac{\zeta_n^-}{N \cdot K - 1}\right\}$$

$$\geq \min\left\{\min\left\{c_{n-1}^{il}, \frac{\zeta_{n-1}^-}{N \cdot K - 1}\right\}, \frac{\zeta_n^-}{N \cdot K - 1}\right\}.$$

If ζ_n^- is monotonically decreasing $\zeta_n^- \downarrow 0$, from the last expression we obtain

$$c_{n+1}^{il} \geq \min\left\{\min\left\{c_{n-1}^{il}, \frac{\zeta_{n-1}^-}{N \cdot K - 1}\right\}, \frac{\zeta_n^-}{N \cdot K - 1}\right\}$$

$$= \min\left\{c_{n-1}^{il}, \frac{\zeta_n^-}{N \cdot K - 1}\right\}$$

$$\geq \cdots \geq \min\left\{c_1^{il}, \frac{\zeta_n^-}{N \cdot K - 1}\right\} = \frac{\zeta_n^-}{N \cdot K - 1} \geq \frac{\zeta_{n+1}^-}{N \cdot K - 1} \equiv \varepsilon_{n+1}.$$

From (2.25) it follows

$$\sum_{i=1}^K \sum_{l=1}^N c_{n+1}^{il} = \sum_{i=1}^K \sum_{l=1}^N c_n^{il} = 1.$$

To conclude, let us notice that in view of (2.24) ζ_n^- is monotically decreasing $\zeta_n^- \downarrow 0$. So, (2.32) is fulfilled. ■

A perquisite meaningful discussion of the usefulness of an iterative procedure is its convergence properties. The main asymptotic properties of this adaptive learning control algorithm are stated in the next section.

2.5 Convergence analysis

In this section we establish the convergence properties of the adaptive learning control algorithm described in the previous section.

To derive this adaptive learning control procedure, we have considered stationarity randomized strategies. We obtained a policy sequence $\{d_n\}$ which according to (2.25), (2.26) (2.27) and (2.28) is essentially nonstationary. As a consequence, we need to prove the convergence of this sequence to the solution $\mathbf{c}^{**}$ (2.21) of the initial optimization problem.

Let us introduce the following Lyapunov function

$$W_n := \left\|\mathbf{c}_n - \mathbf{c}_{\delta_n}^*\right\|^2 + \left\|\boldsymbol{\lambda}_n - \boldsymbol{\lambda}_{\delta_n}^*\right\|^2 \qquad (2.34)$$

starting from $n \geq \inf_{t \geq 1}\left\{t : \left\|\boldsymbol{\lambda}_{\delta_n}^*\right\| \leq \lambda_n^+\right\}$. This Lyapunov function maps the variables $\mathbf{c}_n$ and $\boldsymbol{\lambda}_n$ to real numbers decreasing with time.

The following theorem states the conditions related to the convergence of this adaptive learning control algorithm.

Theorem 1 *Let the controlled Markov chain be ergodic with any fixed distribution of the initial states. Let the loss function Φ_n be given by (2.4). If the control policy generated by the adaptive learning procedure (2.25-2.28) with design parameters $\varepsilon_n, \delta_n, \lambda_n^+, \gamma_n^c$ and γ_n^λ satisfying the following conditions*

$$0 < \delta_n \downarrow 0, \ 0 < \lambda_n^+ \uparrow \infty, \ \gamma_n^c \in (0,1),$$

$$\gamma_n^\lambda = \frac{N \cdot K}{N \cdot K - 1} \gamma_n^c a_n, \quad \sum_{n=1}^{\infty} \gamma_n^c \varepsilon_n \delta_n \left(\lambda_n^+\right)^{-1} = \infty$$

is implemented, then

1. *if*

$$\sum_{n=1}^{\infty} \mu_n < \infty$$

where

$$\mu_n := (\delta_n - \delta_{n+1})^2 \lambda_n^+ (\gamma_n^c \varepsilon_n \delta_n)^{-1} + (\gamma_n^c)^2 + |\delta_n - \delta_{n+1}| \gamma_n^c,$$

then, the control policy (2.25-2.28) converges with probability 1 to the optimal solution, i.e.,

$$W_n \xrightarrow[n \to \infty]{a.s.} 0;$$

2. *if*

$$\frac{\mu_n \lambda_n^+}{\gamma_n^c \varepsilon_n \delta_n} \xrightarrow[n \to \infty]{} 0$$

then, we obtain the convergence in the mean squares sense, i.e.,

$$E\{W_n\} \xrightarrow[n \to \infty]{a.s.} 0.$$

Proof. For $\delta > 0$, the regularized Lagrange function (2.17), is strictly convex. It follows

$$\left(\mathbf{c} - \mathbf{c}_\delta^*\right)^T \nabla_\mathbf{c} L_\delta(\mathbf{c}, \boldsymbol{\lambda}) - \left(\boldsymbol{\lambda} - \boldsymbol{\lambda}_\delta^*\right)^T \nabla_{\boldsymbol{\lambda}} L_\delta(\mathbf{c}, \boldsymbol{\lambda}) \quad (2.35)$$

$$\geq \frac{\delta}{2} \left(\left\|\mathbf{c} - \mathbf{c}_\delta^*\right\|^2 + \left\|\boldsymbol{\lambda} - \boldsymbol{\lambda}_\delta^*\right\|^2 \right)$$

where $\left(\mathbf{c}_\delta^*, \boldsymbol{\lambda}_\delta^*\right)$ is the saddle point of the regularized Lagrange function (2.17), and $\mathbf{c}$, and $\boldsymbol{\lambda}$ are any vectors from the corresponding finite dimensional spaces.

Recall that

$$E\{Z \mid x_n = x(\alpha) \wedge \mathcal{F}_{n-1}\} = E\left\{Z \frac{\chi(x_n = x(\alpha))}{p(\alpha)} \mid \mathcal{F}_{n-1}\right\}.$$

Then, from (2.22) and (2.27) it follows that the gradients (with respect to $c^{\alpha\beta}$ and λ^γ) of the regularized Lagrange function (2.17) can be expressed

2.5. CONVERGENCE ANALYSIS

as a functions of the conditional mathematical expectation of respectively ζ_n and $\boldsymbol{\psi}_n := \left(\psi_n^1, ..., \psi_n^K\right)^T$:

$$E\{\zeta_n \mathbf{e}(x_n \wedge u_n) \mid \mathcal{F}_{n-1}\} \stackrel{a.s.}{=} a_n \frac{\partial}{\partial \mathbf{c}} L_\delta(\mathbf{c}_n, \boldsymbol{\lambda}_n) + b_n \mathbf{e}^{N \cdot K}, \quad (2.36)$$

$$E\{\boldsymbol{\psi}_n \mid \mathcal{F}_{n-1}\} \stackrel{a.s.}{=} \frac{\partial}{\partial \boldsymbol{\lambda}} L_\delta(\mathbf{c}_n, \boldsymbol{\lambda}_n) \quad (2.37)$$

where $\mathbf{e}(x_n \wedge u_n)$ is a vector defined as follows

$$\mathbf{e}(x_n \wedge u_n) := \begin{bmatrix} \chi(x_n = x(1), u_n = u(1)) \\ \ldots \\ \chi(x_n = x(K), u_n = u(1)) \\ \ldots \\ \chi(x_n = x(1), u_n = u(l)) \\ \ldots \\ \chi(x_n = x(K), u_n = u(l)) \\ \ldots \\ \chi(x_n = x(K), u_n = u(N)) \end{bmatrix} \in R^{N \cdot K}. \quad (2.38)$$

Rewriting (2.25), (2.26) and (2.27) in a vector form we obtain

$$\mathbf{c}_{n+1} = \mathbf{c}_n + \gamma_n^c \left(\mathbf{e}(x_n \wedge u_n) - \mathbf{c}_n + \zeta_n \frac{\mathbf{e}^{N \cdot K} - N \cdot K \mathbf{e}(x_n \wedge u_n)}{N \cdot K - 1} \right), \quad (2.39)$$

$$\boldsymbol{\lambda}_{n+1} = \left[\boldsymbol{\lambda}_n + \gamma_n^\lambda \boldsymbol{\psi}_n \right]_{-\lambda_{n+1}^+}^{\lambda_{n+1}^+}, \quad (2.40)$$

$$\boldsymbol{\psi}_n = \mathbf{e}(x_{n+1}) - \left(\sum_{l=1}^N c_n^{\gamma l} + \delta_n \lambda_n^\gamma \right) \mathbf{e}^K. \quad (2.41)$$

where $\mathbf{e}^M$ is a vector defined by

$$\mathbf{e}^M := \big[\underbrace{1, ..., 1}_{M} \big]^T .$$

Substituting (2.39), (2.40) and (2.41), into W_{n+1} (2.34) we derive

$$W_{n+1} \leq \left\| \mathbf{c}_n + \gamma_n^c \left(\mathbf{e}(x_n \wedge u_n) - \mathbf{c}_n + \zeta_n \frac{\mathbf{e}^{N \cdot K} - N \cdot K \mathbf{e}(x_n \wedge u_n)}{N \cdot K - 1} \right) \right.$$

$$\left. - \mathbf{c}_{\delta_{n+1}}^* \right\|^2 + \left\| \boldsymbol{\lambda}_n + \gamma_n^\lambda \boldsymbol{\psi}_n - \boldsymbol{\lambda}_{\delta_{n+1}}^* \right\|^2 = \left\| \left(\mathbf{c}_n - \mathbf{c}_{\delta_n}^* \right) \right.$$

$$+\gamma_n^c \left(\mathbf{e}\left(x_n \wedge u_n\right) - \mathbf{c}_n + \zeta_n \frac{e^{N \cdot K} - N \cdot K e\left(x_n \wedge u_n\right)}{N \cdot K - 1} \right) + \left(\mathbf{c}_{\delta_n}^* - \mathbf{c}_{\delta_{n+1}}^*\right) \Big\|^2$$

$$+ \left\| \left(\boldsymbol{\lambda}_n - \boldsymbol{\lambda}_{\delta_n}^*\right) + \gamma_n^\lambda \psi_n + \left(\boldsymbol{\lambda}_{\delta_n}^* - \boldsymbol{\lambda}_{\delta_{n+1}}^*\right) \right\|^2.$$

Calculating the square of the norms appearing in this inequality, and estimating the resulting terms using the inequality

$$\|\psi_n\| \leq Const \, \gamma_n^\lambda \left(1 + \delta_n \lambda_n^+\right),$$

we obtain

$$W_{n+1} \leq W_n + (\gamma_n^c)^2 \, Const + \left\|\mathbf{c}_{\delta_n}^* - \mathbf{c}_{\delta_{n+1}}^*\right\|^2$$

$$+ 2 \left\|\mathbf{c}_{\delta_n}^* - \mathbf{c}_{\delta_{n+1}}^*\right\| \sqrt{W_n} + 2 \left\|\mathbf{c}_{\delta_n}^* - \mathbf{c}_{\delta_{n+1}}^*\right\| \gamma_n^c Const$$

$$+ \left(\gamma_n^\lambda\right)^2 \left(1 + \delta_n \lambda_n^+\right)^2 Const + \left\|\boldsymbol{\lambda}_{\delta_n}^* - \boldsymbol{\lambda}_{\delta_{n+1}}^*\right\|^2$$

$$+ \gamma_n^\lambda \left(1 + \delta_n \lambda_n^+\right) \left\|\boldsymbol{\lambda}_{\delta_n}^* - \boldsymbol{\lambda}_{\delta_{n+1}}^*\right\| Const + 2 \left\|\boldsymbol{\lambda}_{\delta_n}^* - \boldsymbol{\lambda}_{\delta_{n+1}}^*\right\| \sqrt{W_n}$$

$$+ 2\gamma_n^c \left(\mathbf{c}_n - \mathbf{c}_{\delta_n}^*\right)^T \left(\mathbf{e}\left(x_n \wedge u_n\right) - \mathbf{c}_n + \zeta_n \frac{e^{N \cdot K} - N \cdot K e\left(x_n \wedge u_n\right)}{N \cdot K - 1} \right)$$

$$+ 2\gamma_n^\lambda \left(\boldsymbol{\lambda}_n - \boldsymbol{\lambda}_{\delta_n}^*\right)^T \psi_n$$

where $Const$ is a positive constant.

Combining the terms of the right hand side of this inequality and in view of (2.20), it follows

$$W_{n+1} \leq W_n + Const \cdot \mu_{1,n} \sqrt{W_n} + Const \cdot \mu_{2,n} + w_n \qquad (2.42)$$

where

$$\mu_{1,n} := |\delta_n - \delta_{n+1}|,$$

$$\mu_{2,n} := (\gamma_n^c)^2 + (\delta_n - \delta_{n+1})^2 +$$

$$+ \left(\gamma_n^\lambda\right)^2 \left(1 + \delta_n \lambda_n^+\right)^2 + |\delta_n - \delta_{n+1}| \left(\gamma_n^c + \gamma_n^\lambda \left(1 + \delta_n \lambda_n^+\right)\right),$$

$$w_n := 2\gamma_n^c \left(\mathbf{c}_n - \mathbf{c}_{\delta_n}^*\right)^T \left(\mathbf{e}\left(x_n \wedge u_n\right) - \mathbf{c}_n + \zeta_n \frac{e^{N \cdot K} - N \cdot K e\left(x_n \wedge u_n\right)}{N \cdot K - 1} \right)$$

$$+ 2\gamma_n^\lambda \left(\boldsymbol{\lambda}_n - \boldsymbol{\lambda}_{\delta_n}^*\right)^T \psi_n.$$

Notice that ζ_n (2.23) is a linear function of ξ_n and

$$\left(\mathbf{c}_n - \mathbf{c}_{\delta_n}^*\right)^T \frac{e^{N \cdot K}}{N \cdot K - 1} = 0.$$

2.5. CONVERGENCE ANALYSIS

If (2.36) and (2.37) are used, the following is obtained:

$$E\{w_n \mid \mathcal{F}_{n-1}\} \stackrel{a.s.}{=} -2\frac{N \cdot K}{N \cdot K - 1}\gamma_n^c \left(\mathbf{c}_n - \mathbf{c}_{\delta_n}^*\right)^T E\{\zeta_n \mathbf{e}\left(x_n \wedge u_n\right) \mid \mathcal{F}_{n-1}\}$$

$$+2\gamma_n^\lambda \left(\boldsymbol{\lambda}_n - \boldsymbol{\lambda}_{\delta_n}^*\right)^T E\{\psi_n \mid \mathcal{F}_{n-1}\}$$

$$\stackrel{a.s.}{=} -2\frac{N \cdot K}{N \cdot K - 1}\gamma_n^c a_n \left(\mathbf{c}_n - \mathbf{c}_{\delta_n}^*\right)^T \frac{\partial}{\partial \mathbf{c}} L_\delta \left(\mathbf{c}_n, \boldsymbol{\lambda}_n\right)$$

$$+2\gamma_n^\lambda \left(\boldsymbol{\lambda}_n - \boldsymbol{\lambda}_{\delta_n}^*\right)^T \frac{\partial}{\partial \boldsymbol{\lambda}} L_\delta \left(\mathbf{c}_n, \boldsymbol{\lambda}_n\right).$$

Taking into account the assumptions of this theorem, and the strict convex property (2.35) we deduce

$$E\{w_n \mid \mathcal{F}_{n-1}\} \stackrel{a.s.}{=} -2\frac{N \cdot K}{N \cdot K - 1}\gamma_n^c a_n \left[\left(\mathbf{c}_n - \mathbf{c}_{\delta_n}^*\right)^T \frac{\partial}{\partial \mathbf{c}} L_\delta \left(\mathbf{c}_n, \boldsymbol{\lambda}_n\right)\right.$$

$$\left.- \left(\boldsymbol{\lambda}_n - \boldsymbol{\lambda}_{\delta_n}^*\right)^T \frac{\partial}{\partial \boldsymbol{\lambda}} L_\delta \left(\mathbf{c}_n, \boldsymbol{\lambda}_n\right)\right] \leq -\frac{N \cdot K}{N \cdot K - 1}\gamma_n^c a_n \delta_n W_n.$$

Calculating the conditional mathematical expectation of both sides of (2.42), and in view of the last inequality, we can get

$$E\{W_{n+1} \mid \mathcal{F}_{n-1}\} \stackrel{a.s.}{\leq} \left(1 - \frac{N \cdot K}{N \cdot K - 1}\gamma_n^c a_n \delta_n\right) W_n + Const \left(\mu_{1,n}\sqrt{W_n} + \mu_{2,n}\right). \quad (2.43)$$

In view of

$$2\mu_{1,n}\sqrt{W_n} \leq \mu_{1,n}^2 \rho_n^{-1} + W_n \rho_n$$

(which is valid for any $\rho_n > 0$) for

$$\rho_n := \gamma_n^c a_n \delta_n,$$

it follows

$$2\mu_{1,n}\sqrt{W_n} \leq \mu_{1,n}^2 \left(\gamma_n^c a_n \delta_n\right)^{-1} + W_n \gamma_n^c a_n \delta_n.$$

From this inequality and (2.43), and in view of the following estimation

$$\mu_{1,n}^2 \left(\gamma_n^c a_n \delta_n\right)^{-1} + \mu_{2,n} \leq Const \cdot \mu_n$$

we finally, obtain

$$E\{W_{n+1} \mid \mathcal{F}_{n-1}\} \stackrel{a.s.}{\leq} \left(1 - \frac{1}{N \cdot K - 1}\gamma_n^c a_n \delta_n\right) W_n + Const \cdot \mu_n. \quad (2.44)$$

Observe that
$$a_n = O\left(\frac{\varepsilon_n}{\lambda_n^+}\right).$$

It follows from (2.44) that $\{W_n, \mathcal{F}_n\}$ is a nonnegative quasimartingale.

From the assumptions of this theorem and in view of the Robbins-Siegmund theorem for quasimartingales (see theorem 2 of Appendix A for $x_n = W_n$, etc.) [26, 31], the convergence with probability one follows.

The mean squares convergence follows from (2.44) after applying the operator of conditional mathematical expectation to both sides of this inequality and using lemma A5 given in [26]. ∎

Theorem 1 shows that this adaptive learning control algorithm possesses all the properties that one would desire, i.e., convergence with probability one as well as convergence in the mean squares.

The next corollary states the conditions associated with the sequences $\{\varepsilon_n\}$, $\{\delta_n\}$, $\{\lambda_n^+\}$ and $\{\gamma_n^c\}$.

Corollary 1 *If in theorem 1*

$$\varepsilon_n := \frac{\varepsilon_0}{1 + n^\varepsilon \ln n} \ \left(\varepsilon_0 \in \left[0, (N \cdot K)^{-1}\right), \varepsilon \geq 0\right), \ \delta_n := \frac{\delta_0}{n^\delta} \ (\delta_0, \delta > 0,),$$

$$\lambda_n^+ := \lambda_0^+ \left(1 + n^\lambda \ln n\right) \ \left(\lambda_0^+ > 0, \ \lambda \geq 0\right), \ \gamma_n^c := \frac{\gamma_0}{n^\gamma} \ (\gamma_0 \in (0,1), \ \gamma \geq 0),$$

with

$$\gamma + \varepsilon + \lambda + \delta \leq 1$$

then

1. *the convergence with probability one will take place if*

$$\theta_{a.s.} := \min\{2 - \gamma - \varepsilon - \lambda + \delta; \ 2\gamma\} > 1,$$

2. *the mean squares convergence is guaranteed if*

$$\theta_{m.s.} := (1 - \gamma - \varepsilon - \lambda) > 0.$$

It is easy to check up on these conditions by substituting the parameters given in this corollary in theorem 1 assumptions.

Remark 1 *In the optimization problem related to the regularized Lagrange function $L_\delta(\mathbf{c}, \boldsymbol{\lambda})$ the parameter δ_n must decrease less slowly than any other parameter including ε_n, i.e.,*

$$\delta \leq \varepsilon.$$

2.5. CONVERGENCE ANALYSIS

However, not only is the convergence of a learning scheme important but the convergence speed is also essential. It depends on the number of operations performed by the algorithm during an iteration as well as the number of iterations needed for convergence. The next theorem states the convergence rate of the adaptive learning algorithm described above.

Theorem 2 *Under the conditions of theorem 1 and corollary 1, it follows*

$$W_n \stackrel{a.s.}{=} o\left(\frac{1}{n^\nu}\right)$$

where

$$0 < \nu < \min\{1 - \gamma - \varepsilon - \lambda + \delta;\ 2\gamma - 1; 2\delta\} := \nu^*(\gamma, \varepsilon, \delta).$$

and the positive parameters γ, δ, ε and λ satisfy the following constraints

$$\lambda + \gamma + \varepsilon + \delta \leq 1,\ 1 - \gamma - \varepsilon - \lambda + \delta > 0,\ 2\gamma > 1.$$

Proof. From (2.20), it follows:

$$W_n^* := \|c_n - c^{**}\|^2 + \|\lambda_n - \lambda^{**}\|^2 = \|(c_n - c_n^*) + (c_n^* - c^{**})\|^2 +$$
$$+ \|(\lambda_n - \lambda_n^*) + (\lambda_n^* - \lambda^{**})\|^2 \leq 2\|c_n - c_n^*\|^2 + 2\|c_n^* - c^{**}\|^2 +$$
$$+ 2\|\lambda_n - \lambda_n^*\|^2 + 2\|\lambda_n^* - \lambda^{**}\|^2 \leq 2W_n + C\delta_n^2.$$

Multiplying both sides of the previous inequality by ν_n, we derive

$$\nu_n W_n^* \leq 2\nu_n W_n + \nu_n C \delta_n^2.$$

Selecting $\nu_n = n^\nu$ and in view of lemma 2 [32] and taking into account that

$$\frac{\nu_{n+1} - \nu_n}{\nu_n} = \frac{\nu + o(1)}{n}$$

we obtain

$$0 < \nu < \min\{1 - \gamma - \varepsilon - \lambda + \delta;\ 2\gamma - 1; 2\delta\} := \nu^*(\gamma, \varepsilon, \delta)$$

where the positive parameters γ, δ, ε and λ satisfy the following constraints

$$\lambda + \gamma + \varepsilon + \delta \leq 1,\ 2 - \gamma - \varepsilon - \lambda + \delta > 1,\ 2\gamma > 1.$$

∎

The optimal convergence rate is given by the next corollary.

Corollary 2 *The maximum convergence rate is achieved with the optimal parameters* $\varepsilon^*, \delta^*, \lambda^*, \gamma^*$

$$\varepsilon = \varepsilon^* = \delta = \delta^* = \frac{1}{6}, \ \lambda = \lambda^* = 0, \gamma = \gamma^* = \frac{2}{3}$$

and is equal to

$$\nu^*(\gamma, \varepsilon, \delta) = 2\gamma^* - 1 = \nu^{**} = \frac{1}{3}.$$

Proof. The solution of the linear programming problem

$$\nu^*(\gamma, \varepsilon, \delta) \to \max_{\gamma, \varepsilon, \delta}$$

is given by

$$2\gamma - 1 = 1 + \delta - \varepsilon - \lambda - \gamma = 2\delta$$

or, in equivalent form,

$$\gamma = \frac{2}{3} - \frac{1}{3}(\lambda + \varepsilon - \delta) = \frac{1}{2} + \delta = 1 - \delta - \varepsilon - \lambda.$$

Taking into account that δ_n must decrease less slowly than ε_n (see Remark), we derive

$$\delta \leq \varepsilon$$

and, as a result, the smallest ε maximizing γ is equal to

$$\varepsilon = \delta.$$

Hence

$$\gamma = \frac{2}{3} - \frac{1}{3}\lambda = \frac{1}{2} + \delta = 1 - 2\delta - \lambda.$$

From these relations, we derive

$$\lambda = \frac{1}{2} - 3\delta.$$

Taking into account that $\lambda \geq 0$, we get

$$\delta \leq \frac{1}{6}$$

and, consequently

$$\gamma = \frac{1}{2} + \delta \leq \frac{2}{3}.$$

The optimal parameters are

$$\gamma = \gamma^* = \frac{2}{3}, \ \varepsilon = \varepsilon^* = \delta = \delta^* = \frac{1}{6}, \ \lambda = \lambda^* = 0.$$

The maximum convergence rate is achieved with this choice of parameters and is equal to

$$\nu^*(\gamma, \varepsilon, \delta) = 2\gamma^* - 1 = \nu^{**} = \frac{1}{3}.$$

∎

Before concluding this chapter, let us mention that a set of numerical results are presented in chapter 8, in order to illustrate the performance and the effectiveness of the adaptive control algorithm presented above. The trials presented in chapter 8 show that the algorithm given here as well as the other algorithms based on the Lagrange and the penalty approaches, are computationally efficient and require relatively little storage (memory).

2.6 Conclusions

An adaptive learning control algorithm for finite controlled Markov chains whose transition probabilities are unknown has been presented. The control strategy is designed to achieve the asymptotic minimization of a loss function. To construct this adaptive learning control, a regularized Lagrange function was introduced. The control policy is adjusted using the Bush-Mosteller reinforcement scheme as a stochastic approximation procedure. The convergence properties (convergence with probability one as well as convergence in the mean squares) have been stated. For some class of parameters, we establish that the optimal convergence (learning) rate is equal to $n^{-\frac{1}{3}+\delta}$ (δ is any small positive parameter), and is higher than the convergence rate $\left(n^{-\frac{1}{5}+\delta}\right)$ associated with the stochastic gradient algorithm [30] in which the transition probabilities are estimated and a projection procedure is used.

The algorithm presented in this chapter and its analysis provide the starting-point for a class of adaptive learning controllers for other classes of controlled Markov chains, for example with additional constraints, etc.

2.7 References

1. R. A. Howard, *Dynamic Programming and Markov Processes*, J. Wiley, New York, 1962.

2. S.M. Ross, *Applied Probability Models with Optimization Applications*, Holden-Day, San Francisco, 1970.

3. M. F. Norman, *Markov Processes and Learning Models*, Academic Press, New York, 1976.

4. O. H. Lerma, *Adaptive Markov Control Processes*, Springer-Verlag, London, 1989.

5. O. Hernandez-Lerma and J. B. Lasserre, *Discrete-Time Adaptive Markov Control Processes Basic Optimality Criteria*, Springer-Verlag, Berlin 1996.

6. A. Arapostathis, V. S. Borkar, E. Fernandez-Gaucherand, M. K. Ghosh and S. I. Marcus, Discrete-time controlled Markov processes with average cost criterion: a survey, SIAM Journal of Control and Optimization, vol. 31, pp. 282-344, 1993.

7. D. P. Bertsekas, *Dynamic Programming and Stochastic Control*, Academic Press, New York 1976.

8. J. G. Kemeny and Snell, J. L. *Finite Markov Chains*, Springer-Verlag, Berlin,1976.

9. J. S. Riordon, An adaptive automaton controller for discrete-time Markov processes, *Automatica*, vol. 5, pp. 721-730, 1969.

10. P. Mandl, Estimation and control in Markov chains, *Adv. Appl. Prob.* vol. 6, pp. 40-60, 1974.

11. P. Varaiya, Optimal and suboptimal stationary controls for Markov chains, *IEEE Trans. on Automatic Control*, vol. 23, pp. 388-394, 1978.

12. P. R. Kumar and Woei Lin, Optimal adaptive controllers for unknown Markov chains, *IEEE Trans. on Automatic Control*, vol. 27, pp. 765-774, 1982.

13. B. Doshi and S. E. Shreve, Strong consistency of a modified maximum likelihood estimator for controlled Markov chains, *J. Appl. Prob.*, vol. 17, pp. 726-734, 1980.

14. Y. M. El-Fattah, Gradient approach for recursive estimation and control in finite Markov chains, *Advances in Applied Probability*, vol. 13, pp. 778-803, 1981.

2.7. REFERENCES

15. R. Agrawal, Teneketzis, D. and V. Anantharam, Asymptotically efficient adaptive allocation schemes for controlled Markov chains: finite parameter space, *IEEE Trans. on Automatic Control*, vol. 34, pp. 1249-1259, 1989.

16. R. Agrawal, Minimizing the learning loss in adaptive control of Markov chains under the weak accessibility condition, *J. Appl. Prob.*, vol. 28, pp. 779-790, 1991.

17. M. Duflo, *Random Iterative Models*, Springer-Verlag, Berlin, 1997.

18. A. Benveniste, M. Metivier and P. Priouret, *Stochastic Approximations and Adaptive Algorithms*, Springer-Verlag, Berlin, 1990.

19. H. Kushner and G. G. Yin, *Stochastic Approximation Algorithms*, Springer-Verlag, Berlin, 1997.

20. M. Schäl, Estimation and control in discounted dynamic programming, *Stochastics*, vol 20, pp. 51-71, 1987.

21. M. Sato, and H. Takeda, Learning control of finite Markov chains with unknown transition probabilities, *IEEE Trans. on Automatic Control*, vol. 27, pp. 502-505, 1982.

22. K. Najim, and M. M'Saad, Adaptive control: theory and practical aspects, *Journal of Process Control*, vol. 1, pp. 84-95, 1991.

23. P. R. Kumar, and P. Varaiya, *Stochastic Systems: Estimation, Identification and Adaptive Control*, Prentice-Hall, Englewood Cliffs, 1986.

24. R. Agrawal, and D. Teneketzis, Certainty equivalence control with forcing: revisited, *Syst. Contr. Lett.*, vol. 13, pp. 405-412, 1989.

25. R. Agrawal, Teneketzis D., and V. Anantharam, Asymptotically efficient adaptive allocation schemes for controlled Markov chains: finite parameters space, *IEEE Trans. on Automatic Control*, vol. 34, pp. 1249-1259, 1989.

26. A. S. Poznyak, and K. Najim, *Learning Automata and Stochastic Optimization*, Springer-Verlag, London, 1997.

27. K. Najim, and A. S. Poznyak, *Learning Automata Theory and Applications*, Pergamon Press, London, 1994.

28. R. R. Bush and F. Mosteller, *Stochastic Models for Learning*, J. Wiley, New York, 1955.

29. D. Hall and C. Heyde, *Martingales Limit Theory and its Applications*, Academic Press, New York, 1980.

30. A. V. Nazin and A. S. Poznyak, *Adaptive Choice of Variants*, (in Russian) Nauka, Moscow, 1986.

31. H. Robbins and D. Siegmund, A convergence theorem for nonnegative almost supermartingales and some applications, in *Optimizing Methods in Statistics*, ed. by J. S. Rustagi, Academic Press, New York, pp. 233-257, 1971.

32. A. S. Poznyak and K. Najim, Learning automata with continuous input and changing number of actions, *Int. J. of Systems Science*, vol. 27, pp. 1467-1472, 1996.

Chapter 3

Penalty Function Approach

3.1 Introduction

Markov chains have been widely used to characterize uncertainty in many real-world problems. This chapter presents an adaptive algorithm for the control of unconstrained finite Markov chain. This self-learning control algorithm is based on the penalty function approach [1]. In this control algorithm the transition probabilities of the Markov chain are estimated. The control policy is designed to achieve the minimization of a loss function. The average values of the conditional mathematical expectations of this loss function are assumed to be unknown. Based on the observations of this function, the control policy is adjusted using a stochastic approximation procedure [2-4]. In other words, the adaptive control algorithm is a recursive or on-line method, and can be defined as an algorithm which forms a new estimate (control), incorporating new information (realization, data, etc.), from old estimates using a fixed amount of computations and storage.

3.2 Adaptive learning algorithm

Stochastic approximation techniques are mostly motivated by problems arising in situations when decisions must be made on the basis of, existing or assumed, a priori information about the random on quantities (available data, observations). In this section we shall state an adaptive control algorithm for controlled Markov chains. Consider the following nonlinear programming problem

$$V(d) \to \min_{d \in \mathbf{D}} \qquad (3.1)$$

where the function $V(d)$ and the set $\mathbf{D}$ are defined by (2.5) and (2.7) (see chapter 2).

This nonlinear programming problem can be reduced to the following linear programming problem

$$\widetilde{V}(c) := \sum_{i=1}^{K}\sum_{l=1}^{N} v_{il} c^{il} \to \min_{c \in \mathbf{C}} \qquad (3.2)$$

where the set $\mathbf{C}$ is given by

$$\mathbf{C} = \left\{ c \mid c = \left[c^{il}\right],\ c^{il} \geq 0,\ \sum_{i=1}^{K}\sum_{l=1}^{N} c^{il} = 1, \right. \qquad (3.3)$$

$$\left. \sum_{l=1}^{N} c^{jl} = \sum_{i=1}^{K}\sum_{l=1}^{N} \pi_{ij}^{l} c^{il}\ (i,j=1,...,K;l=1,...,N) \right\}$$

via the standard transformation [5]

$$c^{il} = d^{il} p_i(d) \qquad (3.4)$$

In view of (3.4) we conclude that

$$\sum_{l=1}^{N} c^{il} = p_i(d)\ (i=1,...,K).$$

Notice that for ergodic controlled Markov chains there exists a unique final (stationary) distribution $p_i(d)$ $(i = 1,...,K)$ (irreducibility of its associated transition matrix), and for aperiodic controlled Markov chains it is a nonsingular one:

$$\sum_{l=1}^{N} c^{il} \geq \min_{d \in \mathbf{D}} p_i(d) := c_- > 0$$

and hence, in this case we can define the elements d^{il} of the matrix d as follows

$$d^{il} = c^{il} \left(\sum_{l=1}^{N} c^{il} \right)^{-1}\ (i=1,...,K; l=1,...,N). \qquad (3.5)$$

As a consequence, the solution $c = \left[c^{il}\right]$ of the problem (3.2) would be unique.

The penalty function approach [1] will be used to solve the optimization problem (3.2), (3.3) in which the values v_{il} and π_{ij}^{l} are not a priori known, and the available information at time n corresponds to x_n, u_n, η_n.

3.2. ADAPTIVE LEARNING ALGORITHM

Consider the vector **c**:

$$\mathbf{c}^T := \left(c^{11}, ..., c^{1N}; ...; c^{K1}, ..., c^{KN}\right) \in R^{N \cdot K}$$

and the regularized penalty function defined by

$$\mathcal{P}_{\mu,\delta}(\mathbf{c}) := \mu \sum_{i=1}^{K} \sum_{l=1}^{N} v_{il} c^{il} + \frac{1}{2} \sum_{j=1}^{K} \left(\sum_{l=1}^{N} c^{jl} - \sum_{i=1}^{K} \sum_{l=1}^{N} \pi_{ij}^{l} c^{il} \right)^2 \quad (3.6)$$

$$+ \frac{\delta}{2} \sum_{i=1}^{K} \sum_{l=1}^{N} \left(c^{il}\right)^2, \quad 0 < \mu, \delta \downarrow 0$$

which is given on the simplex S_0^{KN}, defined as follows:

$$S_{\varepsilon}^{KN} := \left\{ \mathbf{c} \mid c^{il} \geq \varepsilon \geq 0, \; \sum_{i=1}^{K} \sum_{l=1}^{N} c^{il} = 1 \; (i=1,...,K; l=1,...,N) \right\}. \quad (3.7)$$

Here μ is the "penalty coefficient" and δ is the "regularizing parameter."
For fixed positive μ and δ, the argument **c** minimizing this regularized penalty function, will be unique and denoted by

$$\mathbf{c}^*_{\mu,\delta} := \arg \min_{\mathbf{c} \in S_0^{KN}} \mathcal{P}_{\mu,\delta}(\mathbf{c}). \quad (3.8)$$

Due to the strict convexity of this penalty function $\mathcal{P}_{\mu,\delta}(\mathbf{c})$ under any $\delta > 0$, this minimum point possesses the Lipshitz property with respect to the parameters δ and μ [1]:

$$\left\| \mathbf{c}^*_{\mu_1,\delta_1} - \mathbf{c}^*_{\mu_2,\delta_2} \right\| \leq C_1 |\mu_1 - \mu_2| + C_2 |\delta_1 - \delta_2|, \; C_1, C_2 \in (0, \infty). \quad (3.9)$$

The next lemma shows that under some conditions related to $\mu = \mu_n$ and $\delta = \delta_n$, the minimum point $\mathbf{c}^*_{\mu,\delta}$ converges to the solution of the optimization problem (3.2) which has the minimal weighted norm (in the regular case this point is unique).

Lemma 1 *If the parameters μ and δ are time-varying, i.e.,*

$$\mu = \mu_n, \; \delta = \delta_n \; (n = 1, 2, ...)$$

such that

$$0 < \mu \downarrow 0, \; 0 < \delta \downarrow 0 \quad (3.10)$$

$$\frac{\delta_n}{\mu_n} \xrightarrow[n \to \infty]{} 0, \; \frac{(\mu_n)^{3/2}}{\delta_n} \xrightarrow[n \to \infty]{} 0 \quad (3.11)$$

then

$$\mathbf{c}^*_{\mu,\delta} \xrightarrow[\mu,\delta \to 0]{} \mathbf{c}^{**} := \arg \min_{\mathbf{c}^*} (\mathbf{c}^*)^T \left(I + V^T V\right) \mathbf{c}^* \qquad (3.12)$$

(the minimization is done over all the solutions $\mathbf{c}^*$ of the linear programming problem (3.2)). Here, the matrix $V \in R^{K \times NK}$ is defined by

$$V := \begin{bmatrix} v_{11} & \mathbf{e}^K(u_2)\Pi_1 & \cdot & \mathbf{e}^K(u_K)\Pi_1 \\ (\mathbf{e}^K(u_1))^T \Pi_2 & v_{22} & \cdot & (\mathbf{e}^K(u_K))^T \Pi_2 \\ \cdot & \cdot & \cdot & \cdot \\ (\mathbf{e}^K(u_1))^T \Pi_K & (\mathbf{e}^K(u_2))^T \Pi_K & \cdot & v_{KK} \end{bmatrix} \qquad (3.13)$$

with

$$v_{ii} = \left(\mathbf{e}^N\right)^T - (\mathbf{e}^K(u_i))^T \Pi_i$$

where the vectors $\mathbf{e}^M$ and $\mathbf{e}^K(u_i)$ are given by

$$\mathbf{e}^M := \underbrace{(1, ..., 1)^T}_{M} \in R^{M \times 1},$$

$$\mathbf{e}^K(u_i) := \left(\underbrace{0, 0, ..., 0, 1, 0, ..., 0}_{i}\right)^T \in R^K$$

and the matrices Π_j ($j = 1, ..., K$) are defined by

$$\Pi_j := \begin{bmatrix} \pi^1_{1j} & \pi^2_{1j} & \cdots & \pi^N_{1j} \\ \pi^1_{2j} & \pi^2_{2j} & \cdots & \pi^N_{2j} \\ \cdot & \cdot & \cdot & \cdot \\ \pi^1_{Kj} & \pi^2_{Kj} & \cdot & \pi^N_{Kj} \end{bmatrix} \in R^{K \times N}. \qquad (3.14)$$

The proof of this lemma is given in [6].

Remark 1 *Notice that the equality constraints*

$$\sum_{l=1}^{N} c^{jl} = \sum_{i=1}^{K} \sum_{l=1}^{N} \pi^l_{ij} c^{il} \ (i, j = 1, ..., K; l = 1, ..., N)$$

can be rewritten in matrix form as follows

$$V\mathbf{c} = 0 \qquad (3.15)$$

with the matrix V satisfying the condition

$$\det V^T V = 0. \qquad (3.16)$$

3.2. ADAPTIVE LEARNING ALGORITHM

Stochastic approximation techniques [4] have a wide range of application in identification, filtering, control, etc.. To find the minimum point $\mathbf{c}^*_{\mu,\delta}$ (3.8) of the function $\mathcal{P}_{\mu,\delta}(\mathbf{c})$ (3.6) when the parameters $\left(v_{il}, \pi^l_{ij}\right)$ are unknown, we will use the stochastic approximation technique to define a recursive procedure

$$\mathbf{c}_{n+1} = \mathbf{c}_{n+1}\left(x_n, u_n, \eta_n, x_{n+1}, \mathbf{c}_n\right)$$

generating the sequence $\{\mathbf{c}_n\}$ which converges in some probability sense to the solution $\mathbf{c}^{**}$ of the initial problem. The recursive structure of our control algorithm is indicated below. It consists of 4 steps.

Step 1. (normalization (projection) procedure): use the available information

$$x_n = x\left(\alpha\right), u_n = u\left(\beta\right), \eta_n, x_{n+1} = x\left(\gamma\right), \eta_n, \mathbf{c}_n\left(c^{il}_n > 0\right)$$

to construct the following function

$$\xi_n := \mu_n \eta_n + \sum_{j=1}^{K}\left(\sum_{l=1}^{N} c_n^{jl} - \sum_{i=1}^{K}\sum_{l=1}^{N}\left(\widehat{\pi}^l_{ij}\right)_n c_n^{il}\right)\left(\chi\left(x_n = x\left(j\right)\right) - \left(\widehat{\pi}^\beta_{\alpha j}\right)_n\right) + \delta_n c_n^{\alpha\beta} \tag{3.17}$$

where $\left(\widehat{\pi}^l_{ij}\right)_n$ represents the estimation at time n of the transition probability π^l_{ij}, and the operator $\chi\left(x(j) = x_{n+1}\right)$ is defined as follows:

$$\chi\left(x(j) = x_{n+1}\right) = \begin{cases} 1 & \text{if } x(j) = x_{n+1} \\ 0 & \text{otherwise} \end{cases} ;$$

and normalize it using the following affine transformation

$$\zeta_n := \frac{a_n \xi_n + b_n}{c_n^{\alpha\beta}}. \tag{3.18}$$

It is interesting to note the similarity between this affine transformation and the normalization (projection) procedure used in the context of stochastic learning automata [1, 7] when the environment response does not belong to the unit segment.

The numerical sequences $\{a_n\}, \{b_n\}$ are given by

$$a_n := \left(2\frac{(\mu_n \sigma + 4)}{\varepsilon_n} + \frac{N \cdot K}{N \cdot K - 1}\delta_n\right)^{-1}, \quad b_n := a_n\left(\mu_n \sigma + 4\right). \tag{3.19}$$

The positive sequences $\{\delta_n\}, \{\varepsilon_n\}$ and $\{\mu_n\}$ will be specified below.

Step 2. (updating procedure): calculate the elements c_{n+1}^{il} using the following recursive algorithm

$$c_{n+1}^{il} = \begin{cases} c_n^{\alpha\beta} + \gamma_n(1 - c_n^{\alpha\beta} - \zeta_n) & i = \alpha \wedge l = \beta \\ c_n^{il} - \gamma_n\left(c_n^{il} - \dfrac{\zeta_n}{N \cdot K - 1}\right) & i \neq \alpha \vee l \neq \beta \end{cases}, \quad (3.20)$$

$$\left(\hat{\pi}_{ij}^l\right)_{n+1} = \left(\hat{\pi}_{ij}^l\right)_n - \qquad (3.21)$$

$$- \begin{cases} \dfrac{1}{\left(S_{\alpha j}^{\beta}\right)_{n+1}} \left[\left(\hat{\pi}_{\alpha j}^l\right)_n - \chi(x_{n+1} = x(j))\right] & i = \alpha \wedge l = \beta \wedge j = \gamma \\ 0 & i \neq \alpha \vee l \neq \beta \vee j = \gamma \end{cases},$$

$$\left(S_{ij}^l\right)_{n+1} = \left(S_{ij}^l\right)_n + \chi\left(x_n = x(i) \wedge u_n = u(l) \wedge x_{n+1} = x(j)\right). \quad (3.22)$$

The deterministic sequence $\{\gamma_n\}$ will be specified below.

Step 3. (new action selection): construct the stochastic matrix

$$d_{n+1}^{il} = c_{n+1}^{il} \left(\sum_{k=1}^{N} c_{n+1}^{ik}\right)^{-1} \quad (i = 1, ..., K, l = 1, ..., N) \qquad (3.23)$$

and according to

$$\Pr\{u_{n+1} = u(l) \mid x_{n+1} = x(\gamma) \wedge \mathcal{F}_n\} = d_{n+1}^{\gamma l}$$

generate randomly a new discrete random variable u_{n+1} as in learning stochastic automata implementation, and get the new observation (realization) η_{n+1} which corresponds to the transition to state x_{n+1}.

Step 4. return to Step 1.

This adaptive control algorithm exhibits some common characteristics with the algorithm presented in chapter 2. It is very close to the algorithms based on learning automata and developed for stochastic optimization purposes [1, 7]

Remark 2 *Practical implementation of recursive algorithms are generally performed with some normalization (projection) which brings the estimated parameter (or some variable) at each instant n into some domain (e.g. prior information on the region where the parameter lies).*

The next lemma shows that the normalized function ζ_n belong to the unit segment $(0, 1)$ and that

$$\mathbf{c}_{n+1} \in S_{\varepsilon_{n+1}}^{KN} \text{ if } \mathbf{c}_n \in S_{\varepsilon_n}^{K \cdot N}.$$

3.2. ADAPTIVE LEARNING ALGORITHM

Lemma 2 *If*

1. *in the sequence (3.18)*
$$c_n^{\alpha\beta} \geq \varepsilon_n, \quad \delta_n \downarrow 0 \tag{3.24}$$

    then
$$\zeta_n \in \left[\zeta_n^-, \zeta_n^+\right] \subset [0,1] \tag{3.25}$$

    where
$$\zeta_n^- = a_n \delta_n, \quad \zeta_n^+ = 1 - \frac{a_n \delta_n}{N \cdot K - 1}; \tag{3.26}$$

2. *in the updating algorithm (3.20)*
$$\zeta_n \in \left[\zeta_n^-, \zeta_n^+\right] \subset [0,1], \quad \gamma_n \in [0,1], \quad \mathbf{c}_n \in S_{\varepsilon_n}^{K \cdot N}$$

    then
$$\mathbf{c}_{n+1} \in S_{\varepsilon_{n+1}}^{K \cdot N}. \tag{3.27}$$

Proof. To prove (3.25) it is enough to notice that

$$\zeta_n$$
$$= \frac{a_n \left(\mu_n \eta_n + \sum_{j=1}^{K} \sum_{l=1}^{N} c_n^{jl} - \sum_{i=1}^{K} \sum_{l=1}^{N} \left(\widehat{\pi}_{ij}^l\right)_n c_n^{il}\right) \left(\chi\left(x_n = x\left(j\right)\right) - \left(\widehat{\pi}_{\alpha j}^{\beta}\right)_n\right)}{c_n^{\alpha\beta}}$$

$$+ \frac{b_n}{c_n^{\alpha\beta}} + a_n \delta_n \geq \frac{-a_n \left(\mu_n \sigma + 4\right) + b_n}{c_n^{\alpha\beta}} + a_n \delta_n = a_n \delta_n = \zeta_n^- > 0$$

and

$$\zeta_n \leq \frac{a_n \left(\mu_n \sigma + 4\right) + b_n}{\varepsilon_n} + a_n \delta_n$$
$$\leq a_n \left(\frac{2}{\varepsilon_n}\left(\mu_n \sigma + 4\right) + \delta_n\right) = \zeta_n^+.$$

Notice that the procedure (3.20) corresponds to Bush-Mosteller reinforcement scheme [1, 7], and simple algebraic calculations demonstrate that (3.25) holds and $\mathbf{c}_{n+1} \in S_0^{KN}$. Indeed, from (3.20) it follows

$$c_{n+1}^{il} \geq (1 - \gamma_n) c_n^{il} + \gamma_n \min\left\{1 - \zeta_n^+; \frac{\zeta_n^-}{N \cdot K - 1}\right\}$$
$$\geq (1 - \gamma_n) c_n^{il} + \gamma_n \min\left\{1 - \zeta_n^+; \frac{\zeta_n^-}{N \cdot K - 1}\right\}.$$

To fulfill the following condition

$$\zeta_n^+ = 1 - \frac{\zeta_n^-}{N \cdot K - 1} < 1. \qquad (3.28)$$

a_n has to be selected as in (3.19).

Taking into account (3.28), the last inequality can be rewritten as follows

$$c_{n+1}^{il} \geq (1 - \gamma_n) \, c_n^{il} + \gamma_n \frac{\zeta_n^-}{N \cdot K - 1} \geq \min\left\{ c_n^{il}, \frac{\zeta_n^-}{N \cdot K - 1} \right\}$$

$$\geq \min\left\{ \min\left\{ c_{n-1}^{il}, \frac{\zeta_{n-1}^-}{N \cdot K - 1} \right\}, \frac{\zeta_n^-}{N \cdot K - 1} \right\}.$$

Notice that in view of (3.19) ζ_n^- is monotically decreasing $\zeta_n^- = a_n \delta_n \downarrow 0$. From the last expression we obtain

$$c_{n+1}^{il} \geq \min\left\{ \min\left\{ c_{n-1}^{il}, \frac{\zeta_{n-1}^-}{N \cdot K - 1} \right\}, \frac{\zeta_n^-}{N \cdot K - 1} \right\}$$

$$= \min\left\{ c_{n-1}^{il}, \frac{\zeta_n^-}{N \cdot K - 1} \right\}$$

$$\geq \cdots \geq \min\left\{ c_1^{il}, \frac{\zeta_n^-}{N \cdot K - 1} \right\} = \frac{\zeta_n^-}{N \cdot K - 1} \geq \frac{\zeta_{n+1}^-}{N \cdot K - 1} \equiv \varepsilon_{n+1}.$$

From (3.20) it follows

$$\sum_{i=1}^{K} \sum_{l=1}^{N} c_{n+1}^{il} = \sum_{i=1}^{K} \sum_{l=1}^{N} c_n^{il} = 1.$$

So, (3.27) is fulfilled. ∎

Now we shall show below that under certain conditions, this adaptive control algorithm has nice asymptotic properties.

3.3 Convergence analysis

In the previous section we have presented and adaptive control algorithm. We now turn to the logical questions: How does this algorithm perform? i.e., what are its properties? In this section we establish the convergence properties of this adaptive control algorithm.

The policy sequence $\{d_n\}$ described by (3.20) and (3.23) is essentially nonstationary. As a consequence, we need to prove the convergence of this sequence to the solution $\mathbf{c}^{**}$ (3.12) of the initial optimization problem.

3.3. CONVERGENCE ANALYSIS

Let us introduce the following Lyapunov function

$$W_n := \left\| \mathbf{c}_n - \mathbf{c}^*_{\delta_n} \right\|^2. \tag{3.29}$$

This choice of Lyapunov function has promise of general applicability. The term "Lyapunov function" has already been used in various contexts. This is always an attractive and efficient function which ensures a certain stabilization [4]. The following theorem states the conditions related to the convergence of this adaptive control algorithm.

Theorem 1 *Let the controlled Markov chain be ergodic with any fixed distribution of the initial states. Let the loss function Φ_n be given by ((2.4) see chapter2). If the control strategy generated by the adaptive algorithm (3.20)-(3.23) with design parameters ε_n, δ_n, μ_n and γ_n satisfying the following conditions*

$$0 < \varepsilon_n, \delta_n, \mu_n \downarrow 0, \ \gamma_n \in (0,1),$$

$$\frac{\delta_n}{\mu_n} \xrightarrow[n\to\infty]{} 0, \ \frac{(\mu_n)^{3/2}}{\delta_n} \xrightarrow[n\to\infty]{} 0, \ \sum_{n=1}^{\infty} \gamma_n \varepsilon_n \delta_n = \infty$$

is used, then

1. *if*

$$\sum_{n=1}^{\infty} \theta_n < \infty$$

   where

$$\theta_n := \gamma_n^2 + (\delta_n - \delta_{n+1})^2 (\gamma_n \varepsilon_n \delta_n)^{-1} + |\delta_n - \delta_{n+1}| \gamma_n + \gamma_n (n\varepsilon_n \delta_n)^{-1},$$

   then, the control strategy (3.20)-(3.23) converges with probability 1 to the optimal solution, i.e.,

$$W_n \xrightarrow[n\to\infty]{a.s.} 0;$$

2. *if*

$$\frac{\theta_n}{\gamma_n \varepsilon_n \delta_n} \xrightarrow[n\to\infty]{} 0$$

   then, we obtain the convergence in the mean squares sense, i.e.,

$$E\{W_n\} \xrightarrow[n\to\infty]{a.s.} 0.$$

Proof. For $\delta > 0$, the regularized penalty function $\mathcal{P}_{\mu,\delta}(\mathbf{c})$ (3.6), is strictly convex. Indeed, for any convex function and for any point $\mathbf{c} \in R^{NK}$ we have

$$\left(\mathbf{c} - \mathbf{c}^*_{\mu,\delta}\right)^T \left(\nabla_\mathbf{c} \mathcal{P}_{\mu,\delta}(\mathbf{c}) - \nabla_\mathbf{c} \mathcal{P}_{\mu,\delta}(\mathbf{c}^*_{\mu,\delta})\right) \geq \left(\mathbf{c} - \mathbf{c}^*_{\mu,\delta}\right)^T \nabla_\mathbf{c}^2 \mathcal{P}_{\mu,\delta}(\mathbf{c}^*_{\mu,\delta}) \left(\mathbf{c} - \mathbf{c}^*_{\mu,\delta}\right).$$

Taking into account (3.13), (3.15) and (3.16) we conclude that the last inequality can be transformed to

$$\left(\mathbf{c} - \mathbf{c}^*_{\mu,\delta}\right)^T \nabla_\mathbf{c} \mathcal{P}_{\mu,\delta}(\mathbf{c}) \geq \left(\mathbf{c} - \mathbf{c}^*_{\mu,\delta}\right)^T \left[V^T V + \delta I\right] \left(\mathbf{c} - \mathbf{c}^*_{\mu,\delta}\right) \qquad (3.30)$$

$$\geq \delta \left\|\mathbf{c} - \mathbf{c}^*_{\mu,\delta}\right\|^2$$

where $\mathbf{c}^*_{\mu,\delta}$ is the minimum point of the regularized penalty function (3.6).

Recall that

$$E\{Z \mid x_n = x(\alpha) \wedge \mathcal{F}_{n-1}\} = E\left\{Z \frac{\chi(x_n = x(\alpha))}{p(\alpha)} \mid \mathcal{F}_{n-1}\right\}$$

and introduce the following notation

$$\varphi_n\left(\widehat{\pi}^l_{ij}\right) := \mu_n \eta_n + \sum_{j=1}^{K} \left(\sum_{l=1}^{N} c_n^{jl} - \sum_{i=1}^{K}\sum_{l=1}^{N} \left(\widehat{\pi}^l_{ij}\right)_n c_n^{il}\right) \left(\chi(x_n = x(j)) - \left(\widehat{\pi}^\beta_{\alpha j}\right)_n\right)$$

$$+ \delta_n c_n^{\alpha\beta}.$$

Then, from (3.20) it follows that the gradient with respect to $\mathbf{c}$ of the regularized penalty function (3.6) can be expressed as a function of the conditional mathematical expectation of ζ_n:

$$E\{\zeta_n \mathbf{e}(x_n \wedge u_n) \mid \mathcal{F}_{n-1}\} = E\left\{\frac{a_n \varphi_n\left(\widehat{\pi}^l_{ij}\right) + b_n}{c_n^{\alpha\beta}} \mathbf{e}(x_n \wedge u_n) \mid \mathcal{F}_{n-1}\right\}$$

$$= E\left\{\frac{a_n \varphi_n\left(\pi^l_{ij}\right) + b_n}{c_n^{\alpha\beta}} \mathbf{e}(x_n \wedge u_n) \mid \mathcal{F}_{n-1}\right\}$$

$$+ E\left\{\frac{a_n}{c_n^{\alpha\beta}} \left[\varphi_n\left(\widehat{\pi}^l_{ij}\right) - \varphi_n\left(\pi^l_{ij}\right)\right] \mathbf{e}(x_n \wedge u_n) \mid \mathcal{F}_{n-1}\right\}$$

$$\stackrel{a.s.}{=} a_n \frac{\partial}{\partial \mathbf{c}} \nabla_\mathbf{c} \mathcal{P}_{\mu_n, \delta_n}(\mathbf{c}_n) + b_n \mathbf{e}^{N \cdot K} + \Delta_n.$$

where

$$\Delta_n := E\left\{\frac{a_n}{c_n^{\alpha\beta}} \left[\varphi_n\left(\widehat{\pi}^l_{ij}\right) - \varphi_n\left(\pi^l_{ij}\right)\right] \mathbf{e}(x_n \wedge u_n) \mid \mathcal{F}_{n-1}\right\}$$

3.3. CONVERGENCE ANALYSIS

and $\mathbf{e}(x_n \wedge u_n)$ is a vector defined as follows

$$\mathbf{e}(x_n \wedge u_n) := \begin{bmatrix} \chi(x_n = x(1), u_n = u(1)) \\ \ldots \\ \chi(x_n = x(K), u_n = u(1)) \\ \ldots \\ \chi(x_n = x(1), u_n = u(l)) \\ \ldots \\ \chi(x_n = x(K), u_n = u(l)) \\ \ldots \\ \chi(x_n = x(K), u_n = u(N)) \end{bmatrix} \in R^{N \cdot K}. \quad (3.31)$$

Notice that under the assumption

$$\sum_{n=1}^{\infty} c_n^{il} \geq \sum_{n=1}^{\infty} \varepsilon_n \geq \sum_{n=1}^{\infty} \gamma_n \varepsilon_n \delta_n = \infty$$

and in view of the Borel-Cantelli lemma [8] and the strong law of large numbers for dependent sequences [9], we derive

$$\|\Delta_n\| = O(\|V_n - V\|) \stackrel{a.s.}{=} o_\omega\left(\frac{1}{\sqrt{n}}\right)$$

where V_n is derived from the matrix V by replacing their elements π_{ij}^l by $\widehat{\pi}_{ij}^l$ and $o_\omega(n^{-1/2})$ is a random sequence tending almost surely to zero, more quickly than $n^{-1/2}$.

Hence

$$E\{\zeta_n \mathbf{e}(x_n \wedge u_n) \mid \mathcal{F}_{n-1}\} \stackrel{a.s.}{=} a_n \frac{\partial}{\partial \mathbf{c}} \nabla_\mathbf{c} \mathcal{P}_{\mu_n, \delta_n}(\mathbf{c}_n) + b_n \mathbf{e}^{N \cdot K} + o_\omega\left(\frac{a_n}{\varepsilon_n \sqrt{n}}\right). \quad (3.32)$$

Rewriting (3.20) in a vector form, we obtain

$$\mathbf{c}_{n+1} = \mathbf{c}_n + \gamma_n \left(\mathbf{e}(x_n \wedge u_n) - \mathbf{c}_n + \zeta_n \frac{\mathbf{e}^{N \cdot K} - N \cdot K \mathbf{e}(x_n \wedge u_n)}{N \cdot K - 1} \right). \quad (3.33)$$

Substituting (3.33) into W_{n+1} (3.29) we derive

$$W_{n+1} \leq \left\| \mathbf{c}_n + \gamma_n \left(\mathbf{e}(x_n \wedge u_n) - \mathbf{c}_n + \zeta_n \frac{\mathbf{e}^{N \cdot K} - N \cdot K \mathbf{e}(x_n \wedge u_n)}{N \cdot K - 1} \right) \right.$$

$$\left. - \mathbf{c}_{\delta_{n+1}}^* \right\|^2 = \left\| (\mathbf{c}_n - \mathbf{c}_{\delta_n}^*) + \right.$$

$$+\gamma_n \left(\mathbf{e}\left(x_n \wedge u_n\right) - \mathbf{c}_n + \zeta_n \frac{e^{N \cdot K} - N \cdot Ke\left(x_n \wedge u_n\right)}{N \cdot K - 1} \right) + \left(\mathbf{c}_{\delta_n}^* - \mathbf{c}_{\delta_{n+1}}^*\right) \bigg\|^2.$$

Calculating the square of the norms appearing in this inequality, and estimating the resulting terms using the inequality

$$\|\zeta_n\| \leq \|\zeta_n^+\| \leq 1,$$

we obtain

$$W_{n+1} \leq W_n + \gamma_n^2 Const + \|\mathbf{c}_{\delta_n}^* - \mathbf{c}_{\delta_{n+1}}^*\|^2$$

$$+ 2 \|\mathbf{c}_{\delta_n}^* - \mathbf{c}_{\delta_{n+1}}^*\| \sqrt{W_n} + 2 \|\mathbf{c}_{\delta_n}^* - \mathbf{c}_{\delta_{n+1}}^*\| \gamma_n Const$$

$$+ 2\gamma_n \left(\mathbf{c}_n - \mathbf{c}_{\delta_n}^*\right)^T \left(\mathbf{e}\left(x_n \wedge u_n\right) - \mathbf{c}_n + \zeta_n \frac{e^{N \cdot K} - N \cdot Ke\left(x_n \wedge u_n\right)}{N \cdot K - 1} \right)$$

where $Const$ is a positive constant.

Combining the terms of the right hand side of this inequality and in view of (3.9), it follows

$$W_{n+1} \leq W_n + Const \cdot \mu_{1,n} \sqrt{W_n} + Const \cdot \mu_{2,n} + w_n \qquad (3.34)$$

where

$$\mu_{1,n} := |\delta_n - \delta_{n+1}|,$$

$$\mu_{2,n} := \gamma_n^2 + (\delta_n - \delta_{n+1})^2 + |\delta_n - \delta_{n+1}|\gamma_n,$$

$$w_n := 2\gamma_n \left(\mathbf{c}_n - \mathbf{c}_{\delta_n}^*\right)^T \left(\mathbf{e}\left(x_n \wedge u_n\right) - \mathbf{c}_n + \zeta_n \frac{e^{N \cdot K} - N \cdot Ke\left(x_n \wedge u_n\right)}{N \cdot K - 1} \right).$$

Notice that

$$\left(\mathbf{c}_n - \mathbf{c}_{\delta_n}^*\right)^T \frac{e^{N \cdot K}}{N \cdot K - 1} = 0.$$

If (3.32) is used, the following is obtained:

$$E\left\{w_n \mid \mathcal{F}_{n-1}\right\} \stackrel{a.s.}{=} -2\frac{N \cdot K}{N \cdot K - 1}\gamma_n \left(\mathbf{c}_n - \mathbf{c}_{\delta_n}^*\right)^T E\left\{\zeta_n \mathbf{e}\left(x_n \wedge u_n\right) \mid \mathcal{F}_{n-1}\right\}$$

$$\stackrel{a.s.}{=} -2\frac{N \cdot K}{N \cdot K - 1}\gamma_n \left(\mathbf{c}_n - \mathbf{c}_{\delta_n}^*\right)^T \left(a_n \frac{\partial}{\partial \mathbf{c}} \nabla_{\mathbf{c}} \mathcal{P}_{\mu_n, \delta_n}(\mathbf{c}_n) + b_n e^{N \cdot K} + o_\omega\left(\frac{a_n}{\varepsilon_n \sqrt{n}}\right) \right)$$

$$= -2\frac{N \cdot K}{N \cdot K - 1}\gamma_n a_n \left(\mathbf{c}_n - \mathbf{c}_{\delta_n}^*\right)^T \frac{\partial}{\partial \mathbf{c}} \nabla_{\mathbf{c}} \mathcal{P}_{\mu_n, \delta_n}(\mathbf{c}_n) + o_\omega\left(\frac{\gamma_n a_n \left(\mathbf{c}_n - \mathbf{c}_{\delta_n}^*\right)}{\varepsilon_n \sqrt{n}}\right).$$

3.3. CONVERGENCE ANALYSIS

Taking into account the assumptions of this theorem, and the strict convexity property (3.30), we deduce

$$E\{w_n \mid \mathcal{F}_{n-1}\} \stackrel{a.s.}{=} -2\frac{N \cdot K}{N \cdot K - 1}\gamma_n a_n \left(\mathbf{c}_n - \mathbf{c}_{\delta_n}^*\right)^T \frac{\partial}{\partial \mathbf{c}}\nabla_{\mathbf{c}}\mathcal{P}_{\mu_n,\delta_n}(\mathbf{c}_n)$$

$$+ o_\omega\left(\frac{\gamma_n a_n}{\varepsilon_n \sqrt{n}}\right)\sqrt{W_n}$$

$$\leq -\frac{N \cdot K}{N \cdot K - 1}\gamma_n a_n \delta_n W_n + o_\omega\left(\frac{\gamma_n a_n}{\varepsilon_n \sqrt{n}}\right)\sqrt{W_n}.$$

Calculating the conditional mathematical expectation of both sides of (3.34), and in view of the last inequality, we can get

$$E\{W_{n+1} \mid \mathcal{F}_{n-1}\} \stackrel{a.s.}{\leq} \left(1 - \frac{N \cdot K}{N \cdot K - 1}\gamma_n a_n \delta_n\right) W_n + Const\left(\mu_{3,n}\sqrt{W_n} + \mu_{2,n}\right), \tag{3.35}$$

where

$$\mu_{3,n} := \mu_{1,n} + o_\omega\left(\frac{\gamma_n a_n}{\varepsilon_n \sqrt{n}}\right).$$

Now we use

$$2\mu_{1,n}\sqrt{W_n} \leq \mu_{1,n}^2 \rho_n^{-1} + W_n \rho_n$$

(which is valid for any $\rho_n > 0$) for

$$\rho_n := \gamma_n a_n \delta_n$$

to conclude

$$2\mu_{1,n}\sqrt{W_n} \leq \mu_{3,n}^2 \left(\gamma_n a_n \delta_n\right)^{-1} + W_n \gamma_n a_n \delta_n.$$

From this inequality and (3.35), and in view of the following estimation

$$\mu_{3,n}^2 \left(\gamma_n a_n \delta_n\right)^{-1} + \mu_{2,n} \leq Const \cdot \theta_n$$

we finally, obtain

$$E\{W_{n+1} \mid \mathcal{F}_{n-1}\} \stackrel{a.s.}{\leq} \left(1 - \frac{1}{N \cdot K - 1}\gamma_n a_n \delta_n\right) W_n + Const \cdot \theta_n. \tag{3.36}$$

Observe that (3.19) leads to

$$a_n = O\left(\varepsilon_n\right).$$

From the assumptions of this theorem and in view of the condition (3.11) and the Robbins-Siegmund theorem (see theorem 2 of Appendix A) [1, 10] which is the key to stochastic approximation, the convergence with probability 1 follows.

The mean squares convergence follows from (3.36) after applying the operator of conditional mathematical expectation to both sides of this inequality and using lemma A5 given in [7]. The theorem is thus constructively established. ∎

Theorem 1 above shows that this adaptive learning control algorithm possess all the properties that one would desire, i.e., convergence with probability 1 as well as convergence in the mean squares.

The conditions related to the sequences $\{\mu_n\}$, $\{\varepsilon_n\}$, $\{\delta_n\}$, and $\{\gamma_n\}$ are now be stated in the following corollary.

Corollary 1 *If in theorem 1*

$$\varepsilon_n := \frac{\varepsilon_0}{1+n^\varepsilon} \left(\varepsilon_0 \in \left[0, (N \cdot K)^{-1}\right), \varepsilon \geq 0\right), \quad \delta_n := \frac{\delta_0}{1+n^\delta \ln n} \left(\delta_0, \delta > 0,\right),$$

$$\mu_n := \frac{\delta_0}{1+n^\mu \ln n} \left(\mu_0, \mu > 0,\right), \quad \gamma_n := \frac{\gamma_0}{n^\gamma} \left(\gamma_0 \in (0,1), \gamma \geq 0\right),$$

with

$$\frac{2}{3}\delta < \mu \leq \delta, \ \gamma + \mu + \varepsilon + \delta \leq 1$$

then

1. *the convergence with probability 1 will take place if*

$$2\gamma > 1,$$

2. *the mean squares convergence is guaranteed if*

$$\varepsilon + \delta < \frac{1}{2}.$$

It is easy to check up on these conditions by substituting the parameters given in this corollary in theorem 1 assumptions.

Remark 3 *In the optimization problem related to the regularized penalty function $\mathcal{P}_{\mu_n,\delta_n}(c_n)$ the parameters μ_n and δ_n must decrease less slowly than ε_n, i.e.,*

$$\mu \leq \varepsilon, \ \delta \leq \varepsilon.$$

In the design of an adaptive controller, we are interested in accuracy as well as in the speed of convergence. The previous theorem addressed the problem of convergence. In what follows, we state the order of convergence rate of the adaptive learning algorithm described above.

3.3. CONVERGENCE ANALYSIS

Theorem 2 *Under the conditions of theorem 1 and corollary 1, it follows*

$$W_n \stackrel{a.s.}{=} o_\omega\left(\frac{1}{n^\nu}\right)$$

where

$$0 < \nu < \min\{2\gamma - 1; 2\delta; 2\mu\} := \nu^*(\gamma, \mu, \delta)$$

and the positive parameters γ, δ, and μ satisfy the constraints

$$\frac{2}{3}\delta < \mu \leq \delta, \ \gamma + \varepsilon + \delta \leq 1, \ 2\gamma > 1, \ \delta \leq \varepsilon, \ \mu \leq \varepsilon.$$

Proof. From (3.9), it follows:

$$W_n^* := \|\mathbf{c}_n - \mathbf{c}^{**}\|^2 = \|(\mathbf{c}_n - \mathbf{c}_n^*) + (\mathbf{c}_n^* - \mathbf{c}^{**})\|^2$$

$$\leq 2\|\mathbf{c}_n - \mathbf{c}_n^*\|^2 + 2\|\mathbf{c}_n^* - \mathbf{c}^{**}\|^2 \leq 2W_n + C\left(\delta_n^2 + \mu_n^2\right).$$

Multiplying both sides of the previous inequality by ν_n, we derive

$$\nu_n W_n^* \leq 2\nu_n W_n + \nu_n C\left(\delta_n^2 + \mu_n^2\right).$$

Selecting $\nu_n = n^\nu$ and in view of lemma 2 [11] and taking into account that

$$\frac{\nu_{n+1} - \nu_n}{\nu_n} = \frac{\nu + o(1)}{n}$$

we obtain

$$0 < \nu < \min\{2\gamma - 1; 2\delta; 2\mu\} := \nu^*(\gamma, \mu, \delta)$$

where the positive parameters γ, δ, ε and μ satisfy the following constraints

$$\frac{2}{3}\delta < \mu \leq \delta, \ \gamma + \varepsilon + \delta \leq 1, \ 2\gamma > 1, \ \delta \leq \varepsilon, \ \mu \leq \varepsilon.$$

∎

The optimal convergence rate is given by the next corollary.

Corollary 2 *The maximum convergence rate is achieved with the optimal parameters $\varepsilon^*, \delta^*, \lambda^*, \gamma^*$*

$$\varepsilon = \varepsilon^* = \mu = \mu^* = \delta = \delta^* = \frac{1}{6}, \ \gamma = \gamma^* = \frac{2}{3}$$

and is equal to

$$\nu^*(\gamma, \varepsilon, \delta) = 2\gamma^* - 1 = \nu^{**} = \frac{1}{3}.$$

Proof. The solution of the linear programming problem

$$\nu^*(\gamma, \mu, \delta) \to \max_{\gamma, \mu, \delta}$$

is given by

$$2\gamma - 1 = 2\delta = 2\mu = 2\delta, \ \gamma + \varepsilon + \delta = 1$$

or, in equivalent form,

$$\gamma = \frac{1}{2} + \delta = 1 - \delta - \varepsilon, \ \mu = \delta.$$

From these equalities it follows that

$$\varepsilon = \frac{1}{2} - 2\delta.$$

Taking into account that δ_n must decrease less slowly than ε_n (see Remark), we derive

$$\delta \leq \varepsilon$$

and, as a result, the smallest ε maximizing γ is equal to

$$\varepsilon = \delta.$$

Hence,

$$\delta = \frac{1}{6}, \ \gamma = \frac{1}{2} + \delta = \frac{2}{3}.$$

So, the optimal parameters are

$$\gamma = \gamma^* = \frac{2}{3}, \ \varepsilon = \varepsilon^* = \delta = \delta^* = \mu = \mu^* = \frac{1}{6}.$$

The maximum convergence rate is achieved with this choice of parameters and is equal to

$$\nu^*(\gamma, \varepsilon, \delta) = 2\gamma^* - 1 = \nu^{**} = \frac{1}{3}.$$

∎

The feasibility and the performance of this adaptive control algorithm is illustrated by two numerical examples presented in the last chapter of this book.

3.4 Conclusions

An adaptive control algorithm for finite controlled Markov chains whose transition probabilities are unknown has been presented. The control strategy is designed to achieve the asymptotic minimization of a loss function. To construct this adaptive control algorithm, a regularized penalty function was introduced. The control policy is adjusted using the Bush-Mosteller reinforcement scheme as a stochastic approximation procedure. The convergence properties (convergence with probability 1 as well as convergence in the mean squares) have been stated. We establish that the optimal convergence rate is equal to $n^{-\frac{1}{3}+\delta}$ (δ is any small positive parameter). The algorithms derived respectively on the basis of Lagrange multipliers and penalty function approaches achieve the same optimal convergence rate. This is a striking result.

3.5 References

1. A. S. Poznyak and K. Najim, *Learning Automata and Stochastic Optimization*, Springer-Verlag, Berlin, 1997.

2. Ya. Z. Tsykin, *Adaptation and Learning in Automatic Systems*, Academic Press, New York, 1971.

3. Ya. Z. Tsykin and A. S. Poznyak, Learning automata, *Journal of Cybernetics and Information Science*, vol. 1, pp. 128-161, 1977.

4. M. Duflo, *Random Iterative Models*, Springer-Verlag, Berlin, 1997.

5. R. A. Howard, *Dynamic Programming and Markov Processes*, J. Wiley, New York, 1962.

6. A. V. Nazin and A. S. Poznyak, *Adaptive Choice of Variants*, (in Russian) Nauka, Moscow, 1986.

7. K. Najim, and A. S. Poznyak, *Learning Automata Theory and Applications*, Pergamon Press, London, 1994.

8. J. L. Doob, *Stochastic Processes*, J. Wiley, New York, 1953.

9. D. Hall and C. Heyde, *Martingales Limit Theory and its Applications*, Academic Press, New York, 1980.

10. H. Robbins and D. Siegmund, A convergence theorem for nonnegative almost supermartingales and some applications, in *Optimizing Methods in Statistics*, ed. by J. S. Rustagi, Academic Press, New York, pp. 233-257, 1971.

11. A. S. Poznyak and K. Najim, Learning automata with continuous input and changing number of actions, *Int. Journal of Systems Science*, vol. 27, pp. 1467-1472, 1996.

Chapter 4

Projection Gradient Method

4.1 Introduction

An adaptive Projection Gradient algorithm will be derived to solve the problem concerning the adaptive control of unconstrained finite Markov chains. This problem has been already stated and solved in chapters 2 and 3, using respectively the Lagrange multipliers and the penalty function approaches. One method of solving this optimization problem which may come to mind is to construct an adaptive algorithm on the basis on the gradient approach. In comparison with these approaches, the algorithm based on gradient and projection techniques involves a procedure for the estimation of the transition matrix. The use of the projection operator makes this algorithm more complex and time consuming than the previous ones. But, from an algorithmic point of view, this approach does not need the implementation of any normalizing procedure. The results given in this chapter are based on the developments presented in [1].

4.2 Control algorithm

To solve the optimization problem

$$\tilde{V}(c) := \sum_{i=1}^{K}\sum_{l=1}^{N} v_{il} c^{il} \to \min_{c \in \mathbf{C}} \qquad (4.1)$$

where the $\mathbf{C} = \mathbf{C}_{\varepsilon=0}$ is given by

$$\mathbf{C}_{\varepsilon} = \left\{ c \mid c = \left[c^{il}\right], c^{il} \geq \varepsilon \geq 0, \sum_{i=1}^{K}\sum_{l=1}^{N} c^{il} = 1, \right. \qquad (4.2)$$

$$\sum_{l=1}^{N} c^{jl} = \sum_{i=1}^{K}\sum_{l=1}^{N} \pi_{ij}^{l} c^{il} \ (i,j=1,...,K; l=1,...,N) \right\}.$$

Based on the *Projection Gradient Method* [2], let us consider the following recurrent procedure:

$$c_{n+1} = \mathcal{P}_{\varepsilon_{n+1}}^{\mathbf{C}}\left\{c_n - \gamma_{n+1}\frac{\partial}{\partial c}\tilde{V}(c_n)\right\} \qquad (4.3)$$

where

$$\frac{\partial}{\partial c}\tilde{V}(c_n) := \left[\frac{\partial}{\partial c^{il}}\tilde{V}(c_n)\right]_{i=1,...,K; l=1,...,N}$$

the nonnegative scalar sequence $\{\gamma_n\}$ is the step gain sequence, and $\mathcal{P}_{\varepsilon_{n+1}}^{\mathbf{C}}(\cdot)$ is the projection operator onto the matrix set $\mathbf{C}_\varepsilon$ given by (4.2).

Notice that this procedure can be implemented only in the case where the information concerning the considered Markov chain is complete. We consider the case where only the realizations of loss function and the state trajectories are available. In this situation, the use of *Stochastic Approximation Techniques* [3-4] seems to be very appropriate. In fact, stochastic approximation techniques have been used to solve many engineering problems [5], and in a resurgence of interest have been considered as learning algorithms for neural networks and neuro-fuzzy systems synthesis [6-8].

Assuming the stationarity of the state distribution of the considered Markov chain and, taking into account:

$$c^{il} = d^{il} p_i(d),$$

for any $i = 1, ..., K; l = 1, ..., N$, we derive

$$\frac{\partial}{\partial c^{il}}\tilde{V}(c_n) = v_{il} = E\left\{\frac{\eta_n \chi(x_n = x(i), u_n = u(l))}{d^{il} p_i(d)}\right\}$$

$$= E\left\{\frac{\eta_n \chi(x_n = x(i), u_n = u(l))}{c^{il}}\right\}.$$

From this formula, we conclude that the realization A_n^{il} of the gradient $\frac{\partial}{\partial c^{il}}\tilde{V}(c_n)$ can be expressed as follows:

$$A_n^{il} = \frac{\eta_n \chi(x_n = x(i), u_n = u(l))}{c^{il}}. \qquad (4.4)$$

Notice that for $\varepsilon_n > 0$ we have

$$c_n^{il} \geq \varepsilon_n > 0. \qquad (4.5)$$

4.2. CONTROL ALGORITHM

To perform the Projection Gradient Scheme, we have to construct the estimates $\left[\left(\widehat{\pi}_{il}^l\right)_n\right]$ of the transition probabilities $\left[\pi_{ij}^l\right]$ involved into the description of the set $\mathbf{C}_\varepsilon$ (4.2). Let us consider the following estimation scheme

$$\left(\widehat{\pi}_{il}^l\right)_{n+1} = \left(\widehat{\pi}_{il}^l\right)_n - \frac{\eta_n \chi\left(x_n = x(i), u_n = u(l)\right)}{s_n^{il}} \left[\left(\widehat{\pi}_{il}^l\right)_n - \chi\left(x_{n+1} = x(j)\right)\right] \quad (4.6)$$

where

$$s_{n+1}^{il} = s_n^{il} + \chi\left(x_n = x(i), u_n = u(l)\right), \; s_0^{il} = 0 \; (i,j = 1,...,K; l = 1,...,N). \quad (4.7)$$

This algorithm corresponds to the recurrent form of the standard arithmetic averaging procedure. This algorithm for transition probabilities updating is arranged so that new estimates replace old ones.

Taking into account the fact that a Markov chain is a dynamic system, it may happen that a change of the control $\left[d_n^{il}\right]$ at each time n may lead to very oscillating and unstable trajectory behaviours. Let us change the control matrix only at a priori fixed instants n_k $(k=1,2,...)$.

Definition 1 *The random strategy $\left[d_n^{il}\right] \in \Sigma$ is said to be "**partially frozen**" if the corresponding matrices $\left[d_n^{il}\right]$ remain unchanged within the given time intervals $[n_k, n_{k+1})$, i.e., for any $n \in [n_k, n_{k+1})$, we have*

$$d_n^{il} = d_{n_k}^{il} \; (k=1,...).$$

Using such a subclass of randomized policy and based on the Stochastic Approximation Approach [3-4], we consider the following recurrent procedure

$$c_{k+1} = \mathcal{P}_{\varepsilon_{k+1}}^{\widehat{\mathbf{C}}} \left\{c_n - \gamma_k A_{n_{k+1}}\right\} k = 1,2,...; n_{k+1} > n_k; n_1 = 1 \quad (4.8)$$

which generate an adaptive control policy for the controlled finite Markov chain, where the elements of the matrix

$$A_{n_{k+1}} := \frac{1}{n_{k+1} - n_k} \sum_{t=n_k}^{n_{k+1}-1} x \eta_t \frac{e(x_t)e^T(u_t)}{e^T(x_t)c_k e(u_t)} \quad (4.9)$$

are estimated by the recursive procedure

$$A_{n+1}^{il} = A_n^{il} - \frac{1}{n_{k+1} - n_k + 1} \left(A_n^{il} - \frac{\eta_n \chi\left(x_n = x(i), u_n = u(l)\right)}{c_n^{il}}\right)$$

defined for each $n \in [n_k, n_{k+1})$ and $i = 1, ..., K; l = 1, ..., N$. The control actions are selected according to the randomized frozen control strategy $\left[d_n^{il}\right] \in \Sigma$ calculated as

$$d_n^{il} = c_k^{il} \left(\sum_{s=1}^{N} c_k^{is}\right)^{-1} \quad n \in [n_k, n_{k+1}) \qquad (4.10)$$

and the projection operator $\mathcal{P}_{\varepsilon_{k+1}}^{\widehat{C}}(\cdot)$ ensures the projection onto the set $\widehat{C}_{\varepsilon_{k+1}}$ defined by

$$\widehat{C}_{\varepsilon_{k+1}} := \left\{ c \mid c = \left[c^{il}\right], c^{il} \geq \varepsilon_{k+1} \geq 0, \sum_{i=1}^{K}\sum_{l=1}^{N} c^{il} = 1, \qquad (4.11)\right.$$

$$\left. \sum_{l=1}^{N} c^{il} = \sum_{i=1}^{K}\sum_{l=1}^{N} \left(\widehat{\pi}_{ij}^{l}\right)_{n_{k+1}-1} c^{il} \; (i,j = 1, ..., K; l = 1, ..., N) \right\}$$

with elements $\left(\widehat{\pi}_{ij}^{l}\right)_{n_{k+1}}$ given by (4.6).

Remark 1 *Projection and related schemes as normalization procedures represent one way to escape the boundedeness issues. The projection operator $\mathcal{P}_{\varepsilon_{k+1}}^{\widehat{C}}(\cdot)$ alters the algorithm (4.8) by projecting the iterates*

$$\left\{c_n - \gamma_k A_{n_{k+1}}\right\}$$

back onto the set $\widehat{C}_{\varepsilon_{k+1}}$.

Remark 2 *For any two points $c^* \in C_\varepsilon \subset R^{K \cdot N}$ and $c \in R^{K \cdot N}$, the main property of the projection operator $\mathcal{P}_\varepsilon^C(\cdot)$ is*

$$\left\|\mathcal{P}_\varepsilon^C(c) - c^*\right\| \leq \|c - c^*\|. \qquad (4.12)$$

For more details see § 2.6.2 [5].

The sequences $\{\gamma_k\}$, $\{\varepsilon_k\}$ and $\{\eta_k\}$ are the design parameters of this adaptive control procedure which have to be selected in such a way as to guarantee the convergence of the matrix sequence $\{c_k\}$ to the solution of the linear programming problem (4.1)-(4.2), in some probability sense. The conditions related to these design parameters are established and discussed in the next sections.

4.3 Estimation of the transition matrix

We begin in this section by recalling an interesting property of estimators. We then present a theorem [1] which states the conditions for which the design parameters of the adaptive procedure (4.8)-(4.11) guarantee the consistency of the transition matrix estimates $\left[\widehat{\pi}_{ij}^l\right]$ (4.6)-(4.7).

Definition 2 *An estimator c_n is said to be a (strongly) consistent estimator of the parameter c if*
$$c_n \xrightarrow{a.s.} c.$$

Definition 3 *An estimator $\widehat{c}_n$ is said to be a weakly consistent estimator of the parameter c if the sequence $\{\widehat{c}_n\}$ converges in probability to c.*

The consistency is intimately associated with the estimation procedures. For example, it is well known that the excitation ensures that the least squares estimator is consistent.

The significance of any definition, of course, resides in its consequences and applications, and so we turn to such question in the next theorem [1].

Theorem 1 *Let us consider a loss sequence $\{\eta_n\}$ satisfying (2.1)-(2.3) and, a regular homogeneous finite Markov chain (see Definition 23 of chapter 1) controlled by the adaptive procedure (4.8)-(4.11). Let us assume also that there exists a numerical nonnegative sequence $\{g_n\}$ which together with the parameters of this adaptive procedure satisfy the following conditions:*

$$\sum_{k=1}^{\infty} (k g_k)^{-1} < \infty, \sum_{k=1}^{\infty} \left(\frac{\Delta n_k}{n_k}\right)^2 < \infty, \quad (4.13)$$

$$\varlimsup_{k \to \infty} \frac{\Delta n_k \sqrt{k g_k}}{\sum_{t=1}^{k} \varepsilon_t \Delta n_t} < \infty, \Delta n_k := (n_{k+1} - n_k) \xrightarrow[k \to \infty]{} \infty. \quad (4.14)$$

Then

1. *for any initial state distribution of this Markov chain and, for any $i = 1, ..., K$ and $l = 1, ..., N$, the following property holds*

$$\lim_{n \to \infty} s_{m+1}^{il} \left(\sum_{t=1}^{t(n)} \varepsilon_t \Delta n_t\right)^{-1} > 0 \quad (4.15)$$

with probability 1, and

2. for any $\rho \in [0,1)$ and $j = 1, ..., K$ we have

$$\left(\widehat{\pi}_{ij}^l\right)_n - \pi_{ij}^l \stackrel{a.s.}{=} o\left(\left(\sum_{t=1}^{t(n)} \varepsilon_t \Delta n_t\right)^{-\frac{\rho}{2}}\right) \qquad (4.16)$$

where the sequences $\left\{s_n^{il}\right\}$ and $\{t(n)\}$ are defined by

$$s_n^{il} = \sum_{\tau=1}^n \chi\left(x_\tau = x(i), u_\tau = u(l)\right), \qquad (4.17)$$

$$t(n) = \{k : n_k \leq n \leq n_{k+1}, k = 1, 2, ...\}. \qquad (4.18)$$

Proof. 1) Let us introduce the sequences

$$S_{n+1}^{il} = \frac{1}{n} s_{n+1}^{il}, \ i = \overline{1,K}, \ l = \overline{1,N}. \qquad (4.19)$$

We shall now state the conditions which guarantee their equivalence with the sequence $\left\{\overline{S}_n^{il}\right\}$ whose elements are given by

$$\overline{S}_n^{il} := \frac{1}{n} \sum_{\tau=1}^n E\left\{\chi\left(x_\tau = x(i), u_\tau = u(l)\right) \mid \widehat{\mathcal{F}}_{t(\tau)}\right\}$$

where

$$\widehat{\mathcal{F}}_{t(\tau)} = \sigma\left(x_{n'(\tau)}; x_s, u_s, \eta_s \mid s = \overline{1, n'(\tau) - 1}\right),$$

and

$$n'(\tau) = \{n_k : n_k \leq \tau \leq n_{k+1}, k = 1, 2, ...\}. \qquad (4.20)$$

In other words, we want to prove that

$$\lim_{n \to \infty} \left(S_n^{il} - \overline{S}_n^{il}\right) = 0. \qquad (4.21)$$

It is clear that

$$S_n^{il} - \overline{S}_n^{il} = \Delta S_n^{il} + O^*\left(\frac{\Delta n_k}{n_k}\right),$$

$$\left|\Delta S_{n+1}^{il}\right| = \frac{n}{n+1}\left|\Delta S_n^{il}\right| \leq \left|\Delta S_n^{il}\right| \ \forall \ (n+1) \neq n'(n+1), \qquad (4.22)$$

where

$$\Delta S_n^{il} := \frac{1}{n} \sum_{t=1}^{t(n)} \sum_{\tau=n_t}^{n_{t+1}-1} \overline{\chi}_\tau^{il}$$

4.3. ESTIMATION OF THE TRANSITION MATRIX

with

$$\overline{\chi}_\tau^{il} = \chi\left(x_\tau = x(i), u_\tau = u(l) - E\left\{\chi\left(x_\tau = x(i), u_\tau = u(l)\right) \mid \widehat{\mathcal{F}}_{t(\tau)}\right\}\right).$$

Here $\{O^*(\delta_n)\}$ is a numerical sequence satisfying

$$0 < \varliminf_{n \to \infty} |O^*(\delta_n)/\delta_n| \leq \varlimsup_{n \to \infty} |O^*(\delta_n)/\delta_n| < \infty. \tag{4.23}$$

The property (4.22) means that the sequence $\{\Delta S_n^{il}\}$ is monotically non-increasing within the interval $\overline{n_k, n_{k+1} - 1}$. Hence to prove (4.21) it is sufficient to state that

$$\Delta S_n^{il} \xrightarrow[n \to \infty]{a.s.} 0.$$

We have

$$\Delta S_{n_k}^{il} = \frac{1}{n_k} \sum_{t=1}^{k} \sum_{\tau=n_t}^{n_{t+1}-1} \overline{\chi}_\tau^{il} = \frac{n_{k-1}}{n_k} \Delta S_{n_{k-1}}^{il} + \frac{1}{n_k} \sum_{\tau=n_k}^{n_{k+1}-1} \overline{\chi}_\tau^{il}.$$

From the last equality, it follows

$$E\left\{\left(\Delta S_{n_k}^{il}\right)^2 \mid \mathcal{F}_{n_{k-1}}\right\} \stackrel{a.s.}{=} \left(1 - \frac{\Delta n_k}{n_k}\right)^2 \left(\Delta S_{n_{k-1}}^{il}\right)^2$$

$$+ E\left\{\left(\frac{1}{n_k} \sum_{\tau=n_k}^{n_{k+1}-1} \overline{\chi}_\tau^{il}\right)^2 \mid \mathcal{F}_{n_{k-1}}\right\}$$

$$\leq \left[1 - 2\frac{\Delta n_k}{n_k}\left(1 + O^*(\frac{\Delta n_k}{n_k})\right)\left(\Delta S_{n_{k-1}}^{il}\right)^2 + \left(\frac{\Delta n_k}{n_k}\right)^2\right]$$

where

$$\mathcal{F}_{n_k} := \widehat{\mathcal{F}}_k = \sigma\left(x_{n'(\tau)}; x_s, u_s, \eta_s \mid s = \overline{1, n'(\tau) - 1}\right).$$

In view of the Robbins-Siegmund theorem [9] (theorem 2 of Appendix A), we obtain the desired result (4.21).

2) Now we can derive the lower bound of $\overline{S}_{n+1}^{il}$:

$$\overline{S}_{n+1}^{il} = \frac{1}{n} \sum_{\tau=1}^{n} E\left\{d_\tau^{il} \chi\left(x_\tau = x(i)\right) \mid \widehat{\mathcal{F}}_{t(\tau)}\right\}$$

$$\geq \frac{1}{n} \sum_{t=1}^{t(n)} d_{n_t}^{il} \sum_{\tau=n_t}^{n_{t+1}-1} E\left\{\chi\left(x_\tau = x(i)\right) \mid \widehat{\mathcal{F}}_t\right\}$$

$$\geq \frac{1}{n} \sum_{t=1}^{t(n)} \varepsilon_t \Delta n_t \theta_t^i, \tag{4.24}$$

where

$$\theta_t^i := \frac{1}{\Delta n_t} \sum_{\tau=n_t}^{n_{t+1}-1} E\left\{\chi\left(x_\tau = x(i)\right) \mid \widehat{\mathcal{F}}_t\right\}. \tag{4.25}$$

Based on the definition of the partially frozen randomized control strategy, the policy d_n remains unchanged within each time interval $\overline{n_k, n_{k+1}-1}$ and, as a result, the given controlled Markov chain can be considered as an aperiodic or regular chain with transition matrix $\Pi(d_n)$. Then, in view of the Rozanov theorem [10] (see chapter 1) (for regular chains, the state probability vector converges exponentially to its stationary value), we derive

$$\left| E\left\{\chi\left(x_\tau = x(i)\right) \mid \widehat{\mathcal{F}}_t\right\} - \sum_{s=1}^{N} c_t^{is} \right|$$

$$\leq o(1) + D_1(d_{n_t}) \exp\left\{-D_2(d_{n_t})(\tau - n_t)\right\}. \tag{4.26}$$

For any $d \in \mathbf{D}$, the regularity of the considered Markov chain leads to

$$D_1(d) \leq \overline{D}_1 < \infty, \quad D_2(d) \geq \overline{D}_2 > 0.$$

Using these estimates in (4.25) and,

$$\sum_{s=1}^{N} c^{is} = p_i(d) \geq \min_{i=1,\dots,K} \min_{d \in \mathbf{D}} p_i(d) := c_- > 0 \tag{4.27}$$

which is valid for all $c \in \mathbf{C}$ associated to a given regular Markov chain, we conclude

$$\theta_t^i = \frac{1}{\Delta n_t} \sum_{\tau=n_t}^{n_{t+1}-1} \left[E\left\{\chi\left(x_\tau = x(i)\right) \mid \widehat{\mathcal{F}}_t\right\} - \sum_{s=1}^{N} c_t^{is} \right] + \sum_{s=1}^{N} c_t^{is}$$

$$\geq c_- - \frac{\overline{D}_1}{\Delta n_t} \sum_{\tau=n_t}^{n_{t+1}-1} \exp\left\{-D_2(\tau - n_t)\right\} + o(1)$$

$$= c_- - \left| O\left(\frac{1}{\Delta n_t}\right) \right| + o(1) \geq \frac{1}{2} c_- > 0 \tag{4.28}$$

for any $t \geq T$, where T is a large enough integer. Taking into account (4.24) and (4.28), we obtain

$$S_{n+1}^{il} \geq \frac{c_-}{2n} \sum_{t=T}^{t(n)} \varepsilon_t \Delta n_t. \tag{4.29}$$

3) Now, we will estimate the rate of convergence to zero of the matrix Δn_t whose elements are defined by

$$\Delta n_t^{il} = S_{n_t}^{il} - \overline{S}_{n_t}^{il}.$$

4.3. ESTIMATION OF THE TRANSITION MATRIX

The following recurrence
$$\Delta n_t^{il} = \frac{k}{n_k}\frac{1}{k}\sum_{t=1}^{k} r_t^{il} = \left(1 - \frac{1}{k}\right)\phi_k \Delta n_{n_{k-1}}^{il} + n_k^{-1} r_k^{il}$$

holds, where
$$\phi_k := \frac{1}{k-1}\frac{n_{k-1}}{n_k}$$

and
$$r_k^{il} := \sum_{\tau=n_t}^{n_{t+1}-1} [\chi(x_\tau = x(i), u_\tau = u(l))$$
$$- P\{x_\tau = x(i), u_\tau = u(l) \mid \widehat{\mathcal{F}}_{t(\tau)}\}].$$

From this recurrence, we derive
$$E\left\{\left(\Delta n_{n_k}^{il}\right)^2 \mid \widehat{\mathcal{F}}_{k-1}\right\} \stackrel{a.s.}{=} \left(1 - \frac{1}{k}\right)^2 \eta_k^2 \left(\Delta n_{n_{k-1}}^{il}\right)$$
$$+ n_k^{-2} E\left\{\left(r_k^{il}\right)^2 \mid \widehat{\mathcal{F}}_{k-1}\right\} \leq \left(1 - \frac{1}{k}\right)^2 \eta_k^2 \left(\Delta n_{n_{k-1}}^{il}\right) + \left(n_k^{-1}\Delta n_k\right)^2$$

or in an equivalent form
$$E\left\{W_{n_k}^{il} \mid \widehat{\mathcal{F}}_{k-1}\right\} \stackrel{a.s.}{\leq} \left(1 - \frac{1}{k}\right)^2 W_{n_{k-1}}^{il} + \left(k^{-1}\Delta n_k\right)^2$$

where
$$W_{n_k}^{il} := k^{-1} n_k \left(\Delta n_{n_k}^{il}\right)^2.$$

By making of use of lemma 5 (Appendix A), we obtain
$$W_{n_k}^{il} \stackrel{a.s.}{=} o(v_k^{-1})$$

and
$$\Delta n_{n_k}^{il} \stackrel{a.s.}{=} o(n_k^{-1}\sqrt{kg_k\Delta n_k}). \tag{4.30}$$

Based on (4.20), (4.29) and (4.30), we derive
$$s_{\tau+1}^{il} = \sum_{n=1}^{\tau} \chi(x_\tau = x(i), u_\tau = u(l))$$
$$\geq \sum_{n=1}^{n'(\tau)} \chi(x_n = x(i), u_n = u(l)) = n'(\tau) S_{n'(\tau)}^{il}$$
$$= n'(\tau)\left(\overline{S}_{n'(\tau)}^{il} + \Delta_{n'(\tau)}^{il}\right)$$

$$\overset{a.s.}{\geq} n'(\tau)\left[\frac{c_-}{2n'(\tau)}\sum_{t=1}^{t(\tau)}\varepsilon_t\Delta_{n_t} + o\left(\frac{\sqrt{t(\tau g_{t(\tau)})}}{n'(\tau)}\left(n''(\tau) - n'(\tau)\right)\right)\right]$$

where
$$n''(\tau) = \{n_{k+1}: n_k \leq \tau < n_{k+1}, k = 1, 2, ...\}.$$

From the last inequality and in view of the assumptions of this theorem, we conclude that

$$s_{\tau+1}^{il} \overset{a.s.}{\geq} \sum_{t=1}^{t(\tau)}\varepsilon_t\Delta_{n_t}\left(\frac{1}{2}c_- + o(1)\right) \underset{\tau\to\infty}{\to} \infty. \quad (4.31)$$

From this inequality we state (4.15).

4) Let us now state assertion (4.16). From (4.31) we have

$$s_n^{il} \underset{n\to\infty}{\overset{a.s.}{\to}} \infty,$$

and, hence, starting from some integer $n_0 = n_0(\omega)$, for any $n \geq n_0$

$$s_n^{il} \overset{a.s.}{\geq} 0$$

and

$$\left(\Delta\pi_{ij}^l\right)_n = \left(\widehat{\pi}_{ij}^l\right)_n - \pi_{ij}^l$$

$$= \frac{1}{s_n^{il}}\sum_{t=1}^n \chi\left(x_t = x(i), u_t = u(l)\right)\left[\chi\left(x_{t+1} = x(j)\right) - \pi_{ij}^l\right]$$

$$= \left(1 - \frac{\chi\left(x_n = x(i), u_n = u(l)\right)}{s_n^{il}}\right)\left(\Delta\pi_{ij}^l\right)_{n-1}$$

$$+ \frac{\chi\left(x_n = x(i), u_n = u(l)\right)}{s_n^{il}}\left[\chi\left(x_{n+1} = x(j)\right) - \pi_{ij}^l\right].$$

Let us consider the σ-algebra $\widehat{\mathcal{F}}_{n-1}$ defined as

$$\widehat{\mathcal{F}}_n := \sigma\left(x_t, u_t \mid t = 1, ..., n\right).$$

s_n^{il} and $\left(\widehat{\pi}_{ij}^l\right)_{n-1}$ are $\widehat{\mathcal{F}}_{n-1}$-measurable. We derive

$$E\left\{\left(s_n^{il}\right)\rho\left(\Delta\pi_{ij}^l\right)_n^2 \mid \widehat{\mathcal{F}}_{n-1}\right\}$$

$$\overset{a.s.}{\leq} \left[1 - \frac{2-\rho}{s_n^{il}}\chi\left(x_n = x(i), u_n = u(l)\right)\right]\left[\left(s_{n-1}^{il}\right)\rho\left(\Delta\pi_{ij}^l\right)_{n-1}^2\right]$$

4.3. ESTIMATION OF THE TRANSITION MATRIX

$$+\frac{2\chi\left(x_n = x(i), u_n = u(l)\right)}{\left(s_n^{il}\right)^{2-\rho}}.$$

For any $\rho \in [0,1)$, we have

$$\sum_{n=n_0}^{\infty} \frac{\chi\left(x_n = x(i), u_n = u(l)\right)}{s_n^{il}} = \sum_{k=1}^{\infty} \frac{1}{k} = \infty$$

and

$$\sum_{n=n_0}^{\infty} \frac{\chi\left(x_n = x(i), u_n = u(l)\right)}{\left(s_n^{il}\right)^{2-\rho}} = \sum_{k=1}^{\infty} \frac{1}{k^{2-\rho}} < \infty$$

with probability 1.

From these two relations and in view of the Robbins-Siegmund theorem [9] (theorem 2 of Appendix A), we conclude

$$\left(s_n^{il}\right) \rho \left(\Delta \pi_{ij}^l\right)_n^2 \xrightarrow[n \to \infty]{a.s.} 0.$$

This expression is equivalent to (4.16). The theorem is proved. ∎

Corollary 1 *If in the previous theorem the sequences $\{\varepsilon_k\}$ and $\{n_k\}$ are selected as follows*

$$\varepsilon_k \sim k^{-\beta}, \quad n_k \sim k^{\kappa}$$

with

$$\kappa > \beta, \; 0 \leq \beta < \frac{1}{2},$$

then, for any $i = 1, ..., K$ and $l = 1, ..., N$

$$\lim_{n \to \infty} \frac{s_n^{il}}{n^{1-\frac{\beta}{\kappa}}} > 0 \tag{4.32}$$

with probability 1.

Proof. From (4.16) it follows

$$\sum_{t=1}^{k} \varepsilon_t \Delta n_t \sim \int_1^k t^{-\beta} t^{\kappa-1} dt \sim k^{\kappa-\beta} \xrightarrow[k \to \infty]{} \infty.$$

Notice that, for example, conditions (4.13) and (4.14) of the theorem are fulfilled for

$$g_k = k^\rho$$

with the parameter ρ satisfying

$$0 < \rho \leq 1 - 2\beta.$$

Taking into account that
$$t(n) \sim n^{\frac{1}{\kappa}}$$
we obtain
$$\sum_{t=1}^{t(n)} \varepsilon_t \Delta n_t \sim n^{1-\frac{\beta}{\kappa}}$$
and, as a result, we have (4.32). The corollary is proved. ∎

The theory of self-learning (adaptive) systems time is able to solve many problems arising in practice. The devised adaptive algorithm can, in very wide conditions of indeterminacy ensure the achievement of the desired control goal.

The key to being able to analyse the behaviour of the adaptive control algorithm described above is presented in the next section.

4.4 Convergence analysis

The next theorem states the convergence of the loss function Φ_n to its minimal value Φ_* (see lemma 1 of chapter 2).

Theorem 2 *If the loss function satisfies (2.1)-(2.3) and, the considered regular homogeneous finite Markov chain (see Definition 23 of chapter 1) is controlled by the adaptive procedure (4.8)-(4.11) with the parameters $\{\gamma_k\}$, $\{\varepsilon_k\}$ and $\{n_k\}$ satisfying*

$$0 < \varepsilon_k \underset{k\to\infty}{\to} 0, \; \gamma_k > 0, \; k n_k^{-1} \underset{k\to\infty}{\to} 0, \tag{4.33}$$

$$\Delta n_k = n_{k+1} - n_k \underset{k\to\infty}{\to} \infty, \tag{4.34}$$

$$\lim_{k\to\infty} \gamma_k^{-1} \left[n_k^{-1} \Delta n_k + |\varepsilon_{k+1} - \varepsilon_k| + \Delta n_k \left(\sum_{t=1}^{k} \varepsilon_t \Delta n_t \right)^{-1} \right] = 0, \tag{4.35}$$

$$\gamma_{k+1}^{-1} \Delta n_{k+1} \geq \gamma_k^{-1} \Delta n_k, \; \sum_{k=1}^{\infty} \frac{\gamma_k \Delta n_k}{\varepsilon_k n_k} + \left(\frac{\Delta n_k}{n_k} \right)^2 < \infty \tag{4.36}$$

and there exists a nonnegative sequence $\{h_k\}$ such that

$$\sum_{k=1}^{\infty} h_k^2 < \infty, \; \overline{\lim_{k\to\infty}} \, \Delta n_k \left(h_k \sum_{t=1}^{k} \varepsilon_t \Delta n_t \right)^{-1} < \infty \tag{4.37}$$

then, for any initial value $c_1 \in \widehat{C}_{\varepsilon_1}$, the loss sequence $\{\Phi_n\}$ converges to its minimal value Φ_*, with probability 1, i.e.,

$$\Phi_n \underset{n\to\infty}{\overset{a.s.}{\to}} \Phi_*$$

4.4. CONVERGENCE ANALYSIS

Proof. Let us consider any point $\bar{c} \in \mathbf{C}_{\varepsilon=0}$ as well as its projection $\tilde{c}_k$ onto the set $\widehat{\mathbf{C}}_{\varepsilon_k}$. Then, using the property (4.12) of the projection operator $\mathcal{P}^{\widehat{\mathbf{C}}}_{\varepsilon_k}\{\cdot\}$ and, the uniform boundedness of the set $\widehat{\mathbf{C}}_{\varepsilon_k}$ with respect to n and $\omega \in \Omega$, we get

$$\|c_{k+1} - \tilde{c}_{k+1}\|^2 = \sum_{i,l}\left(c_{k+1}^{il} - \tilde{c}_{k+1}^{il}\right)^2$$

$$\leq \sum_{i,l}\left(c_k^{il} - \gamma_k A_{n_{k+1}}^{il} - \tilde{c}_{k+1}^{il}\right)^2$$

$$= \|c_k - \tilde{c}_{k+1}\|^2 - 2\gamma_k \sum_{i,l}\left(c_k^{il} - \tilde{c}_{k+1}^{il}\right)A_{n_{k+1}}^{il} + \gamma_k^2 \|A_{n_{k+1}}\|^2$$

$$\leq \|c_k - \tilde{c}_k\|^2 - 2\gamma_k \sum_{i,l}\left(c_k^{il} - \tilde{c}_k^{il}\right)A_{n_{k+1}}^{il}$$

$$+\gamma_k^2 \|A_{n_{k+1}}\|^2 + K_1\|\tilde{c}_k - \tilde{c}_{k+1}\|\left(1 + \gamma_k\|A_{n_{k+1}}\|\right)$$

where

$$k = 1, 2, ...;\quad K_1 = const \in (0, \infty).$$

Weighting these inequalities with $\Delta n_t/(2\gamma_t)$ and, summing up them over $t = 1$ up to time $t = k$, we obtain

$$\frac{1}{n_{k+1} - 1}\sum_{t=1}^{k}\Delta n_t \sum_{i,l}\left(c_t^{il} - \tilde{c}_t^{il}\right)A_{n_{t+1}}^{il}$$

$$\leq \frac{1}{2(n_{k+1} - 1)}\sum_{t=1}^{k}\gamma_t^{-1}\Delta n_t\left(\|c_t - \tilde{c}_t\|^2 - \|c_{t+1} - \tilde{c}_{t+1}\|^2\right)$$

$$+\frac{1}{2(n_{k+1} - 1)}\sum_{t=1}^{k}\gamma_t^{-1}\Delta n_t\left[\gamma_t^2\|A_{n_{t+1}}^{il}\|^2\right.$$

$$\left. + K_1\|c_t - \tilde{c}_{t+1}\|\left(1 + \gamma_k\|A_{n_{t+1}}\|\right)\right].$$

Based on the definition of the matrix $A_{n_{t+1}}$ (4.9), the previous inequality leads to

$$\frac{1}{n_{k+1} - 1}\sum_{t=1}^{k}\sum_{i,l}\left(c_t^{il} - \tilde{c}_t^{il}\right)\sum_{s=n_t}^{n_{t+1}-1}\frac{\eta_s\chi\left(x_s = x(i), u_s = u(l)\right)}{e^T(x_s)c_t e(x_s)}$$

$$= \frac{1}{n_{k+1} - 1}\sum_{\tau=1}^{n_{k+1}-1}\theta_\tau$$

with

$$\theta_\tau := \sum_{i,l} \left(c^{il}_{t(\tau)} - \widetilde{c}^{il}_{t(\tau)} \right) \sum_{s=n_t}^{n_{t+1}-1} \frac{\eta_\tau \chi\left(x_\tau = x(i), u_\tau = u(l)\right)}{e^T(x_\tau) c_t e(x_\tau)}$$

and $t(\tau)$ is defined by (4.18).

Notice that the arithmetic average of the random variables θ_τ is asymptotically equal to the arithmetic average of their conditional mathematical expectations $E\{\theta_\tau \mid \mathcal{F}_\tau\}$, i.e.,

$$\frac{1}{n} \sum_{\tau=1}^n \theta_\tau \stackrel{a.s.}{=} \frac{1}{n} \sum_{\tau=1}^n E\{\theta_\tau \mid \mathcal{F}_\tau\} + o(1) \tag{4.38}$$

where

$$\mathcal{F}_\tau = \sigma\left(x_\tau, \eta_s, x_s, u_s \mid s = \overline{1, \tau-1}\right).$$

Taking into account the assumptions of this theorem, (4.18) follows directly from lemma 3 (see Appendix A). Indeed,

$$\sum_{\tau=1}^\infty \tau^{-2} \sum_{i,l} E\left\{ \left(c^{il}_{t(\tau)} - \widetilde{c}^{il}_{t(\tau)} \right) \frac{\eta_\tau^2 \chi\left(x_\tau = x(i), u_\tau = u(l)\right)}{\left(c^{il}_{t(\tau)}\right)^2} \mid \mathcal{F}_\tau \right\}$$

$$\leq \text{const} \sum_{\tau=1}^\infty \tau^{-2} \sum_{i,l} \left(c^{il}_{t(\tau)} \sum_{s=1}^N c^{is}_{t(\tau)} \right)^{-1}$$

$$\leq C(\omega) \sum_{\tau=1}^\infty \tau^{-2} \varepsilon_{t(\tau)}^{-1} \leq C(\omega) \sum_{t=1}^\infty \sum_{s=n_t}^{n_{t+1}-1} s^{-2} \varepsilon_t^{-1}$$

$$\leq C(\omega) \sum_{t=1}^\infty \varepsilon_t^{-1} \int_{n_t-1}^{n_{t+1}-1} x^{-2} dx$$

$$= C(\omega) \sum_{t=1}^\infty \varepsilon_t^{-1} \left(\frac{1}{n_t - 1} - \frac{1}{n_{t+1} - 1} \right) < \infty. \tag{4.39}$$

This series converges because

$$\sum_{k=1}^\infty \gamma_k \Delta n_k \varepsilon_k^{-1} n_k^{-1} < \infty$$

and

$$n_{k+1} \gamma_k \underset{k \to \infty}{\to} \infty.$$

Taking into account the property of the loss function

$$E\left\{\eta_\tau \chi\left(x_\tau = x(i)\right) \mid \widehat{\mathcal{F}}_\tau \wedge x_\tau = x(j)\right\} \stackrel{a.s.}{=} v_{ij},$$

4.4. CONVERGENCE ANALYSIS

in the calculation of the conditional mathematical expectation $E\{\theta_\tau \mid \mathcal{F}_\tau\}$, we get

$$E\{\theta_\tau \mid \mathcal{F}_\tau\} \stackrel{a.s.}{=} \sum_{i,l} \left(c^{il}_{t(\tau)} - \tilde{c}^{il}_{t(\tau)}\right) \left(\sum_{s=1}^{N} c^{is}_{t(\tau)}\right)^{-1} \chi\left(x_\tau = x(i)\right) \cdot$$

$$\cdot E\{\eta_\tau \mid \widehat{\mathcal{F}}_\tau \wedge x_\tau = x(j)\}$$

$$= \sum_{i,l} \left(c^{il}_{t(\tau)} - \tilde{c}^{il}_{t(\tau)}\right) \left(\sum_{s=1}^{N} c^{is}_{t(\tau)}\right)^{-1} \chi\left(x_\tau = x(i)\right) v_{ij}.$$

From this relation, we derive

$$\frac{1}{n}\sum_{\tau=1}^{n} E\{\theta_\tau \mid \mathcal{F}_\tau\} \stackrel{a.s.}{=} \frac{1}{n}\sum_{\tau=1}^{n}\sum_{i,l} v_{ij}\chi\left(x_\tau = x(i)\right) d^{ij}_{n'(\tau)} -$$

$$- \sum_{i,l} v_{ij}\tilde{c}^{il} - r_{1n}$$

where

$$r_{1n} := \sum_{i,l} v_{ij} \left[\frac{1}{n}\sum_{\tau=1}^{n} \tilde{c}^{il}_{t(\tau)}\chi\left(x_\tau = x(i)\right) \left(\sum_{s=1}^{N} c^{is}_{t(\tau)}\right)^{-1} - \tilde{c}^{il}\right]. \quad (4.40)$$

Finally, combining all the previous inequalities, we conclude:

$$\frac{1}{n_{k+1} - 1} \sum_{\tau=1}^{n_{k+1}-1} \sum_{i,l} v_{ij}\chi\left(x_\tau = x(i)\right) d^{ij}_{n'(\tau)} - \sum_{i,l} v_{ij}\tilde{c}^{il}$$

$$\leq \tilde{r}_{1n_k} + r_{2n_k} + r_{3n_k} + o(1) \quad (4.41)$$

where

$$\tilde{r}_{1n_k} := r_{1n}\mid_{n=n_{k+1}-1}, \quad (4.42)$$

$$r_{2n_k} := \frac{1}{2(n_{k+1}-1)} \sum_{t=1}^{k} \Delta n_t \gamma_t^{-1} \left(\|c_t - \tilde{c}_t\|^2 - \|c_{t+1} - \tilde{c}_{t+1}\|^2\right), \quad (4.43)$$

$$r_{3n_k} := \frac{1}{2(n_{k+1}-1)} \sum_{t=1}^{k} \Delta n_t \gamma_t^{-1} \left[\gamma_t^2 \left\|A_{n_{t+1}}\right\|^2\right.$$

$$\left. + K_1 \|c_t - \tilde{c}_{t+1}\| \left(1 + \gamma_t \left\|A_{n_{t+1}}\right\|\right)\right].$$

Let us now show that, when $k \to \infty$, the right hand side of (4.41) tends to zero, with probability 1.

a) To show that
$$\tilde{r}_{1n_k} \xrightarrow[k\to\infty]{a.s.} 0,$$
let us decompose r_{1n} (4.40) into the sum of two terms:
$$r_{1n} = r'_{1n} + r''_{1n} \tag{4.44}$$
where
$$r'_{1n} := \sum_{i,l} v_{ij} \frac{1}{n} \sum_{\tau=1}^{n} \left(\tilde{c}^{il}_{t(\tau)} - \tilde{c}^{il} \right) \chi \left(x_\tau = x(i) \right) \left(\sum_{s=1}^{N} c^{is}_{t(\tau)} \right)^{-1} \tag{4.45}$$
and
$$r''_{1n} := \sum_{i,l} v_{ij} \tilde{c}^{il} \frac{1}{n} \sum_{\tau=1}^{n} \left[\chi \left(x_\tau = x(i) \right) \left(\sum_{s=1}^{N} c^{is}_{t(\tau)} \right)^{-1} - 1 \right]. \tag{4.46}$$

Selecting h_k as follows
$$h_k := \frac{1}{\sqrt{k g_k}}$$
we satisfy the conditions of theorem 1, and, hence, we obtain the consistency of the estimates of the transition probability matrix, i.e.,
$$\left(\tilde{\pi}^l_{ij} \right)_n \xrightarrow[n\to\infty]{a.s.} \pi^l_{ij} \quad \left(i,j = \overline{1,K}, l = \overline{1,N} \right).$$

Based on (2.15) (see chapter 2) and
$$\tilde{c}^{il}_k \xrightarrow[k\to\infty]{} \tilde{c}^{il} \text{ if } \varepsilon_k \xrightarrow[k\to\infty]{} 0$$
we conclude:
$$r''_{1n} \xrightarrow[k\to\infty]{a.s.} 0.$$

Let us now prove that
$$r''_{1n_k} = r''_{1n} \mid_{n=n_{k+1}-1} \xrightarrow[k\to\infty]{a.s.} 0.$$

We have
$$\frac{1}{n_{k+1}-1} \sum_{\tau=1}^{n_{k+1}-1} \chi \left(x_\tau = x(i) \right) \left(\sum_{s=1}^{N} c^{is}_{t(\tau)} \right)^{-1}$$
$$= \frac{1}{n_{k+1}-1} \sum_{t=1}^{k} \left(\sum_{s=1}^{N} c^{is}_t \right)^{-1} \sum_{\tau=1}^{n_{k+1}-1} \chi \left(x_\tau = x(i) \right) \stackrel{a.s.}{=}$$
$$\stackrel{a.s.}{=} \frac{1}{n_{k+1}-1} \sum_{t=1}^{k} \left(\sum_{s=1}^{N} c^{is}_t \right)^{-1} \sum_{\tau=1}^{n_{k+1}-1} P \left(x_\tau = x(i) \mid \mathcal{F}_{n_t} \right) + o(1). \tag{4.47}$$

4.4. CONVERGENCE ANALYSIS

This equality follows directly from lemma 3 (see Appendix A), the assumptions of this theorem and, from the convergence of the series

$$\sum_{t=1}^{\infty} n_t^{-2} E\left\{ \left(\sum_{s=1}^{N} c_t^{is}\right)^{-2} \left[\sum_{\tau=1}^{n_{k+1}-1} \chi\left(x_\tau = x(i)\right)\right.\right.$$

$$\left.\left. - P\left(x_\tau = x(i) \mid \mathcal{F}_{n_t}\right)\right]^2 \mid \mathcal{F}_{n_t}\right\}$$

$$\stackrel{a.s.}{\leq} C(\omega) \sum_{t=1}^{\infty} n_t^{-2} (\Delta n_t)^2 \stackrel{a.s.}{<} \infty.$$

From (4.47), it follows that

$$\tilde{r}_{1n_k}'' \stackrel{a.s.}{=} \sum_{i,l} v_{ij} \tilde{c}^{il} \frac{1}{n_{k+1}-1} \sum_{t=1}^{k} \left[\left(\sum_{s=1}^{N} c_t^{is}\right)^{-1}\right.$$

$$\left. \cdot \sum_{\tau=n_t}^{n_{t+1}-1} P\left(x_\tau = x(i) \mid \mathcal{F}_{n_t}\right) - \Delta n_t\right] + o(1).$$

Using (4.26), we derive

$$\tilde{r}_{1n_k}'' \stackrel{a.s.}{\leq} C(\omega) \frac{1}{n_{k+1}-1} \sum_{t=1}^{k} \sum_{\tau=n_t}^{n_{t+1}-1} \exp\left\{-\overline{D}_2 (\tau - n_t)\right\} o(1)$$

$$\leq C_1(\omega) k (n_{k+1} - 1)^{-1} + o(1) \stackrel{a.s.}{\underset{k\to\infty}{\to}} 0.$$

So, we prove that

$$\tilde{r}_{1n_k} \stackrel{a.s.}{\underset{k\to\infty}{\to}} 0.$$

b) Let us now consider the term r_{2n_k}. Using the estimation

$$\left|\sum_{t=1}^{k} \left(\|c_t - \tilde{c}_t\|^2 - \|c_{t+1} - \tilde{c}_{t+1}\|^2\right)\right|$$

$$= \|c_1 - \tilde{c}_1\|^2 - \|c_{k+1} - \tilde{c}_{k+1}\|^2 \leq const < \infty$$

and applying lemma 7 (see Appendix A), we prove that

$$\lim_{k\to\infty} r_{2n_k} \stackrel{a.s.}{=} 0.$$

c) Let us consider the term r_{3n_k}. In view of theorem 1, we get

$$\left|\left(\hat{\pi}_{ij}^l\right)_n - \left(\hat{\pi}_{ij}^l\right)_{n+1}\right| \leq \left(s_{n+1}^{ij}\right)^{-1} \stackrel{a.s.}{\leq} C(\omega) \left(\sum_{t=1}^{t(n)} \varepsilon_t \Delta n_t\right)^{-1}$$

with
$$C(\omega) \stackrel{a.s.}{\in} (0, \infty).$$

It is easy to demonstrate that
$$\|c_t - \tilde{c}_{t+1}\| \leq K1 \left(|\varepsilon_t - \varepsilon_{t+1}| + \|\hat{\pi}_{n_t} - \hat{\pi}_{n_{t+1}}\| \right)$$

for some $K_1 \in (0, \infty)$. From the previous inequality, the following relation follows:
$$\|c_k - \tilde{c}_{k+1}\| \stackrel{a.s.}{\leq} C(\omega) \left[|\varepsilon_k - \varepsilon_{k+1}| + \Delta n_k \left(\sum_{t=1}^{k} \varepsilon_t \Delta n_t \right)^{-1} \right].$$

Based on the following inequalities
$$2ab \leq a^2 + b^2, \quad \left(\sum_{i=1}^{m} a_i \right)^2 \leq m \sum_{i=1}^{m} a_i^2,$$

which are valid for any a and b, we conclude
$$r_{3n_k} \leq \frac{K_2}{n_{k+1} - 1} \sum_{t=1}^{k} \Delta n_t \gamma_t^{-1} \left(\gamma_t^2 \|A_{n_{t+1}}\|^2 + \|c_t - \tilde{c}_{t+1}\| \right)$$
$$\leq \frac{K_2}{n_{k+1} - 1} \sum_{t=1}^{k} \gamma_t \left[\sum_{s=n_t}^{n_{t+1}-1} \frac{\eta_s^2}{[e^T(x_s) c_s e(u_s)]^2} \right.$$
$$\left. + C(\omega) \Delta n_t \gamma_t^{-1} \left(|\varepsilon_t - \varepsilon_{t+1}| + \Delta n_t \left(\sum_{s=1}^{t} \varepsilon_s \Delta n_s \right)^{-1} \right) \right]$$
$$= \frac{K_2}{n_{k+1} - 1} \sum_{\tau=1}^{n_{k+1}-1} \theta'_\tau + \frac{K_2(\omega)}{n_{k+1} - 1} \sum_{t=1}^{k} \theta''_\tau$$

where
$$\theta'_\tau := \gamma_{t(\tau)} \eta_\tau^2 \left[e^T(x_\tau) c_\tau e(u_\tau) \right]^{-2}$$

and
$$\theta''_\tau := \Delta n_t \gamma_t^{-1} \left(|\varepsilon_t - \varepsilon_{t+1}| + \Delta n_t \left(\sum_{s=1}^{t} \varepsilon_s \Delta n_s \right)^{-1} \right).$$

Observe that
$$\sum_{\tau=1}^{\infty} \tau^{-1} E \left\{ \theta'_\tau \mid \mathcal{F}_{n'(\tau)} \right\} \stackrel{a.s.}{\leq} C(\omega) \sum_{\tau=1}^{\infty} \tau^{-1} \gamma_{t(\tau)} \varepsilon_{t(\tau)}^{-1}$$

4.4. CONVERGENCE ANALYSIS

$$\leq C(\omega) \sum_{k=1}^{\infty} \gamma_k \varepsilon_k^{-1} \sum_{\tau=n_k}^{n_{k+1}-1} \tau^{-1}$$

$$\leq C(\omega) \sum_{k=1}^{\infty} \gamma_k \varepsilon_k^{-1} \frac{n_{k+1} - n_k}{n_k} < \infty,$$

we conclude

$$\sum_{\tau=1}^{\infty} \tau^{-1} \theta'_\tau \stackrel{a.s.}{<} \infty,$$

and hence (law of large numbers),

$$\frac{1}{n} \sum_{\tau=1}^{\infty} \tau^{-1} \stackrel{a.s.}{\underset{k \to \infty}{\to}} 0.$$

In view of the Toeplitz lemma (lemma 8 of Appendix A), we derive

$$\frac{1}{n_{k+1} - 1} \sum_{t=1}^{k} \theta''_\tau \stackrel{a.s.}{\underset{k \to \infty}{\to}} 0.$$

Hence,

$$r_{3n_k} \stackrel{a.s.}{\underset{k \to \infty}{\to}} 0.$$

d) To finish the proof of this theorem, we have to calculate the limits in (4.41):

$$\overline{\lim_{k \to \infty}} \frac{1}{n_{k+1} - 1} \sum_{\tau=1}^{n_{k+1}-1} \sum_{i,l} v_{ij} \chi(x_\tau = x(i)) d^{ij}_{n'(\tau)}$$

$$\stackrel{a.s.}{\leq} \sum_{i,l} v_{ij} \tilde{c}^{il}. \tag{4.48}$$

According to lemma 3 (see Appendix A), we derive

$$\Phi_n \stackrel{a.s.}{=} \overline{\Phi}_n + o(1)$$

where

$$\overline{\Phi}_n := \frac{1}{n} E\{\eta_\tau \mid \mathcal{F}_\tau\}$$

$$= \sum_{\tau=1}^{n} \sum_{i,l} v_{ij} \chi(x_\tau = x(i)) d^{ij}_{n'(\tau)}. \tag{4.49}$$

Indeed, to apply this lemma, it is sufficient to demonstrate that

$$\sum_{\tau=1}^{\infty} \tau^{-2} E\{\eta_\tau^2 \mid \mathcal{F}_\tau\} \leq C(\omega) \sum_{\tau=1}^{\infty} \tau^{-2} \sum_{i,l} \left(c^{ij}_{t(\tau)} \sum_{s,\tau} c^{s\tau}_{t(\tau)} \right)^{-1}$$

$$\overset{a.s.}{\le} C(\omega) \sum_{\tau=1}^{\infty} \tau^{-2} \varepsilon_{t(\tau)}^{-1} \le C(\omega) \sum_{k=1}^{\infty} \varepsilon_k^{-1} \frac{\Delta n_k}{n_k n_{k+1}} < \infty.$$

Taking into account that

$$\overline{\Phi}_n = \frac{n_k}{n} \overline{\Phi}_{n_k} + \frac{1}{n} \sum_{\tau=n_k}^{n} \sum_{i,l} v_{ij} \chi\left(x_\tau = x(i)\right) d_{n_k}^{ij}$$

with

$$k = t(n), \ n_k = n'(n),$$

and based on the assumptions of this theorem, we get

$$\left|\overline{\Phi}_n - \overline{\Phi}_{n_k}\right| \le \frac{n - n_k}{n} \overline{\Phi}_{n_k} + const \frac{\Delta n_k}{n}$$

$$\le const \frac{\Delta n_k}{n} \to 0$$

when $n \to \infty$. Hence, from this relation and, inequality (4.48), we finally obtain

$$\varlimsup_{n \to \infty} \Phi_n = \varlimsup_{n \to \infty} \overline{\Phi}_{n_k} \le \sum_{i,l} v_{ij} \tilde{c}^{ij} = V(\tilde{d})$$

where

$$\tilde{d}^{ij} := \tilde{c}^{ij} \left(\sum_{s=1}^{N} \tilde{c}^{is}\right)^{-1}, i = \overline{1,K}, \ j = \overline{1,N}$$

and

$$\tilde{d} = \left\|\tilde{d}^{ij}\right\|_{\substack{i=\overline{1,K} \\ j=\overline{1,N}}}.$$

The last inequality is valid for any $\tilde{d} \in \mathbf{D}$. The theorem is proved. ∎

The conditions (4.33)-(4.37) of theorem 2 give the class of the design parameters $\{\gamma_k\}, \{\varepsilon_k\}$ and $\{n_k\}$ of the adaptive control algorithm (4.8)-(4.11) which guarantee the convergence of the loss sequence $\{\Phi_n\}$ to its minimal value Φ_*.

Corollary 2 *For the special (but commonly used) class of the parameters*

$$\gamma_k = \gamma k^{-\nu}, \ \varepsilon_k = \varepsilon k^{-\theta}, \ n_k = [k^\kappa] \ (\gamma, \kappa > 0) \tag{4.50}$$

the corresponding convergence conditions (4.33)-(4.37) can be transformed into the following simple form

$$0 < \theta < \nu < 1 - \theta, \ 1 < \kappa. \tag{4.51}$$

The convergence of an adaptive scheme is important but the convergence speed is also essential. It depends on the number of operations performed by the algorithm during an iteration as well as the number of iterations needed for convergence. The next section is dedicated to the speed of the adaptive control algorithm (4.8)-(4.11).

4.5 Rate of adaptation and its optimization

The next theorem gives the estimation of the convergence rate (or adaptation rate) for the convergence of the sequence $\{\Phi_n\}$ to its minimal value Φ_*.

Theorem 3 *Let us consider the adaptive control algorithm (4.8)-(4.11) for which the parameters $\{\gamma_k\}$, $\{\varepsilon_k\}$ and $\{n_k\}$ are given by (4.50)-(4.51) Then, for any positive δ and any initial values $c_1 \in \widehat{\mathbf{C}}_{\varepsilon_1}$ and $x_1 \in X$, we have*

$$\Phi_n - \Phi_* \leq o(n^{\delta - \varphi_1}) + O^*(n^{-\varphi_2})$$

with probability 1, where

$$\varphi_1 := \kappa^{-1} \min\left\{\frac{\kappa - \theta}{2}, \nu - \theta\right\}$$

and

$$\varphi_2 := \kappa^{-1} \min\{1 - \theta - \nu,\ \theta,\ \kappa - 1\}.$$

Proof. Let us observe that, by means of (4.38), we can express $o(1)$ as

$$o(1) = \frac{1}{n_{k+1} - 1} \sum_{\tau=1}^{n_{k+1}-1} (\theta_\tau - E\{\theta_\tau \mid \mathcal{F}_\tau\}) := r_{4n_k}.$$

Applying lemma 4 (see Appendix A), we obtain

$$r_{4n_k} \stackrel{a.s.}{=} o\left(n_k^{\delta - \frac{\kappa - \theta}{2\kappa}}\right).$$

for any $\delta > 0$.
From

$$\psi_n := n^{1 - 2\delta - \frac{\theta}{\kappa}}$$

and the inequality

$$\lim_{n \to \infty} \left(\frac{\psi_n}{\psi_{n-1}} - 1\right) n = 1 - 2\delta - \frac{\theta}{\kappa} < 1,$$

it follows

$$\sum_{n=1}^{\infty} n^{-2} \psi_n E\left\{(\theta_n - E\{\theta_n \mid \mathcal{F}_n\})^2 \mid \mathcal{F}_n\right\}$$

$$\stackrel{a.s.}{\leq} \sum_{n=1}^{\infty} n^{-2} \psi_n E\{\theta_n^2 \mid \mathcal{F}_n\}$$

$$= \sum_{n=1}^{\infty} n_n^{-2} \psi_n \sum_{i,j} \chi\left(x_n = x(i)\right) \left(c_{t(n)}^{ij} - \tilde{c}_{t(n)}^{ij}\right)^2 \cdot$$

$$\cdot \left(c_{t(n)}^{ij} \sum_{s=1}^{N} c_{t(n)}^{is}\right)^{-1} E\left\{\eta_n^2\left(x(i), u(j), \omega\right)\right\}$$

$$\leq C_1(\omega) \sum_{n=1}^{\infty} n_n^{-2} \psi_n \varepsilon_{t(n)}^{-1}$$

$$\leq C_2(\omega) \sum_{n=1}^{\infty} n_n^{-1-\delta} < \infty$$

with probability 1. In these calculations, we take into account the following relations

$$\varepsilon_{t(n)} = \varepsilon \left(1 + n'(n)\right)^{-\frac{\theta}{\kappa}} \geq \varepsilon(1+n)^{-\frac{\theta}{\kappa}}.$$

a) We proceed to the estimation of the random variable r'_{1n}. Using relation (4.44) and theorem 1, we get

$$r'_{1n} \stackrel{a.s.}{\leq} C_3 n^{-1} \sum_{\tau=1}^{n} \left\|\tilde{c}_{t(\tau)} - \tilde{c}\right\|$$

$$\leq C_4 n^{-1} \sum_{k=1}^{t(n)} \Delta n_k \left(\varepsilon_k + \|\tilde{\pi}_{n_k} - \pi\|\right)$$

$$\leq C_4 n^{-\frac{\theta}{\kappa}} + o(n^{\delta - \frac{\kappa-\theta}{2\kappa}}) \; \forall \delta > 0.$$

Here C_i $(i = 1, ...)$ are nonnegative random variables bounded with probability 1.

To estimate r''_{1n_k}, it is enough to apply lemma 4 (see Appendix A) as before:

$$\left|r''_{1n_k}\right| \stackrel{a.s.}{\leq} C_6 k^{1-\kappa} + o(k^{\delta - \frac{1}{2}}).$$

Indeed, to apply this lemma we have to prove the convergence of the following series with $\psi_k := k^{3-2\kappa-2\delta}$:

$$\sum_{t=1}^{\infty} t^{-2} \psi_t E\left\{\left(\sum_{l=1}^{N} c_t^{il}\right)^{-2} \left[\sum_{\tau=n_t}^{n_{t+1}-1} (\chi\left(x_\tau = x(i)\right)\right.\right.$$

$$\left.\left. - P\left\{x_\tau = x(i) \mid \mathcal{F}_{n_t}\right\}\right)\right]^2 \mid \mathcal{F}_{n_t}\right\}$$

$$\stackrel{a.s.}{\leq} C_\tau(\omega) \sum_{t=1}^{\infty} t^{-(1+2\delta)} < \infty$$

4.5. RATE OF ADAPTATION AND ITS OPTIMIZATION

and to estimate the limit
$$\lim_{k \to \infty} \left(\frac{\psi_k}{\psi_k - 1} - 1 \right) k = 3 - 2\kappa - 2\delta < 2.$$

Taking into account that
$$\left| r''_{1n_k} \right| \overset{a.s.}{\leq} C_6 n_k^{\frac{1-\kappa}{\kappa}} + o(n_k^{(\delta - \frac{1}{2})/\kappa}).$$

Hence, for any $\delta > 0$ we argue as before to conclude
$$|\tilde{r}_{1n_k}| \overset{a.s.}{\leq} o(n_k^{-\frac{\kappa-\delta}{2\kappa+\delta}}) + C_5 n_k^{\frac{1-\kappa}{\kappa}} + C_6 n_k^{\frac{1-\kappa}{\kappa}}.$$

b) To estimate the rate of decreasing of the terms r_{2n_k} to zero, it is enough to apply lemma 7 (see Appendix A):
$$|r_{2n_k}| \overset{a.s.}{\leq} \frac{C_\tau}{n_{k+1} - 1} \cdot \frac{\Delta n_k}{\gamma_k} \leq C_8 n_k^{\frac{1-\nu}{\kappa}}.$$

c) Finally, for the terms r_{3n_k}, after applying lemma 5 (see Appendix A), we obtain
$$|r_{3n_k}| \overset{a.s.}{\leq} C_9 n_k^{\frac{\theta + \kappa - 1}{2}} + o(n_k^{-\frac{(\nu - \theta)}{\kappa + \delta}}).$$

This estimate follows from the facts that
$$\sum_{n=1}^{\infty} n^{-1} n^{\frac{\nu - \theta}{\kappa}} E \left\{ \gamma_{t(n)} \eta_n^2 \left[e^T(x_n) c_n e(u_x) \right]^{-2} \mid \mathcal{F}_n \right\}$$
$$\overset{a.s.}{\leq} C_{10} \sum_{n=1}^{\infty} n^{-1-\delta} < \infty$$

and
$$\lim_{n \to \infty} \left[\frac{(n+1)^{\frac{\nu}{\kappa} - \delta}}{n^{\frac{\nu}{\kappa} - \delta}} - 1 \right] n = \frac{\nu}{\kappa} - \delta < 1.$$

Combining all the estimates obtained above, we derive:
$$\frac{1}{n_{k+1} - 1} \sum_{\tau=1}^{n_{k+1}-1} \sum_{i,j} v_{ij} \chi(x_\tau = x(i)) d_{n'(\tau)}^{ij} - \sum_{i,j} v_{ij} c^{ij}$$
$$\overset{a.s.}{\leq} o(n_k^{\delta - \frac{\kappa - \theta}{2\kappa}}) + C_6 n_k^{\frac{1-\kappa}{\kappa}} + C_9 n_k^{\frac{\theta + \nu - 1}{\nu}} + o(n_k^{\delta - \frac{\nu - \theta}{2}}).$$

Using the homomorphism between the sets $\mathbf{C}\mid_{\varepsilon=0}$ and $\mathbf{D}$, we get the estimate
$$\sum_{i,j} v_{ij} \tilde{c}^{ij} = \tilde{V}(\tilde{c}) \geq \min_{d \in \mathbf{D}} V(d) = \Phi^*.$$

Thus we are able to apply the key technical lemma 4 (see Appendix A) to derive the following relation:

$$\overline{\Phi}_n := \frac{1}{n} \sum_{\tau=1}^{n} \sum_{i,j} v_{ij} \chi\left(x_\tau = x(i)\right) d_{n'(\tau)}^{ij}$$

$$\stackrel{a.s.}{=} \Phi_n + o(n^{\delta-\frac{1}{2}})$$

for any $\delta > 0$.

As it was shown in the proof of theorem 2

$$\left|\overline{\Phi}_n - \Phi_n\right| \leq const \frac{\Delta n_k}{n_k},$$

can be transformed into

$$\left|\overline{\Phi}_n - \Phi_{n_k}\right| \leq O^*(n^{\frac{1}{\kappa}}), \; k = t(n).$$

Taking into account the expressions (4.50) and (4.51) of the parameters, we finally obtain

$$|\Phi_n - \Phi^*| \leq C_9 n^{-\frac{1-\theta-\kappa}{\kappa}} + o(n^{\delta-\frac{\kappa-\theta}{2\kappa}}) + C_6 n^{-\frac{\kappa-1}{\kappa}} + C_5 n^{-\frac{\theta}{\kappa}}$$

$$+ o(n^{\delta-\frac{\nu-\theta}{\kappa}}) + o(n^{\delta-\frac{1}{2}}).$$

The theorem is proved. ∎

Now, we are ready to determine the optimal parameters of the adaptive control procedure studied in this chapter.

Theorem 4 *The best order of the adaptation process induced by the adaptive control algorithm (4.8)-(4.11) within the class of parameters sequences (4.50) and (4.51), is equal to $o(n^{\delta-\frac{1}{5}})$, i.e.,*

$$\Phi_n - \Phi^* \stackrel{a.s.}{\leq} o(n^{\delta-\frac{1}{5}}) \; \forall \delta > 0$$

and can be achieved if

$$\nu = \nu^* := \frac{1}{2}, \; \theta = \theta^* := \frac{1}{4}, \; \kappa = \kappa^* := \frac{5}{4}$$

As it follows from this theorem, the best adaptation rate is close to $n^{-\frac{1}{5}}$.

The next section will be concerned with the difference between the adaptation rates corresponding respectively to the situations with complete and not complete information.

4.6 On the cost of uncertainty

Tsypkin [11] said: "Lack of knowledge is overcome by learning. The smaller the *a priori* knowledge, the longer is the period necessary for learning. This is a natural cost of ignorance." The theorem given below estimates the order of the adaptation process in the case of complete information on the transition matrices and the average values of the loss function for the controlled Markov chains.

As it was shown in chapter 2, in the case of complete a priori information, the solution of the problem related to the control of Markov chains is equivalent to the solution of the corresponding Linear Programming Problem (see (2.10) of chapter 2). In turn, this problem was shown to be equivalent to another Linear Programming Problem (2.11)-(2.13) of chapter 2. This problem may have a nonunique solution $c^* \in \mathbf{C}$, but each solution generates a stationary random strategy d^* according to the formula

$$(d^*)^{il} = \frac{(c^*)^{il}}{\sum_{s=1}^{N}(c^*)^{is}} \quad (i=1,...,K; l=1,...,N). \tag{4.52}$$

Theorem 5 *For any regular Markov chain controlled by the stationary randomized strategy $d^* \in \mathbf{D}$ (4.52), the corresponding sequence $\{\Phi_n\}$ of loss functions converges to its limit Φ^* with the rate having an order equal to $o(n^{\delta-\frac{1}{2}})$ i.e.,*

$$\Phi_n \stackrel{a.s.}{=} \Phi^* + o(n^{\delta-\frac{1}{2}}) \tag{4.53}$$

Proof. Using lemma 4 (see Appendix A), it is easy to show that

$$\Phi_n \stackrel{a.s.}{=} \sum_{i=1}^{K}\sum_{j=1}^{N} v_{ij}\frac{1}{n}\chi(x_t = x(i), u_t = u(j)) + o(n^{\delta-\frac{1}{2}})$$

$$\stackrel{a.s.}{=} \sum_{i=1}^{K}\sum_{j=1}^{N} v_{ij}(d^*)^{ij} S_n(i) + o(n^{\delta-\frac{1}{2}}) \tag{4.54}$$

where

$$S_n(i) := \frac{1}{n}\sum_{t=1}^{n}\sum_{j=1}^{K}\chi(x_{t-1} = x(j))\sum_{l=1}^{N}\pi_{ji}^{l}(d^*)^{il} + o(n^{\delta-\frac{1}{2}}).$$

Based on the expression (4.54), we can state the following recurrent equation:

$$S_n = \Pi^T(d^*)S_{n-1} + o(n^{\delta-\frac{1}{2}})\mathbf{e}^K,$$

for the random vectors
$$S_n := (S_n(1), ..., S_n(K))^T.$$

In the case of regular Markov chains, the system of linear equations
$$S = \Pi^T(d^*)S, \ \left(\mathbf{e}^K\right)^T S = 1$$

has a unique solution $p(d^*)$. We conclude that there exists $\lambda \in (0,1)$ such that
$$\left[\Pi^T(d^*)\right]^n S_1 = p(d^*) + O(\lambda^n)\mathbf{e}^K.$$

Using this formula, we derive
$$S_n = p(d^*) + \left[\Pi^T(d^*)\right]^{n-1} S_1 - p(d^*)$$
$$+ \sum_{t=1}^{n-1} \left[\Pi^T(d^*)\right]^{n-1-t} \mathbf{e}^K o(n^{\delta-\frac{1}{2}})$$
$$= p(d^*) + \left[O(\lambda^n) + O^*(\sum_{t=1}^{n} \lambda^{n-t} t^{\delta-\frac{1}{2}})\right]\mathbf{e}^K$$
$$= p(d^*) + o(n^{\delta-\frac{1}{2}})\mathbf{e}^K.$$

Applying this result to (4.54) and taking into account that (see (2.4) of chapter 2)
$$\Phi_* \stackrel{a.s.}{=} \min_{d \in \mathbf{D}} V(d)$$

we finally obtain
$$\Phi_n \stackrel{a.s.}{=} \Phi_* + o(n^{\delta-\frac{1}{2}}).$$

The theorem is proved. ∎

4.7 Conclusions

An adaptive algorithm for the control of finite Markov chains has been presented. This algorithm is based on the Projection Gradient approach. Results concerning the convergence and the convergence rate have been stated. Based on the last theorem and the results of chapters 2 and 3 we also conclude that:

- the Lagrange Multipliers and Penalty function approaches lead to an adaptation rate equal to $o(n^{\delta-\frac{1}{2}})$;

- the algorithm based on the Projection Gradient approach used for adaptive control of regular Markov chains, provides an adaptation rate equal to $o(n^{\delta-\frac{1}{5}})$. It works more slowly than the adaptive control algorithms described in the previous chapters.

- for the case of complete information, the adaptation is equal to $o(n^{\delta-\frac{1}{2}})$.

Hitherto, we have been concerned with adaptive control of unconstrained finite Markov chains. A number of new and potentially useful results have been presented in this part.

The next part of this book will be dedicated to the adaptive control of constrained finite Markov chains.

4.8 References

1. A. V. Nazin and A. S. Poznyak, *Adaptive Choice of Variants*, (in Russian) Nauka, Moscow, 1986.

2. B. T. Polyak, *Introduction to Optimization*, Optimization Software, Publication Division, New York, 1987.

3. H. Kushner and G. G. Yin, *Stochastic Approximation Algoritms*, Springer-Verlag, New York, 1997.

4. M. Duflo, *Random Iterative Models*, Springer-Verlag, Berlin, 1997.

5. A. S. Poznyak and K. Najim, *Learning Automata and Stochastic Optimization*, Springer-Verlag, Berlin, 1997.

6. K. Najim and A. S. Poznyak, Neural networks synthesis based on stochastic approximation algorithm, *Int. J. of Systems Science*, vol. 25, pp. 1219-1222, 1994.

7. A. S. Poznyak, K. Najim and M. Chtourou, Use of recursive stochastic algorithm for neural networks synthesis, *Appl. Math. Modelling*, vol. 17, pp. 444-448, 1993.

8. K. Najim and E. Ikonen, Distributed logic processor trained under constraints using stochastic approximation techniques, *IEEE Trans. On Systems, Man, and Cybernetics, Part A*, vol. 29, pp. 421,426, 1999.

9. H. Robbins and D. Siegmund, A convergence theorem for nonnegative almost supermartingales and some applications, in *Optimizing Methods in Statistics*, ed. by J. S. Rustagi, Academic Press, New York, pp. 233-257, 1971.

10. Yu. A. Rozanov, *Random Processes*, (in Russian) Nauka, Moscow, 1973.

11. Ya. Z. Tsypkin, *Foundations of the Theory of Learning Systems*, Academic Press, New York, 1973.

Part II

Constrained Markov Chains

Chapter 5

Lagrange Multipliers Approach

5.1 Introduction

The theory and algorithms of the previous part were concerned with the adaptive control of unconstrained Markov chains [1]. Many engineering problems require algorithms which handle constraints because of physical limitations. Constrained control of Markov chains has been considered by several authors [2-5]. There exist several studies dealing with adaptive control of constrained Markov chains.

In the case of single constraint, the recursive estimation of the bias needed for a simple randomization between two policies to steer a long-range average cost to a given value has been introduced in [6-8]. This recursive estimation belongs to the stochastic approximation techniques. The optimal adaptive control of Markov chains, under average-cost constraints has been studied by Altman and Shwartz [9-12]. In these studies, the transition probabilities have been considered as the unknown parameters and, two classes (action time-sharing and time sharing) of policies have been introduced to provide strong consistency of different adaptive control schemes using some recursive procedures for the estimation of the transition probabilities and the expectation of the instantaneous costs. They suggested a class of Asymptotically Stationary (AS) strategies and showed that, under each of the cost criteria, the cost of these AS-strategies depend only on its limiting points. These authors also discuss the sensitivity of optimal policies and optimal values to small changes in the transition matrix and in the instantaneous cost functions. They establish the convergence of the optimal value for the discount constrained finite horizon problem to the optimal value of the corresponding infinite horizon problem.

A novel adaptive control algorithm[1] for the control of constrained Markov chains whose transition probabilities are unknown is presented in this chapter [13]. A finite set of algebraic constraints is considered. This adaptive control algorithm is based on the Lagrange multipliers approach [14] with an additional regularizing term providing the continuity of the corresponding saddle point with respect to the transition probability matrix and the conditional expectation values of the loss and constrained functions. In this control algorithm the transition probabilities of the Markov chain are not estimated. The control policy uses only the observations of the realizations of the loss functions and the constraints. This control law is adapted using the Bush-Mosteller reinforcement scheme [14-15] which is related to stochastic approximation procedures [16-18]. The Bush-Mosteller reinforcement scheme [19] is commonly used in the design of stochastic learning automata to solve many engineering problems. Learning deals with the ability of systems to improve their responses based on past experience. Controlling a Markov chain may be reduced to the design of a control policy which achieves some optimality of the control strategy under (or without) some constraints. In this study the optimality is associated with the minimization of a loss function which is assumed to be bounded under a set of algebraic constraints. So, the main features of this adaptive algorithm are:

- *the use of the Stochastic Learning Automata approach* to construct a recursive procedure to generate the asymptotically optimal control policy;

- *the use of a Modified Lagrange Function including a regularizing term* to guarantee the continuity in the parameters of the corresponding linear programming problem whose solution is connected with the optimal values of the main loss function under the given constraints;

- *the estimation of the adaptation rate and its optimization* within a class of the design parameters involved in the suggested adaptive procedure.

5.2 System description

In general, the behaviour of a controlled Markov chain is similar to the behaviour of a controlled dynamic system and can be described as follows. At each time n the system is observed to be in one state x_n. Whenever the system is in the state x_n one decision u_n (control action) is chosen according to some rule to achieve the desired control objective. In other words, the

---

[1]Reprinted from Automatica, vol. 35, A. S. Poznyak and K. Najim, Adaptive control of constrained finite Markov chains, Copyright (1998), with permission from Elsevier Science.

5.2. SYSTEM DESCRIPTION

decision is selected to guarantee that the resulting state process performs satisfactorily. Then, at the next time $n+1$ the system goes to the state x_{n+1}. In the case when the state and action sets are finite, and the transition from one state to another is random according to a fixed distribution, we deal with *Finite Markov Chains*.

Definition 1 *For $\{x_n\}$ and $\{u_n\}$ of a given controlled Markov chain, the sequence $\{\Phi_n^0\}$ of **loss functions** Φ_n^0 is defined as follows*

$$\Phi_n^0 := \frac{1}{n} \sum_{t=1}^{n} \eta_t^0 \qquad (5.1)$$

and the sequences $\{\Phi_n^m\}$ $(m = 1, ..., M)$ are given by

$$\Phi_n^m := \frac{1}{n} \sum_{t=1}^{n} \eta_t^m.$$

*They are involved in the **constraints***

$$\Phi_n^m \overset{a.s.}{\leq} 0 + o_\omega(1) \qquad (5.2)$$

where $o_\omega(1)$ is a random sequence tending to zero with probability 1 as $n \to \infty$, i.e.,

$$\overline{\lim_{n \to \infty}} \Phi_n^m \overset{a.s.}{\leq} 0, m = 0, ..., M.$$

Generally, for any nonstationary random control strategy $\{d_n\} \in \Sigma$ the sequences $\{\Phi_n^m\}$ $(m = 0, ..., M)$ (5.1) may have no limits (in any probability sense). Nevertheless, there exist a lower and an upper bounds for their partial limit points. So, a low bound have been calculated in [11] and [20]. These partial limit points belong to the intervals $[(\Phi^m)_*, (\Phi^m)^*]$ which are given by the following lemma.

Lemma 1 *For any ergodic controlled Markov chain with any distribution of initial states*

$$(\Phi^m)_* := \min_{d \in \mathbf{D}} V_m(d) \overset{a.s.}{\leq} \underline{\lim_{n \to \infty}} \Phi_n^m \leq \overline{\lim_{n \to \infty}} \Phi_n^m \overset{a.s.}{\leq} \max_{d \in \mathbf{D}} V_m(d) := (\Phi^m)^* \qquad (5.3)$$

where

$$V_m(d) = \sum_{i=1}^{K} \sum_{l=1}^{N} v_{il}^m d^{il} p_i(d) \quad (m = 0, ..., M) \qquad (5.4)$$

the set $\mathbf{D}$ of stochastic matrices d is defined by

$$d \in \mathbf{D} := \left\{ d \mid d^{il} \geq 0, \sum_{l=1}^{N} d^{il} = 1 \ (i = 1, ..., K; l = 1, ..., N) \right\}. \qquad (5.5)$$

and the vector $p^T(d) = (p_1(d), ..., p_K(d))$ satisfies the following linear algebraic equation

$$p(d) = \Pi^T(d) p(d) \tag{5.6}$$

Proof. Let us rewrite Φ_n^m ($m = 0, ..., M$) as follows:

$$\Phi_n^m = \sum_{i=1}^{K} \sum_{l=1}^{N} \left(\frac{1}{n} \sum_{t=1}^{n} \eta_t^m \right) \chi(x_n = x(i), u_n = u(l)) = \sum_{i=1}^{K} \sum_{l=1}^{N} \widehat{v}_n^{ilm} \widehat{\theta}_n^{il}$$

where

$$\widehat{v}_n^{ilm} := \frac{\sum_{t=1}^{n} \eta_t^m \chi(x_t = x(i), u_t = u(l))}{1 + \sum_{t=1}^{n} \chi(x_t = x(i), u_t = u(l))},$$

$$\widehat{\theta}_n^{il} := \frac{1}{n} \left[1 + \sum_{t=1}^{n} \chi(x_t = x(i), u_t = u(l)) \right].$$

If for a random realization $\omega \in \Omega$ we have

$$\sum_{t=1}^{\infty} \chi(x_t = x(i), u_t = u(l)) < \infty$$

then $\widehat{\theta}_n^{il}(\omega) \to 0$.

Consider now the realizations $\omega \in \Omega$ for which

$$\sum_{t=1}^{\infty} \chi(x_t = x(i), u_t = u(l)) = \infty.$$

Using the properties of the controlled Markov chain and its associated loss sequence, and applying the law of large number for dependent sequences [21] we, for all indexes obtain

$$\left(\widehat{v}_n^{ilm} - v_{il}^m \right) \xrightarrow[n \to \infty]{a.s.} 0,$$

$$\left(\frac{\sum_{t=1}^{n} \chi(x_t = x(i), x_{t+1} = x(j))}{1 + \sum_{t=1}^{n} \chi(x_t = x(i))} - \pi_{ij}\left(\widehat{d}_n^{il}\right) \right) \xrightarrow[n \to \infty]{} 0$$

where

$$\widehat{d}_n^{il} := \frac{\sum_{t=1}^{n} \chi(x_t = x(i), u_t = u(l))}{1 + \sum_{t=1}^{n} \chi(x_t = x(i))}.$$

Therefore, we have

$$\widehat{\theta}_n^{il} = \frac{1}{n}\left[1 + \frac{\sum_{t=1}^n \chi(x_t = x(i), u_t = u(l))}{1 + \sum_{t=1}^n \chi(x_t = x(i))}\left(1 + \sum_{t=1}^n \chi(x_t = x(i))\right)\right]$$

$$= \widehat{d}_n^{il}\frac{1}{n}\sum_{t=1}^n \chi(x_t = x(i)) + O\left(\frac{1}{n}\right) = \widehat{d}_n^{il} p_i\left(\widehat{d}_n^{il}\right) + O\left(\frac{1}{n}\right)$$

where the components $p_i(\cdot)$ satisfies (5.6).

The previous equalities leads to

$$\Phi_n^m - \sum_{i=1}^K \sum_{l=1}^N v_{il}^m \widehat{d}_n^{il} p_i\left(\widehat{d}_n^{il}\right) = \Phi_n^m - V\left(\widehat{d}_n^{il}\right) \overset{a.s.}{\underset{n\to\infty}{\to}} 0.$$

Consequently (5.3) holds. ∎

The problem to be solved is stated in the next section.

5.3 Problem formulation

The problem of adaptive control of finite Markov chains is considered and formulated in this section.

According to lemma 1, the minimal point

$$\lim_{n\to\infty} \Phi_n^0 \overset{a.s.}{=} \min_{d\in \mathbf{D}} V_0(d)$$

under the given constraints

$$\overline{\lim}_{n\to\infty} \Phi_n^m \overset{a.s.}{\leq} 0 \quad (m=1,...,M)$$

can be reached within the simple class of stationary strategies $\{d\} \in \Sigma_s$. The problem related to the minimization of the asymptotic realization of the loss function Φ_n^0 within the class Σ of random strategies $\{d_n\}$

$$\overline{\lim}_{n\to\infty} \Phi_n^0 \to \inf_{\{d_n\}\in\Sigma} \quad (a.s.) \qquad (5.7)$$

subject to the constraints

$$\overline{\lim}_{n\to\infty} \Phi_n^m \leq 0 \quad (m=1,...,M) \qquad (5.8)$$

can be solved in the class Σ_s of stationary strategies, and the minimal value $\Phi^0_{\min}$ of the asymptotic realization of the loss function $\varlimsup_{n\to\infty} \Phi^0_n$ is given by the solution of the following nonlinear optimization problem

$$V_0(d) \to \min_{d\in \mathbf{D}} \qquad (5.9)$$

subject to

$$p(d) = \Pi^T(d)p(d), \ V_m(d) \leq 0 \ (m=1,...,M).$$

The adaptive control problem will be stated as follows:

Based on the available observations

$$\left(x_1, u_1, \eta_1; ....; x_{n-1}, u_{n-1}, \eta_{n-1}; x_n\right)$$

develop a sequence $\{d_n\}$ of random matrix

$$d_n = d_n\left(x_1, u_1, \eta_1; ....; x_{n-1}, u_{n-1}, \eta_{n-1}; x_n\right)$$

such that the sequence $\{\Phi^0_n\}$ of loss functions Φ^0_n reaches its minimal value $\min_{d\in\mathbf{D}} V_0(d)$ *under the constraints (5.8).*

This constrained optimization problem can be tackled in many ways, but here we only note that by, for example, the Lagrange multipliers, the optimal solution can be determined.

An adaptive control algorithm for solving this problem is described in the next section.

5.4 Adaptive learning algorithm

In this section we shall design an adaptive control algorithm related to the control problem formulated in the previous section.

Let us consider the following nonlinear programming problem

$$V(d) \to \min_{d\in \mathbf{D}} \qquad (5.10)$$

$$p(d) = \Pi^T(d)p(d) \qquad (5.11)$$

$$V_m(d) \leq 0 \ (m=1,...,M), \qquad (5.12)$$

5.4. ADAPTIVE LEARNING ALGORITHM

where the functions $V_m(d)$ $(m = 0, ..., M)$ and the set $\mathbf{D}$ are defined by (5.4) and (5.5).

The transformation [22]

$$c^{il} = d^{il} p_i(d) \tag{5.13}$$

converts this problem into the following linear programming problem

$$\tilde{V}_0(c) := \sum_{i=1}^{K} \sum_{l=1}^{N} v_{il}^0 c^{il} \to \min_{c \in \mathbf{C}} \tag{5.14}$$

where the set $\mathbf{C}$ is given by

$$\mathbf{C} = \left\{ c \mid c = \left[c^{il}\right], \ c^{il} \geq 0, \ \sum_{i=1}^{K} \sum_{l=1}^{N} c^{il} = 1, \right. \tag{5.15}$$

$$\sum_{l=1}^{N} c^{jl} = \sum_{i=1}^{K} \sum_{l=1}^{N} \pi_{ij}^{l} c^{il} \ (i,j = 1, ..., K; l = 1, ..., N)$$

$$\left. \tilde{V}_m(c) := \sum_{i=1}^{K} \sum_{l=1}^{N} v_{il}^m c^{il} \leq 0 \ (m = 1, ..., M) \right\}.$$

In view of (5.13), it follows

$$\sum_{l=1}^{N} c^{il} = p_i(d) \ (i = 1, ..., K).$$

For ergodic controlled Markov chains there exists a unique final distribution $p_i(d)$ $(i = 1, ..., K)$ (irreducibility of its associated transition matrix), and for aperiodic controlled Markov chains it is a nonsingular one:

$$\sum_{l=1}^{N} c^{il} \geq \min_{d \in \mathbf{D}} p_i(d) := c_- > 0$$

it follows that in this case, the elements d^{il} of the matrix d can be defined as follows

$$d^{il} = c^{il} \left(\sum_{l=1}^{N} c^{il} \right)^{-1} (i = 1, ..., K; l = 1, ..., N). \tag{5.16}$$

In general, the solution $c = \left[c^{il}\right]$ of this linear programming problem (5.14), (5.15) is not unique. That's why, later we introduce a regularization term into the corresponding Lagrange function to ensure the uniqueness of the solution.

To solve this optimization problem (5.14), (5.15) in which the values v_{il}^m ($m = 1, ..., M$) and π_{ij}^l are not a priori known, and the available information at time n corresponds to x_n, u_n, η_n, the Lagrange multipliers approach [14] will be used.

To solve the optimization problem stated before, let us consider the following regularized (augmented) Lagrange function

$$L_\delta(\mathbf{c}, \boldsymbol{\lambda}) := \sum_{i=1}^{K} \sum_{l=1}^{N} v_{il}^0 c^{il} - \sum_{j=1}^{K} \lambda_j^0 \left[\sum_{l=1}^{N} c^{jl} - \sum_{i=1}^{K} \sum_{l=1}^{N} \pi_{ij}^l c^{il} \right] \quad (5.17)$$

$$+ \sum_{m=1}^{M} \lambda^m \sum_{i=1}^{K} \sum_{l=1}^{N} v_{il}^m c^{il} + \frac{\delta}{2} \left(\|\mathbf{c}\|^2 - \|\boldsymbol{\lambda}\|^2 \right), \quad \delta > 0$$

which is given on the set $S_0^{K \cdot N} \times R^K \times R_+^M$, where the simplex $S_0^{K \cdot N}$ is defined as follows:

$$S_\varepsilon^{K \cdot N} := \left\{ \mathbf{c} \mid c^{il} \geq \varepsilon \geq 0, \ \sum_{i=1}^{K} \sum_{l=1}^{N} c^{il} = 1 \ (i = 1, ..., K; l = 1, ..., N) \right\}. \quad (5.18)$$

and the vectors $\mathbf{c}$ and $\boldsymbol{\lambda}$

$$\mathbf{c}^T \ : \ = \left(c^{11}, ..., c^{1N}; ...; c^{K1}, ..., c^{KN} \right) \in R^{K \cdot N},$$
$$\boldsymbol{\lambda}^{0T} \ : \ = \left(\lambda_1^0, ..., \lambda_K^0 \right) \in R^K,$$
$$\boldsymbol{\lambda}_c^T \ : \ = \left(\lambda^1, ..., \lambda^M \right) \in R_+^M,$$
$$R_+^M \ : \ = \left\{ \boldsymbol{\lambda}_c \in R^M \mid \lambda^i \geq 0 \ i = 1, ..., M \right\}.$$

The vectors $\boldsymbol{\lambda}^0 \in R^K$ and $\boldsymbol{\lambda}_c \in R_+^M$ are the Lagrange multipliers associated respectively to the given equality and inequalities constraints. The saddle point of this regularized Lagrange function will be denoted by

$$(\mathbf{c}_\delta^*, \boldsymbol{\lambda}_\delta^*) := \arg \min_{\mathbf{c} \in S_0^{KN}} \max_{\boldsymbol{\lambda} \in R^K \times R_+^M} L_\delta(\mathbf{c}, \boldsymbol{\lambda}). \quad (5.19)$$

The function $L_\delta(\mathbf{c}, \boldsymbol{\lambda})$ ($\delta > 0$) is strictly convex. Then, this saddle point is unique and possesses the Lipshitz property with respect to parameter δ [14]:

$$\left\| \mathbf{c}_{\delta_1}^* - \mathbf{c}_{\delta_2}^* \right\| + \left\| \boldsymbol{\lambda}_{\delta_1}^* - \boldsymbol{\lambda}_{\delta_2}^* \right\| \leq Const \left| \delta_1 - \delta_2 \right|. \quad (5.20)$$

It has been shown in [14] that if $\delta \to 0$ the saddle point $(\mathbf{c}_\delta^*, \boldsymbol{\lambda}_\delta^*)$ converges to the solution of the optimization problem (5.14) which has the minimal norm (in the regular case this point is unique):

$$(\mathbf{c}_\delta^*, \boldsymbol{\lambda}_\delta^*) \underset{\delta \to 0}{\to} (\mathbf{c}^{**}, \boldsymbol{\lambda}^{**}) := \arg \min_{c^*, \lambda^*} \left(\|\mathbf{c}^*\|^2 + \|\boldsymbol{\lambda}^*\|^2 \right) \quad (5.21)$$

5.4. ADAPTIVE LEARNING ALGORITHM

(the minimization is done over all saddle points of the nonregularized Lagrange functions).

The remainder of this section will now focus on the development of an algorithm realizing the adaptive control of constrained finite Markov chains.

Stochastic approximation techniques [16-18] which are applicable to a wide class of stochastic optimization problems, will be used to find the saddle point $(\mathbf{c}_\delta^*, \boldsymbol{\lambda}_\delta^*)$ (5.19) of the function $L_\delta(\mathbf{c}, \boldsymbol{\lambda})$ (5.17) when the parameters $\left(v_{il}^m, \pi_{ij}^l\right)$ are unknown. A recursive procedure

$$\mathbf{c}_{n+1} = \mathbf{c}_{n+1}(x_n, u_n, \boldsymbol{\eta}_n, x_{n+1}, \mathbf{c}_n, \boldsymbol{\lambda}_n),$$
$$\boldsymbol{\lambda}_{n+1} = \boldsymbol{\lambda}_{n+1}(x_n, u_n, \boldsymbol{\eta}_n, x_{n+1}, \mathbf{c}_n, \boldsymbol{\lambda}_n)$$

which generates the sequences $\{\mathbf{c}_n\}$ and $\{\boldsymbol{\lambda}_n\}$ will be developed. This procedure has to converge in some probability sense to the solution $(\mathbf{c}^{**}, \boldsymbol{\lambda}^{**})$ of the initial problem. Let us make a formal statement of this adaptive control algorithm.

Step 1. (*normalization procedure*): at each time n, use the available information (information gathering)

$$x_n = x(\alpha), u_n = u(\beta), \boldsymbol{\eta}_n, x_{n+1} = x(\gamma), \boldsymbol{\eta}_n, \mathbf{c}_n\left(c_n^{il} > 0\right), \boldsymbol{\lambda}_n$$

to build the following function

$$\xi_n := \eta_n^0 - \left(\left(\lambda_\alpha^0\right)_n - \left(\lambda_\gamma^0\right)_n\right) + \sum_{m=1}^M \lambda_n^m \eta_n^m + \delta_n c_n^{\alpha\beta} \qquad (5.22)$$

and normalize (scale) it using the following affine transformation

$$\zeta_n := \frac{a_n \xi_n + b_n}{c_n^{\alpha\beta}} \qquad (5.23)$$

where $\{a_n\}, \{b_n\}$ are given by

$$a_n := \left(2\frac{\left(\sigma^0 + \left(2 + \sum_{m=1}^M \sigma^m\right)\lambda_n^+\right)}{\varepsilon_n} + \frac{N \cdot K}{N \cdot K - 1}\delta_n\right)^{-1}, \qquad (5.24)$$

$$b_n := \left(\sigma^0 + \left(2 + \sum_{m=1}^M \sigma^m\right)\lambda_n^+\right) a_n.$$

The positive sequences $\{\varepsilon_n\}, \{\delta_n\}$ and $\{\lambda_n^+\}$ will be specified below.

Notice that: i) different normalization procedures have been used in the context of learning automata when the environment emits arbitrary responses between 0 and 1 [14-15]. ii) The presented normalization procedure can be interpreted as a projection operator.

Step 2. (adaptive procedure): calculate the elements c_{n+1}^{il} using the following iterative algorithm

$$c_{n+1}^{il} = \begin{cases} c_n^{\alpha\beta} + \gamma_n^c(1 - c_n^{\alpha\beta} - \zeta_n) & i = \alpha \wedge l = \beta \\ c_n^{il} - \gamma_n^c \left(c_n^{il} - \dfrac{\zeta_n}{N \cdot K - 1} \right) & i \neq \alpha \vee l \neq \beta \end{cases}, \quad (5.25)$$

$$\left(\lambda_j^0\right)_{n+1} = \left[\left(\lambda_j^0\right)_n + \gamma_n^\lambda \left(\psi_j^0\right)_n\right]_{-\lambda_{n+1}^+}^{\lambda_{n+1}^+}, \quad j = 1, ..., K, \quad (5.26)$$

$$\lambda_{n+1}^m = \left[\lambda_n^m + \gamma_n^\lambda \psi_n^m\right]_0^{\lambda_{n+1}^+}, \quad m = 1, ..., M, \quad (5.27)$$

where

$$\left(\psi_j^0\right)_n = \sum_{l=1}^N c_n^{jl} - \delta_n \left(\lambda_j^0\right)_n - \chi\left(x(j) = x_{n+1}\right), \quad (5.28)$$

$$\psi_n^m = \eta_n^m - \delta_n \lambda_n^m. \quad (5.29)$$

The operators $[y]_{-\lambda_{n+1}^+}^{\lambda_{n+1}^+}$ and $\chi(x(j) = x_{n+1})$ are defined as follows:

$$[y]_{s_1}^{s_2} = \begin{cases} y & \text{if } y \in [s_1, s_2] \\ s_2 & \text{if } y > s_2 \\ s_1 & \text{if } y < s_1 \end{cases}$$

and

$$\chi(x(j) = x_{n+1}) = \begin{cases} 1 & \text{if } x(j) = x_{n+1} \\ 0 & \text{otherwise} \end{cases}.$$

The deterministic sequences $\{\gamma_n^c\}$ and $\{\gamma_n^\lambda\}$ will be specified below. These recursions constitute the heart of the adaptive control algorithm.

Step 3. (selection of a new action): construct the stochastic matrix

$$d_{n+1}^{il} = c_{n+1}^{il} \left(\sum_{k=1}^N c_{n+1}^{ik}\right)^{-1} \quad (i = 1, ..., K, l = 1, ..., N) \quad (5.30)$$

and according to

$$\Pr\{u_{n+1} = u(l) \mid x_{n+1} = x(\gamma) \wedge \mathcal{F}_n\} = d_{n+1}^{\gamma l}$$

5.4. ADAPTIVE LEARNING ALGORITHM

generate randomly a new discrete random variable u_{n+1} (this procedure can be done as in learning stochastic automata implementation [14-15], and getting the new observation (realization) η_{n+1} which corresponds to the transition to state x_{n+1}).

Step 4. return to Step 1.

Figure 5.1 shows how the key blocks, normalization, learning algorithm and new action selection, are used in the Matlab^{TM} mechanization of this adaptive control algorithm (see Appendix B).

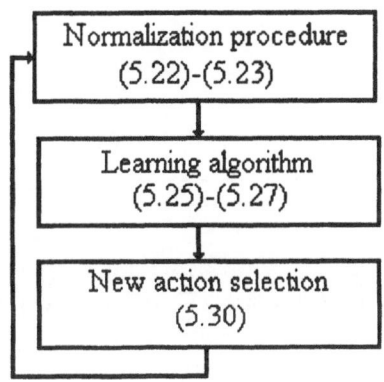

Figure 5.1: Diagram of the adaptive control algorithm.

The next lemma shows that the normalized function ζ_n belong to the unit segment $(0, 1)$ and that

$$\mathbf{c}_{n+1} \in S_{\varepsilon_{n+1}}^{K \cdot N} \text{ if } \mathbf{c}_n \in S_{\varepsilon_n}^{K \cdot N}.$$

Lemma 2 *If*

1. *in the recurrence (5.23)*

$$c_n^{\alpha\beta} \geq \varepsilon_n, \quad \delta_n \downarrow 0 \tag{5.31}$$

then

$$\zeta_n \in \left[\zeta_n^-, \zeta_n^+\right] \subset [0, 1] \tag{5.32}$$

where

$$\zeta_n^- = a_n \delta_n, \quad \zeta_n^+ = 1 - \frac{a_n \delta_n}{N \cdot K - 1}; \tag{5.33}$$

2. in the algorithm (5.25)
$$\zeta_n \in \left[\zeta_n^-, \zeta_n^+\right] \subset [0,1], \; \gamma_n^c \in [0,1], \; \mathbf{c}_n \in S_{\varepsilon_n}^{KN}$$
then
$$\mathbf{c}_{n+1} \in S_{\varepsilon_{n+1}}^{KN}. \tag{5.34}$$

Proof. To prove the statement (5.32), let us notice that

$$\zeta_n = \frac{a_n \left(\eta_n^0 - \left((\lambda_\alpha^0)_n - (\lambda_\gamma^0)_n\right) - \sum_{m=1}^{M} \lambda_n^m \eta_n^m + \delta_n c_n^{\alpha\beta}\right) + b_n}{c_n^{\alpha\beta}} + a_n \delta_n$$

$$\geq \frac{-a_n \left(\sigma^0 + \left(2 + \sum_{m=1}^{M} \sigma^m\right) \lambda_n^+\right) + b_n}{c_n^{\alpha\beta}} + a_n \delta_n = a_n \delta_n = \zeta_n^- > 0$$

and

$$\zeta_n \leq \frac{a_n \left(\sigma^0 + \left(2 + \sum_{m=1}^{M} \sigma^m\right) \lambda_n^+\right) + b_n}{\varepsilon_n} + a_n \delta_n$$

$$\leq a_n \left(\frac{2}{\varepsilon_n} \left(\sigma^0 + \left(2 + \sum_{m=1}^{M} \sigma^m\right) \lambda_n^+\right) + \delta_n\right) = \zeta_n^+.$$

The procedure (5.25) corresponds to Bush-Mosteller reinforcement scheme [14-15], and simple algebraic calculations demonstrate that (5.32) holds and $\mathbf{c}_{n+1} \in S_0^{K \cdot N}$. Clearly, from (5.25) it follows

$$c_{n+1}^{il} \geq (1 - \gamma_n^c) c_n^{il} + \gamma_n^c \min \left\{1 - \zeta_n^+; \frac{\zeta_n^-}{N \cdot K - 1}\right\}$$

$$\geq (1 - \gamma_n^c) c_n^{il} + \gamma_n^c \min \left\{1 - \zeta_n^+; \frac{\zeta_n^-}{N \cdot K - 1}\right\}.$$

To fulfill the following condition

$$\zeta_n^+ = 1 - \frac{\zeta_n^-}{N \cdot K - 1} < 1. \tag{5.35}$$

a_n has to be selected as in (5.24).

Taking into account relation (5.35), the last inequality can be rewritten as follows

$$c_{n+1}^{il} \geq (1 - \gamma_n^c) c_n^{il} + \gamma_n^c \frac{\zeta_n^-}{N \cdot K - 1} \geq \min \left\{c_n^{il}, \frac{\zeta_n^-}{N \cdot K - 1}\right\}$$

5.5. CONVERGENCE ANALYSIS

$$\geq \min\left\{\min\left\{c_{n-1}^{il}, \frac{\zeta_{n-1}^-}{N \cdot K - 1}\right\}, \frac{\zeta_n^-}{N \cdot K - 1}\right\}.$$

If ζ_n^- is monotically decreasing $\zeta_n^- \downarrow 0$, from the last expression we obtain

$$c_{n+1}^{il} \geq \min\left\{\min\left\{c_{n-1}^{il}, \frac{\zeta_{n-1}^-}{N \cdot K - 1}\right\}, \frac{\zeta_n^-}{N \cdot K - 1}\right\} = \min\left\{c_{n-1}^{il}, \frac{\zeta_n^-}{N \cdot K - 1}\right\}$$

$$\geq \cdots \geq \min\left\{c_1^{il}, \frac{\zeta_n^-}{N \cdot K - 1}\right\} = \frac{\zeta_n^-}{N \cdot K - 1} \geq \frac{\zeta_{n+1}^-}{N \cdot K - 1} \equiv \varepsilon_{n+1}.$$

From (5.25) it follows

$$\sum_{i=1}^{K}\sum_{l=1}^{N} c_{n+1}^{il} = \sum_{i=1}^{K}\sum_{l=1}^{N} c_n^{il} = 1.$$

In view of (5.24) ζ_n^- is monotically decreasing $\zeta_n^- \downarrow 0$. So, (5.34) is fulfilled.
∎

The convergence analysis of this adaptive control algorithm is carried out in the remainder of this text.

5.5 Convergence analysis

Hitherto, we have presented an adaptive control algorithm for constrained Markov chains. Our first order of business in this section is to analyze the asymptotic behaviour of this control algorithm.

In the development of this adaptive learning control procedure, we have considered stationarity randomized strategies. We obtained a policy sequence $\{d_n\}$ which according to (5.25), (5.26), (5.28), (5.29) and (5.30) is essentially nonstationary. As a consequence, we need to prove the convergence of this sequence to the solution $(\mathbf{c}^{**}, \boldsymbol{\lambda}^{**})$ (5.21) of the initial optimization problem.

Consider the following Lyapunov function

$$W_n := \left\|\mathbf{c}_n - \mathbf{c}_{\delta_n}^*\right\|^2 + \left\|\boldsymbol{\lambda}_n - \boldsymbol{\lambda}_{\delta_n}^*\right\|^2 \tag{5.36}$$

starting from $n \geq \inf_{t \geq 1}\left\{t : \left\|\boldsymbol{\lambda}_{\delta_n}^*\right\| \leq \lambda_n^+\right\}$.

Now we shall show below that under certain conditions, the algorithm described in the previous section, has nice asymptotic properties.

Theorem 1 *Let the controlled Markov chain be ergodic with any fixed distribution of the initial states. Let the loss function Φ_n be given by (5.1). If the* control policy generated by the adaptive learning procedure (5.25-5.30) with design parameters ε_n, δ_n, λ_n^+, γ_n^c and γ_n^λ satisfying the following conditions

$$0 < \delta_n \downarrow 0,\ 0 < \lambda_n^+ \uparrow \infty,\ \gamma_n^c \in (0,1),$$

$$\gamma_n^\lambda = \frac{N \cdot K}{N \cdot K - 1} \gamma_n^c a_n,\ \sum_{n=1}^{\infty} \gamma_n^c \varepsilon_n \delta_n \left(\lambda_n^+\right)^{-1} = \infty \qquad (5.37)$$

is used, then

1. *if*

$$\sum_{n=1}^{\infty} \mu_n < \infty$$

where

$$\mu_n := (\delta_n - \delta_{n+1})^2 \lambda_n^+ (\gamma_n^c \varepsilon_n \delta_n)^{-1} + (\gamma_n^c)^2 + |\delta_n - \delta_{n+1}| \gamma_n^c,$$

then, the control policy (5.25)-(5.30) converges with probability 1 to the optimal solution, i.e.,

$$W_n \underset{n \to \infty}{\overset{a.s.}{\to}} 0;$$

2. *if*

$$\frac{\mu_n \lambda_n^+}{\gamma_n^c \varepsilon_n \delta_n} \underset{n \to \infty}{\to} 0$$

then, we obtain the convergence in the mean squares sense, i.e.,

$$E\{W_n\} \underset{n \to \infty}{\overset{a.s.}{\to}} 0.$$

Proof. For $\delta > 0$, the regularized Lagrange function (5.17), is strictly convex. It follows

$$(\mathbf{c} - \mathbf{c}_\delta^*)^T \nabla_{\mathbf{c}} L_\delta(\mathbf{c}, \boldsymbol{\lambda}) - (\boldsymbol{\lambda} - \boldsymbol{\lambda}_\delta^*)^T \nabla_{\boldsymbol{\lambda}} L_\delta(\mathbf{c}, \boldsymbol{\lambda}) \qquad (5.38)$$

$$\geq \frac{\delta}{2} \left(\|\mathbf{c} - \mathbf{c}_\delta^*\|^2 + \|\boldsymbol{\lambda} - \boldsymbol{\lambda}_\delta^*\|^2 \right)$$

where $(\mathbf{c}_\delta^*, \boldsymbol{\lambda}_\delta^*)$ is the saddle point of the regularized Lagrange function (5.17), and $\mathbf{c}$, and $\boldsymbol{\lambda}$ are any vectors from the corresponding finite dimensional spaces.

Recall that

$$E\{Z \mid x_n = x(\alpha) \wedge \mathcal{F}_{n-1}\} = E\left\{ Z \frac{\chi(x_n = x(\alpha))}{p(\alpha)} \;\Big|\; \mathcal{F}_{n-1} \right\}.$$

5.5. CONVERGENCE ANALYSIS

Then, from (5.22), (5.28) and (5.29) it follows that the gradients (with respect to $c^{\alpha\beta}$ and λ^{γ}) of the regularized Lagrange function (5.17) can be expressed as a function of the conditional mathematical expectation of respectively ζ_n and $\boldsymbol{\psi}_n := \left((\psi_1^0)_n, ..., (\psi_K^0)_n, \psi_n^1, ..., \psi_n^M\right)^T$:

$$E\{\zeta_n \mathbf{e}(x_n \wedge u_n) \mid \mathcal{F}_{n-1}\} \stackrel{a.s.}{=} a_n \frac{\partial}{\partial \mathbf{c}} L_\delta(\mathbf{c}_n, \boldsymbol{\lambda}_n) + b_n \mathbf{e}^{N \cdot K}, \tag{5.39}$$

$$E\{\boldsymbol{\psi}_n \mid \mathcal{F}_{n-1}\} \stackrel{a.s.}{=} \frac{\partial}{\partial \boldsymbol{\lambda}} L_\delta(\mathbf{c}_n, \boldsymbol{\lambda}_n) \tag{5.40}$$

where $\mathbf{e}(x_n \wedge u_n)$ is a vector defined as follows

$$\mathbf{e}(x_n \wedge u_n) := \begin{bmatrix} \chi(x_n = x(1), u_n = u(1)) \\ \dots \\ \chi(x_n = x(K), u_n = u(1)) \\ \dots \\ \chi(x_n = x(1), u_n = u(l)) \\ \dots \\ \chi(x_n = x(K), u_n = u(l)) \\ \dots \\ \chi(x_n = x(K), u_n = u(N)) \end{bmatrix} \in R^{N \cdot K}. \tag{5.41}$$

Rewriting (5.25), (5.26), (5.28) and (5.29) in a vector form we obtain

$$\mathbf{c}_{n+1} = \mathbf{c}_n + \gamma_n^c \left(\mathbf{e}(x_n \wedge u_n) - \mathbf{c}_n + \zeta_n \frac{\mathbf{e}^{N \cdot K} - N \cdot K \mathbf{e}(x_n \wedge u_n)}{N \cdot K - 1} \right), \tag{5.42}$$

$$\boldsymbol{\lambda}_{n+1} = \left[\boldsymbol{\lambda}_n + \gamma_n^\lambda \boldsymbol{\psi}_n\right]_{s_{n+1}}^{\lambda_{n+1}^+}, \tag{5.43}$$

$$\boldsymbol{\psi}_n = \begin{cases} \left(\sum_{l=1}^N c_n^{\gamma l} - \delta_n \lambda_n^\gamma\right) \mathbf{e}^K - \mathbf{e}(x_{n+1}) & K \\ \psi_n^m = \eta_n^m - \delta_n \lambda_n^m\}_{m=1,...,M} & M \end{cases} \in R^{K+M} \tag{5.44}$$

where $\mathbf{e}^M$ is a vector defined by

$$\mathbf{e}^M := \underbrace{[1, ..., 1]}_{M}{}^T$$

and

$$s_{n+1} = \begin{cases} -\lambda_{n+1}^+ & \text{for the first } K \text{ components of } \boldsymbol{\lambda}_{n+1} \\ 0 & \text{for the other components of } \boldsymbol{\lambda}_{n+1} \end{cases}$$

Substituting (5.42), (5.43) and (5.44), into W_{n+1} (5.36) we derive

$$W_{n+1} \leq \left\| \mathbf{c}_n + \gamma_n^c \left(\mathbf{e}\left(x_n \wedge u_n\right) - \mathbf{c}_n + \zeta_n \frac{e^{N \cdot K} - N \cdot K e\left(x_n \wedge u_n\right)}{N \cdot K - 1} \right) \right.$$

$$\left. - \mathbf{c}_{\delta_{n+1}}^* \right\|^2 + \left\| \boldsymbol{\lambda}_n + \gamma_n^\lambda \boldsymbol{\psi}_n - \boldsymbol{\lambda}_{\delta_{n+1}}^* \right\|^2 = \left\| \left(\mathbf{c}_n - \mathbf{c}_{\delta_n}^* \right) \right.$$

$$\left. + \gamma_n^c \left(\mathbf{e}\left(x_n \wedge u_n\right) - \mathbf{c}_n + \zeta_n \frac{e^{N \cdot K} - N \cdot K e\left(x_n \wedge u_n\right)}{N \cdot K - 1} \right) + \left(\mathbf{c}_{\delta_n}^* - \mathbf{c}_{\delta_{n+1}}^* \right) \right\|^2$$

$$+ \left\| \left(\boldsymbol{\lambda}_n - \boldsymbol{\lambda}_{\delta_n}^* \right) + \gamma_n^\lambda \boldsymbol{\psi}_n + \left(\boldsymbol{\lambda}_{\delta_n}^* - \boldsymbol{\lambda}_{\delta_{n+1}}^* \right) \right\|^2.$$

Calculating the square of the norms appearing in this inequality, and estimating the resulting terms using the inequality

$$\|\boldsymbol{\psi}_n\| \leq Const \, \gamma_n^\lambda \left(1 + \delta_n \lambda_n^+ \right),$$

we obtain

$$W_{n+1} \leq W_n + (\gamma_n^c)^2 \, Const + \left\| \mathbf{c}_{\delta_n}^* - \mathbf{c}_{\delta_{n+1}}^* \right\|^2$$

$$+ 2 \left\| \mathbf{c}_{\delta_n}^* - \mathbf{c}_{\delta_{n+1}}^* \right\| \sqrt{W_n} + 2 \left\| \mathbf{c}_{\delta_n}^* - \mathbf{c}_{\delta_{n+1}}^* \right\| \gamma_n^c Const$$

$$+ \left(\gamma_n^\lambda \right)^2 \left(1 + \delta_n \lambda_n^+ \right)^2 Const + \left\| \boldsymbol{\lambda}_{\delta_n}^* - \boldsymbol{\lambda}_{\delta_{n+1}}^* \right\|^2$$

$$+ \gamma_n^\lambda \left(1 + \delta_n \lambda_n^+ \right) \left\| \boldsymbol{\lambda}_{\delta_n}^* - \boldsymbol{\lambda}_{\delta_{n+1}}^* \right\| Const + 2 \left\| \boldsymbol{\lambda}_{\delta_n}^* - \boldsymbol{\lambda}_{\delta_{n+1}}^* \right\| \sqrt{W_n}$$

$$+ 2\gamma_n^c \left(\mathbf{c}_n - \mathbf{c}_{\delta_n}^* \right)^T \left(\mathbf{e}\left(x_n \wedge u_n\right) - \mathbf{c}_n + \zeta_n \frac{e^{N \cdot K} - N \cdot K e\left(x_n \wedge u_n\right)}{N \cdot K - 1} \right)$$

$$+ 2\gamma_n^\lambda \left(\boldsymbol{\lambda}_n - \boldsymbol{\lambda}_{\delta_n}^* \right)^T \boldsymbol{\psi}_n,$$

where $Const$ is a positive constant.

Combining the terms of the right hand side of this inequality and in view of (5.20), it follows

$$W_{n+1} \leq W_n + Const \cdot \mu_{1,n} \sqrt{W_n} + Const \cdot \mu_{2,n} + w_n \qquad (5.45)$$

where

$$\mu_{1,n} := |\delta_n - \delta_{n+1}|,$$

$$\mu_{2,n} := (\gamma_n^c)^2 + (\delta_n - \delta_{n+1})^2$$

$$+ \left(\gamma_n^\lambda \right)^2 \left(1 + \delta_n \lambda_n^+ \right)^2 + |\delta_n - \delta_{n+1}| \left(\gamma_n^c + \gamma_n^\lambda \left(1 + \delta_n \lambda_n^+ \right) \right),$$

$$w_n := 2\gamma_n^c \left(\mathbf{c}_n - \mathbf{c}_{\delta_n}^* \right)^T \left(\mathbf{e}\left(x_n \wedge u_n\right) - \mathbf{c}_n + \zeta_n \frac{e^{N \cdot K} - N \cdot K e\left(x_n \wedge u_n\right)}{N \cdot K - 1} \right)$$

5.5. CONVERGENCE ANALYSIS

$$+2\gamma_n^\lambda \left(\boldsymbol{\lambda}_n - \boldsymbol{\lambda}_{\delta_n}^*\right)^T \boldsymbol{\psi}_n.$$

Notice that ζ_n (5.23) is a linear function of ξ_n and

$$\left(\mathbf{c}_n - \mathbf{c}_{\delta_n}^*\right)^T \frac{\mathbf{e}^{N \cdot K}}{N \cdot K - 1} = 0.$$

If (5.39) and (5.40) are used, the following is obtained:

$$E\{w_n \mid \mathcal{F}_{n-1}\} \stackrel{a.s.}{=} -2\frac{N \cdot K}{N \cdot K - 1}\gamma_n^c \left(\mathbf{c}_n - \mathbf{c}_{\delta_n}^*\right)^T E\{\zeta_n \mathbf{e}\,(x_n \wedge u_n) \mid \mathcal{F}_{n-1}\}$$

$$+2\gamma_n^\lambda \left(\boldsymbol{\lambda}_n - \boldsymbol{\lambda}_{\delta_n}^*\right)^T E\{\boldsymbol{\psi}_n \mid \mathcal{F}_{n-1}\}$$

$$\stackrel{a.s.}{=} -2\frac{N \cdot K}{N \cdot K - 1}\gamma_n^c a_n \left(\mathbf{c}_n - \mathbf{c}_{\delta_n}^*\right)^T \frac{\partial}{\partial \mathbf{c}} L_\delta\left(\mathbf{c}_n, \boldsymbol{\lambda}_n\right)$$

$$+2\gamma_n^\lambda \left(\boldsymbol{\lambda}_n - \boldsymbol{\lambda}_{\delta_n}^*\right)^T \frac{\partial}{\partial \boldsymbol{\lambda}} L_\delta\left(\mathbf{c}_n, \boldsymbol{\lambda}_n\right).$$

Taking into account the assumptions of this theorem, and the strict convexity property (5.38) we deduce

$$E\{w_n \mid \mathcal{F}_{n-1}\} \stackrel{a.s.}{=} -2\frac{N \cdot K}{N \cdot K - 1}\gamma_n^c a_n \left[\left(\mathbf{c}_n - \mathbf{c}_{\delta_n}^*\right)^T \frac{\partial}{\partial \mathbf{c}} L_\delta\left(\mathbf{c}_n, \boldsymbol{\lambda}_n\right)\right.$$

$$\left. -\left(\boldsymbol{\lambda}_n - \boldsymbol{\lambda}_{\delta_n}^*\right)^T \frac{\partial}{\partial \boldsymbol{\lambda}} L_\delta\left(\mathbf{c}_n, \boldsymbol{\lambda}_n\right)\right] \leq -\frac{N \cdot K}{N \cdot K - 1}\gamma_n^c a_n \delta_n W_n.$$

Calculating the conditional mathematical expectation of both sides of (5.45), and in view of the last inequality, we can get

$$E\{W_{n+1} \mid \mathcal{F}_{n-1}\} \stackrel{a.s.}{\leq} \left(1 - \frac{N \cdot K}{N \cdot K - 1}\gamma_n^c a_n \delta_n\right) W_n + Const\left(\mu_{1,n}\sqrt{W_n} + \mu_{2,n}\right). \tag{5.46}$$

If we use the easily verified fact that

$$2\mu_{1,n}\sqrt{W_n} \leq \mu_{1,n}^2 \rho_n^{-1} + W_n \rho_n$$

(which is valid for any $\rho_n > 0$) for

$$\rho_n := \gamma_n^c a_n \delta_n$$

it follows

$$2\mu_{1,n}\sqrt{W_n} \leq \mu_{1,n}^2 \left(\gamma_n^c a_n \delta_n\right)^{-1} + W_n \gamma_n^c a_n \delta_n.$$

From this inequality and (5.46), and in view of the following estimation

$$\mu_{1,n}^2 \left(\gamma_n^c a_n \delta_n\right)^{-1} + \mu_{2,n} \leq Const \cdot \mu_n$$

we finally, obtain

$$E\{W_{n+1} \mid \mathcal{F}_{n-1}\} \stackrel{a.s.}{\leq} \left(1 - \frac{1}{N \cdot K - 1}\gamma_n^c a_n \delta_n\right) W_n + Const \cdot \mu_n. \quad (5.47)$$

From (5.47) we conclude that $\{W_n, \mathcal{F}_n\}$ is a nonnegative quasimartingale [23] (see Appendix A).

Observe that

$$a_n = O\left(\frac{\varepsilon_n}{\lambda_n^+}\right).$$

From the assumptions of this theorem and in view of Robbins-Siegmund theorem for quasimartingales [23] (see Appendix A), the convergence with probability 1 follows.

The mean squares convergence follows from (5.47) after applying the operator of mathematical expectation to both sides of this inequality and using lemma A5 given in [15]. ∎

The term $\gamma_n^c \varepsilon_n \delta_n (\lambda_n^+)^{-1}$ can be interpreted as a *"generalized adaptation gain"* of the adaptive control procedure in the extended vector space $R^{N \cdot K + K + M}$. To reach any point belonging to this space from any initial point, $\sum_{n=1}^{\infty} \gamma_n^c \varepsilon_n \delta_n (\lambda_n^+)^{-1}$ must diverge (see (5.37)) because we do not know how far is the optimal solution point $(\mathbf{c}^{**}, \boldsymbol{\lambda}^{**})$ from the starting point $(\mathbf{c}_1, \boldsymbol{\lambda}_1)$.

Theorem 1 shows that this adaptive learning control algorithm possess all the properties that one would desire, i.e., convergence with probability 1 as well as convergence in the mean squares.

The conditions associated with the sequences $\{\varepsilon_n\}$, $\{\delta_n\}$, $\{\lambda_n^+\}$ and $\{\gamma_n^c\}$ are stated in the next corollary.

Corollary 1 *If in theorem 1*

$$\varepsilon_n := \frac{\varepsilon_0}{1 + n^\varepsilon \ln n} \left(\varepsilon_0 \in \left[0, (N \cdot K)^{-1}\right), \varepsilon \geq 0\right), \quad \delta_n := \frac{\delta_0}{n^\delta} \ (\delta_0, \delta > 0,),$$

$$\lambda_n^+ := \lambda_0^+ \left(1 + n^\lambda \ln n\right) \ \left(\lambda_0^+ > 0, \ \lambda \geq 0\right), \quad \gamma_n^c := \frac{\gamma_0}{n^\gamma} \ (\gamma_0 \in (0,1), \ \gamma \geq 0),$$

with

$$\gamma + \varepsilon + \lambda + \delta \leq 1$$

1. *the convergence with probability 1 will take place if*

$$\theta_{a.s.} := \min\{2 - \gamma - \varepsilon - \lambda + \delta; \ 2\gamma\} > 1,$$

5.5. CONVERGENCE ANALYSIS

2. the mean squares convergence is guaranteed if

$$\theta_{m.s.} := (1 - \gamma - \varepsilon - \lambda) > 0.$$

It is easy to check up on these conditions by substituting the parameters given in this corollary in theorem 1 assumptions.

Remark 1 *In the optimization problem related to the regularized Lagrange function $L_\delta(\mathbf{c}, \boldsymbol{\lambda})$ the parameter δ_n must decrease less slowly than any other parameter including ε_n, i.e.,*

$$\delta \leq \varepsilon.$$

However, not only is the analysis of the convergence of an iterative scheme important, but the convergence speed is also essential. The next theorem states the convergence rate of the adaptive learning algorithm described above.

Theorem 2 *Under the conditions of theorem 1 and corollary 1, it follows*

$$W_n \stackrel{a.s.}{=} o\left(\frac{1}{n^\nu}\right)$$

where

$$0 < \nu < \min\{1 - \gamma - \varepsilon - \lambda + \delta;\ 2\gamma - 1;\ 2\delta\} := \nu^*(\gamma, \varepsilon, \delta).$$

and the positive parameters γ, δ, ε and λ satisfy the following constraints

$$\lambda + \gamma + \varepsilon + \delta \leq 1,\ 1 - \gamma - \varepsilon - \lambda + \delta > 0,\ 2\gamma > 1.$$

Proof. From (5.20), it follows:

$$W_n^* := \|p_n - p^{**}\|^2 + \|\lambda_n - \lambda^{**}\|^2 = \|(p_n - p_n^*) + (p_n^* - p^{**})\|^2$$
$$+ \|(\lambda_n - \lambda_n^*) + (\lambda_n^* - \lambda^{**})\|^2 \leq 2\|p_n - p_n^*\|^2 + 2\|p_n^* - p^{**}\|^2$$
$$+ 2\|\lambda_n - \lambda_n^*\|^2 + 2\|\lambda_n^* - \lambda^{**}\|^2 \leq 2W_n + C\delta_n^2$$

Multiplying both sides of the previous inequality by ν_n, we derive

$$\nu_n W_n^* \leq 2\nu_n W_n + \nu_n C \delta_n^2$$

Selecting $\nu_n = n^\nu$ and in view of lemma 2 [24] and taking into account that

$$\frac{\nu_{n+1} - \nu_n}{\nu_n} = \frac{\nu + o(1)}{n}$$

we obtain

$$0 < \nu < \min\{1 - \gamma - \varepsilon - \lambda + \delta;\ 2\gamma - 1;\ 2\delta\} := \nu^*(\gamma, \varepsilon, \delta)$$

where the positive parameters γ, δ, ε and λ satisfy the following constraints

$$\lambda + \gamma + \varepsilon + \delta \leq 1,\ 2 - \gamma - \varepsilon - \lambda + \delta > 1,\ 2\gamma > 1.$$

∎

The basic result concerning the optimal convergence rate can now be stated in the following corollary.

Corollary 2 *The maximum convergence rate is achieved with the optimal parameters ε^*, δ^*, λ^* and γ^**

$$\varepsilon = \varepsilon^* = \delta = \delta^* = \frac{1}{6},\ \lambda = \lambda^* = 0, \gamma = \gamma^* = \frac{2}{3}$$

and is equal to

$$\nu^*(\gamma, \varepsilon, \delta) = 2\gamma^* - 1 = \nu^{**} = \frac{1}{3}.$$

Proof. The solution of the linear programming problem

$$\nu^*(\gamma, \varepsilon, \delta) \to \max_{\gamma, \varepsilon, \delta}$$

is given by

$$2\gamma - 1 = 1 + \delta - \varepsilon - \lambda - \gamma = 2\delta$$

or, in equivalent form,

$$\gamma = \frac{2}{3} - \frac{1}{3}(\lambda + \varepsilon - \delta) = \frac{1}{2} + \delta = 1 - \delta - \varepsilon - \lambda.$$

Taking into account that δ_n must decrease less slowly than ε_n (see Remark 1), we derive

$$\delta \leq \varepsilon$$

and, as a result, the smallest ε maximizing γ is equal to

$$\varepsilon = \delta.$$

Hence

$$\gamma = \frac{2}{3} - \frac{1}{3}\lambda = \frac{1}{2} + \delta = 1 - 2\delta - \lambda.$$

5.6. CONCLUSIONS

From these relations, we derive

$$\lambda = \frac{1}{2} - 3\delta.$$

Taking into account that $\lambda \geq 0$, we get

$$\delta \leq \frac{1}{6}$$

and, consequently

$$\gamma = \frac{1}{2} + \delta \leq \frac{2}{3}.$$

The optimal parameters are

$$\gamma = \gamma^* = \frac{2}{3}, \; \varepsilon = \varepsilon^* = \delta = \delta^* = \frac{1}{6}, \; \lambda = \lambda^* = 0.$$

The maximum convergence rate is achieved with this choice of parameters and is equal to

$$\nu^*(\gamma, \varepsilon, \delta) = 2\gamma^* - 1 = \nu^{**} = \frac{1}{3}.$$

∎

Several numerical simulations are presented in the last chapter of this book, in order to illustrate the performance of the adaptive control algorithm given above, and to get a feeling for the properties of the adaptive control algorithm based on the Lagrange multipliers approach. Some practical aspects are also examined in the last chapter. A Matlab program dealing with this adaptive control algorithm is given in Appendix B.

5.6 Conclusions

The main contribution of this chapter is the development of an adaptive learning algorithm for Constrained Controlled Finite Markov Chains. The Stochastic Learning Automata approach is used to construct a recursive procedure to generate the asymptotically optimal control policy. This control algorithm uses a Modified Lagrange Function including a regularizing term to guarantee the continuity in the parameters of the corresponding linear programming problem whose solution is connected with the optimal values of the main loss function under the given constraints. In this algorithm, the control policy is adjusted using the Bush-Mosteller reinforcement scheme which is related to stochastic approximation techniques. The convergence with probability 1 as well as the convergence in the mean squares have been stated. It has been demonstrated that the optimal convergence rate is equal to $n^{-\frac{1}{3}+\delta}$ (δ is any small positive parameter). The next chapter will be concerned with the development of an adaptive control algorithm on the basis of the Penalty Function Approach which is commonly used for solving constrained optimization problems.

5.7 References

1. A. Arapostathis, V. S. Borkar, E. Fernandez-Gaucherand, M. K. Ghosh and S. I. Marcus, Discrete-time controlled Markov processes with average cost criterion: a survey, *SIAM Journal of Control and Optimization*, vol. 31, pp. 282-344, 1993.

2. M. Haviv, On constrained Markov decision processes, *Operations Research Letters*, vol. 19, pp. 25-28, 1996.

3. E. A. Feinberg and A. Shwartz, Constrained discounted dynamic programming, *Mathematics of Operation Research*, vol. 21, pp. 922-944, 1996.

4. A. Altman, A. Hordijk and L. C. M. Kallenberg, On the value function in constrained control of Markov chains, *Mathematical Methods of Operations Research*, vol. 44, pp. 387-399, 1996.

5. A. Altman, Constrained Markov decision processes with total cost criteria: occupation measures and primal LP, *Mathematical Methods of Operations Research*, vol. 43, pp. 45-72, 1996.

6. A. M. Makowski and A. Shwartz, Implementation issues for Markov decision processes. In *Stochastic Differential Systems, Stochastic Control Theory*, Springer-Verlag, IMA Volumes in Mathematics & Its Applications, n°.10, ed. by W. Fleming and P. L. Lions, pp. 323-337, 1988.

7. D. -J. Ma and A. M. Makowski, A class of steering policies under a recurrent condition, *Proceedings of the 27-th IEEE Conference on Decision and Control*, Austin (TX) USA, December, pp. 1192-1197, 1988.

8. D. J-. Ma and A.M. Makowski, A class of two-dimensional stochastic approximation and steering policies for Markov decision processes, *Proceedings of the 31-st IEEE Conference on Decision and Control*, Tucson (Arizona) USA, December, pp. 3344-3349, 1992.

5.7. REFERENCES

9. E. Altman and A.Shwartz, Optimal priority assignment: A time sharing approach, *IEEE Transactions on Automatic Control*, vol. 34, pp. 1089-1102, 1989.

10. E. Altman and A.Shwartz, Adaptive control of constrained Markov chains, *IEEE Transactions on Automatic Control*, vol. 36, pp. 454-462, 1991.

11. E. Altman and A.Shwartz, Adaptive control of constrained Markov chains: criteria and policies, *Annals of Operation Research* vol.28, pp. 101-134, 1991.

12. E. Altman and A.Shwartz, Sensitivity of constrained Markov decision processes, *Annals of Operation Research*, vol. 32, pp. 1-22, 1991.

13. A. S. Poznyak and K. Najim, Adaptive control of constrained finite Markov chains, *Automatica*, vol. 35, pp. 777-789, 1999.

14. A. S. Poznyak and K. Najim, *Learning Automata and Stochastic Optimization*, Springer-Verlag, London, 1997.

15. K. Najim and A. S. Poznyak, *Learning Automata Theory and Applications*, Pergamon Press, London, 1994.

16. M. T. Wasan, *Stochastic Approximation*, Cambridge University Press, 1969.

17. H. Kushner and G. G. Yin, *Stochastic Approximation Algorithms and Applications*, Springer-Verlag, New York, 1997.

18. M. Duflo, *Random Iterative Models*, Springer-Verlag, Berlin, 1997.

19. R. R. Bush and F. Mosteller, *Stochastic Models for Learning*, J. Wiley, New York, 1955.

20. C. Derman, *Finite State Markovian Decision Processes*, Academic Press, New York, 1970.

21. D. Hall and C. Heyde, *Martingales Limit Theory and its Applications*, Academic Press, New York, 1980.

22. R. A. Howard, *Dynamic Programming and Markov Processes*, J. Wiley, New York, 1962.

23. H. Robbins and D. Siegmund, A convergence theorem for nonnegative almost supermartingales and some applications, in *Optimizing Methods in Statistics*, ed. by J. S. Rustagi, Academic Press, New York, pp. 233-257, 1971.

24. A. S. Poznyak and K. Najim, Learning automata with continuous input and changing number of actions, *Int. J. of Systems Science*, vol. 27, 1467-1472, 1996.

Chapter 6

Penalty Function Approach

6.1 Introduction

The theory of stochastic algorithms has attracted much interest during the last few decades. The main contribution of this chapter consists of the development of a new adaptive control algorithm [1-4] for constrained finite Markov chains, on the basis of the *Penalty Function Approach* [5] (without slack variables). In this control algorithm[1] [6] the transition probabilities of the Markov chain are estimated as well as the average values of the constraints. The control policy uses only the observations of the loss functions and the constraints (the available information). This control law is adapted using the Bush-Mosteller reinforcement scheme [5, 7] which is related to stochastic approximation procedures [4, 8-11]. Using a stochastic approximation procedure [12], Najim et al. [13] have developed a constrained long-range predictive control algorithm based on neural networks. The Bush-Mosteller reinforcement scheme (with constant or time varying correction factor) [7] is commonly used in the design of stochastic learning automata to solve many engineering problems. In has been used by Najim and Poznyak [5, 7] for the development of optimization algorithms which can be regarded as stochastic approximation procedures. The main results are stated and proved and have easy-to-verify assumptions. The proofs of intermediate lemma and theorems are given in Appendix A. This control algorithm and its asymptotic analysis differ deeply from the usually developed results in the literature which are principally dedicated to the adaptive control of Markov chains in which the transition probability and the initial distribution depend upon an unknown parameter. These results concern the consistency of the estimator.

---

[1]Penalty function and adaptive control of constrained finite Markov chains, K. Najim and A. S. Poznyak. Copyright John Wiley & Sons Limited. Reproduced with permission.

6.2 System description and problem formulation

The design of an adaptive learning control algorithm for controlled Markov chains will be based on the minimization of a loss function subject to some algebraic constraints. Let us first introduce some definitions concerning the controlled Markov chains, the loss function, and the constraints to be considered in this study.

In general, the behaviour of a controlled Markov chain is similar to the behaviour of a controlled dynamic system and can be described as follows. At each time n the system is observed to be in one state x_n. Whenever the system is in the state x_n one decision u_n (control action) is chosen according to some rule to achieve the desired control objective. In other words, the decision is selected to guarantee that the resulting state process performs satisfactorily. Then, at the next time $n+1$ the system goes to the state x_{n+1}. In the case when the state and action sets are finite, and the transition from one state to another is random according to a fixed distribution, we deal with *Finite Markov Chains*.

Some definitions related to these systems are given below.

For the states $\{x_n\}$ and the control actions $\{u_n\}$ of a given controlled Markov chain, the sequence $\{\Phi_n^0\}$ of **loss functions** Φ_n^0 can be defined as follows:

$$\Phi_n^0 := \frac{1}{n} \sum_{t=1}^{n} \eta_t^0 \qquad (6.1)$$

and the sequences $\{\Phi_n^m\}$ $(m = 1, ..., M)$ are given by

$$\Phi_n^m := \frac{1}{n} \sum_{t=1}^{n} \eta_t^m.$$

They are involved in the **constraints**

$$\Phi_n^m \stackrel{a.s.}{\leq} 0 + o_\omega(1) \qquad (6.2)$$

where $o_\omega(1)$ is a random sequence tending to zero with probability 1 as $n \to \infty$.

For unconstrained control problems, the loss functions define the goal of the control strategy to be developed. In other words, the minimization of the performance index Φ_n^0 represents the control objective. In some situations (resource allocation, communication, etc.), the behaviour of a given stochastic process is subject to some limitations. These limitations are in this chapter, modelled by the constraints $\{\Phi_n^m\}$ $(m = 1, ..., M)$.

6.2. SYSTEM DESCRIPTION AND PROBLEM FORMULATION

It is well known [1-3] that the limit in the long-run expected average cost per unit time cannot be guaranteed and, therefore, it is customary to replace the limit by lim sup or lim inf. Here, we are concerned with a similar problem. In fact, generally, for any nonstationary random control strategy $\{d_n\} \in \Sigma$ the sequences $\{\Phi_n^m\}$ ($m = 0, ..., M$) (6.1) may have no limits (in any probability sense). Nevertheless, there exist a lower and an upper bounds for their partial limit points. These points belong to the intervals $[(\Phi^m)_*, (\Phi^m)^*]$ (see lemma 1 of chapter 5). As a result, the problem related to the minimization of the performance index or control objective, i.e., the minimization of the asymptotic realization of the loss function Φ_n^0 (6.1) within the class Σ of random strategies $\{d_n\}$

$$\overline{\lim_{n \to \infty}} \Phi_n^0 \to \inf_{\{d_n\} \in \Sigma} (a.s.) \tag{6.3}$$

subject to the constraints (6.2)

$$\overline{\lim_{n \to \infty}} \Phi_n^m \leq 0 \ (m = 1, ..., M) \tag{6.4}$$

can be solved in the class Σ_s of stationary strategies, and the minimal value $\Phi_{\min}^0$ of the asymptotic realization of the loss function $\overline{\lim_{n \to \infty}} \Phi_n^0$ is given by the solution of the following nonlinear optimization problem

$$V_0(d) \to \min_{d \in \mathbf{D}} \tag{6.5}$$

subject to

$$p(d) = \Pi^T(d) p(d), \ V_m(d) \leq 0 \ (m = 1, ..., M).$$

where

$$V_m(d) = \sum_{i=1}^{K} \sum_{l=1}^{N} v_{il}^m d^{il} p_i(d) \ (m = 0, ..., M) \tag{6.6}$$

and, the set $\mathbf{D}$ of stochastic matrices d is defined by

$$d \in \mathbf{D} := \left\{ d \mid d^{il} \geq 0, \sum_{l=1}^{N} d^{il} = 1 \ (i = 1, ..., K; l = 1, ..., N) \right\}. \tag{6.7}$$

We are now in the position to state the problem of interest in this chapter. The adaptive control problem will be stated as follows:

Based on the available observations (history)

$$\left(x_1, u_1, \eta_1; ....; x_{n-1}, u_{n-1}, \eta_{n-1}; x_n \right)$$

develop a sequence $\{d_n\}$ of random matrix

$$d_n = d_n\left(x_1, u_1, \eta_1; ....; x_{n-1}, u_{n-1}, \eta_{n-1}; x_n\right)$$

such that the sequence $\{\Phi_n^0\}$ of loss functions Φ_n^0 reaches its minimal value $\min_{d\in \mathbf{D}} V_0(d)$ under the constraints (6.4).

In the previous chapter, we presented an adaptive control (self-learning control) algorithm based on the Lagrange multipliers optimization approach. A new adaptive control algorithm [6] for solving this problem is described in the next section.

6.3 Adaptive learning algorithm

In this section we shall be concerned with the development of an adaptive learning control algorithm on the basis of the penalty function approach [6]. Many engineering problems can be formulated as constrained optimization problems.

The control problem of interest in this chapter is thus the minimization of

$$V(d) \to \min_{d \in \mathbf{D}} \tag{6.8}$$

under the constraints:

$$p(d) = \Pi^T(d)\, p(d) \tag{6.9}$$

$$V_m(d) \leq 0 \quad (m = 1, ..., M). \tag{6.10}$$

where the functions $V_m(d)$ $(m = 0, ..., M)$ and the set $\mathbf{D}$ are defined by (6.6) and (6.7).

This constrained optimization problem constitutes a nonlinear programming problem. We confront a situation in which there exists a real need for development of a control algorithm involving learning.

Our main purpose here is to set the scene for our subsequent development of an adaptive (self-learning) control algorithm. Using the transformation [14]

$$c^{il} = d^{il} p_i(d), \tag{6.11}$$

the previous nonlinear programming problem will be converted into the following linear programming problem

$$\tilde{V}_0(c) := \sum_{i=1}^{K} \sum_{l=1}^{N} v_{il}^0 c^{il} \to \min_{c \in \mathbf{C}} \tag{6.12}$$

6.3. ADAPTIVE LEARNING ALGORITHM

where the set $\mathbf{C}$ is given by

$$\mathbf{C} = \left\{ c \mid c = \left[c^{il}\right], \ c^{il} \geq 0, \ \sum_{i=1}^{K}\sum_{l=1}^{N} c^{il} = 1, \right. \tag{6.13}$$

$$\sum_{l=1}^{N} c^{jl} = \sum_{i=1}^{K}\sum_{l=1}^{N} \pi_{ij}^{l} c^{il} \ (i,j = 1, ..., K; l = 1, ..., N) \tag{6.14}$$

$$\left. \tilde{V}_m(c) := \sum_{i=1}^{K}\sum_{l=1}^{N} v_{il}^m c^{il} \leq 0 \ (m = 1, ..., M) \right\}. \tag{6.15}$$

Based on (6.11), it follows

$$\sum_{l=1}^{N} c^{il} = p_i(d) \ (i = 1, ..., K).$$

Let us recall that:

i) an ergodic class (set) is a collection X of recurrent states with the probability that, when starting from one of the states in X, all states will be visited with probability 1;

ii) an ergodic Markov chain has only one class, and that class is ergodic;

iii) if a Markov chain is ergodic then there are limiting probabilities for all states belonging to X. In other words, for ergodic controlled Markov chains there exists a unique final distribution $p_i(d)$ $(i = 1, ..., K)$ (irreducibility of its associated transition matrix), and for aperiodic controlled Markov chains, $p_i(d)$ $(i = 1, ..., K)$ are nonsingular:

$$\sum_{l=1}^{N} c^{il} \geq \min_{d \in \mathbf{D}} p_i(d) := c_- > 0.$$

It follows that in this case, the elements d^{il} of the matrix d can be defined as follows

$$d^{il} = c^{il} \left(\sum_{l=1}^{N} c^{il}\right)^{-1} \ (i = 1, ..., K; l = 1, ..., N). \tag{6.16}$$

Now, it is certainly true that the solution $c = \left[c^{il}\right]$ of the problem (6.12) would be unique.

Two remarks are in order at this point.

Remark 1 *The attention given currently to the class of ergodic controlled Markov chains is due to their very interesting intrinsic properties.*

Remark 2 *Notice that the equality constraints (6.3) can be rewritten in compact form, i.e., a matrix form as follows*

$$Vc = 0 \qquad (6.17)$$

with the matrix V satisfying the condition

$$\det V^T V = 0. \qquad (6.18)$$

Here, the matrix $V \in R^{K \times NK}$ is defined by

$$V := \begin{bmatrix} v_{11} & \mathbf{e}^K(u_2)\Pi_1 & . & \mathbf{e}^K(u_K)\Pi_1 \\ (\mathbf{e}^K(u_1))^T \Pi_2 & v_{22} & . & \mathbf{e}^K(u_K)\Pi_2 \\ . & . & . & . \\ (\mathbf{e}^K(u_1))^T \Pi_K & (\mathbf{e}^K(u_2))^T \Pi_K & . & v_{KK} \end{bmatrix} \qquad (6.19)$$

with

$$v_{ii} = \left(\mathbf{e}^N\right)^T - (\mathbf{e}^K(u_i))^T \Pi_i$$

where the vectors $\mathbf{e}^M$ and $\mathbf{e}^K(u_i)$ are given by

$$\mathbf{e}^M := \underbrace{(1,...,1)^T}_{M} \in R^{M \times 1},$$

$$\mathbf{e}^K(u_i) := \Big(\underbrace{0,0,...,0,1,0,...,0}_{i}\Big)^T \in R^K$$

and the matrices Π_j $(j = 1,...,K)$ are defined by

$$\Pi_j := \begin{bmatrix} \pi^1_{1j} & \pi^2_{1j} & \cdots & \pi^N_{1j} \\ \pi^1_{2j} & \pi^2_{2j} & \cdots & \pi^N_{2j} \\ . & . & . & . \\ \pi^1_{Kj} & \pi^2_{Kj} & . & \pi^N_{Kj} \end{bmatrix} \in R^{K \times N}. \qquad (6.20)$$

Remark 3 *The functions $\tilde{V}_m(\mathbf{c})$ $(m = 0, ..., M)$ defined by (6.12) and (6.15) can also be rewritten in the following matrix (compact) form:*

$$\tilde{V}_m(\mathbf{c}) = \left(\mathbf{e}^K\right)^T \Xi_m \mathbf{c} \quad (m = 0, ..., M) \qquad (6.21)$$

where the matrices Ξ_m are defined as follows

$$\Xi_m := \begin{bmatrix} v^m_{11} v^m_{12} \cdots v^m_{1N} & 0 & 0 & 0 \\ 0 & v^m_{21} v^m_{22} \cdots v^m_{2N} & 0 & 0 \\ . & & . & \\ 0 & 0 & 0 & v^m_{K1} v^m_{K2} \cdots v^m_{KN} \end{bmatrix} \in R^{K \times NK}$$

$$(6.22)$$

6.3. ADAPTIVE LEARNING ALGORITHM

Based on these matrix representations (6.19) and (6.21) we can rewrite the linear programming problem (6.12) in the following equivalent matrix form

$$\tilde{V}_0(c) := \left(\mathbf{e}^K\right)^T \Xi_0 \mathbf{c} \to \min_{c \in \mathbf{C}} \qquad (6.23)$$

where

$$\mathbf{C} = \Big\{ c \mid c = \left[c^{il}\right], \ c^{il} \geq 0, \ \sum_{i=1}^{K}\sum_{l=1}^{N} c^{il} = 1,$$

$$V\mathbf{c} = 0$$

$$\left(\mathbf{e}^K\right)^T \Xi_m \mathbf{c} \leq 0 \ (m = 1, ..., M) \Big\}. \qquad (6.24)$$

The penalty function approach is the simplest and best known approach for solving constrained optimization problems. It consists of transforming the initial problem into an unconstrained optimization one. The penalty function (auxiliary function) is chosen so such that it coincides with the objective function (criterion) to be minimized in the admissible domain defined by the constraints, and it increases rapidly outside the admissible domain. This optimization approach [5] will be used to solve the optimization problem (6.12), (6.3) in which the values v_{il} and π^l_{ij} are not a priori known, and the available information at time n corresponds to x_n, u_n and η_n.

Consider the vector $\mathbf{c}$:

$$\mathbf{c}^T := \left(c^{11}, ..., c^{1N}; ...; c^{K1}, ..., c^{KN}\right) \in R^{N \cdot K}$$

and the regularized penalty function given by

$$\mathcal{P}_{\mu,\delta}(\mathbf{c}) := \mu \tilde{V}_0(\mathbf{c}) + \frac{1}{2} \|V\mathbf{c}\|^2 + \frac{1}{2} \sum_{m=1}^{M} \left(\left[\tilde{V}_m(\mathbf{c})\right]^+\right)^2 + \frac{\delta}{2} \|\mathbf{c}\|^2 = \qquad (6.25)$$

$$= \mu \left(\mathbf{e}^K\right)^T \Xi_0 \mathbf{c} + \frac{1}{2} \|V\mathbf{c}\|^2 + \frac{1}{2} \sum_{m=1}^{M} \left(\left[\left(\mathbf{e}^K\right)^T \Xi_m \mathbf{c}\right]^+\right)^2 + \frac{\delta}{2} \|\mathbf{c}\|^2, \ 0 < \mu, \delta \downarrow 0$$

where

$$[z]^+ := \begin{cases} z & \text{if } z \geq 0 \\ 0 & \text{if } z < 0 \end{cases}. \qquad (6.26)$$

The regularized penalty function (6.25) is given on the simplex S_0^{KN}, defined as follows:

$$S_\varepsilon^{KN} := \Big\{ \mathbf{c} \mid c^{il} \geq \varepsilon \geq 0, \ \left(\mathbf{e}^{NK}\right)^T \mathbf{c} = 1 \ (i = 1, ..., K; l = 1, ..., N) \Big\}. \qquad (6.27)$$

Here μ is the "penalty coefficient" and δ is the "regularizing parameter."

Notice that the function

$$\frac{1}{2}\sum_{m=1}^{M}\left[\left(\mathbf{e}^{K}\right)^{T}\Xi_{m}\mathbf{c}\right]^{+}$$

is not differentiable on $\mathbf{c}$ in the point $\mathbf{c}=0$, but the quadratic term

$$\frac{1}{2}\sum_{m=1}^{M}\left(\left[\left(\mathbf{e}^{K}\right)^{T}\Xi_{m}\mathbf{c}\right]^{+}\right)^{2}$$

involved in the penalty function (6.25) is differentiable everywhere in $R^{N \cdot K}$. The penalty function approach reduces the solution of the considered constrained optimization problem to the solution of an unconstrained optimization problem. The penalty functions have been used to solve optimization problems well before they gained mathematical respectability. They appear in many forms, and are often called by special names (restraint, etc.) in various applications [15-16].

For fixed positive μ and δ, the argument $\mathbf{c}$ minimizing this regularized penalty function, will be unique and denoted by

$$\mathbf{c}^{*}_{\mu,\delta} := \arg\min_{\mathbf{c}\in S_0^{KN}} \mathcal{P}_{\mu,\delta}(\mathbf{c}). \qquad (6.28)$$

Due to the strict convexity of this penalty function $\mathcal{P}_{\mu,\delta}(\mathbf{c})$ under any $\delta > 0$, this minimum point possesses the Lipshitz property with respect to the parameters δ and μ [5]:

$$\left\|\mathbf{c}^{*}_{\mu_1,\delta_1} - \mathbf{c}^{*}_{\mu_2,\delta_2}\right\| \leq C_1 \left|\mu_1 - \mu_2\right| + C_2 \left|\delta_1 - \delta_2\right|, \quad C_1, C_2 \in (0,\infty). \qquad (6.29)$$

We see in light of what has been said above that for this extremization problem there is no ready made machinery that we can put our constrained optimization problem into and produce an algorithm by turning the crank.

The next lemma shows that under some conditions related to $\mu = \mu_n$ and $\delta = \delta_n$, the minimum point $\mathbf{c}^{*}_{\mu,\delta}$ converges to the solution of the optimization problem (6.12) which has the minimal weighted norm (in the regular case this point is unique)

Lemma 1 *If the parameters μ and δ are time-varying, i.e.,*

$$\mu = \mu_n, \; \delta = \delta_n \; (n = 1, 2, ...)$$

such that

$$0 < \mu \downarrow 0, \; 0 < \delta \downarrow 0 \qquad (6.30)$$

6.3. ADAPTIVE LEARNING ALGORITHM

$$\frac{\delta_n}{\mu_n} \xrightarrow[n\to\infty]{} 0, \quad \frac{(\mu_n)^{3/2}}{\delta_n} \xrightarrow[n\to\infty]{} 0 \qquad (6.31)$$

then the following claim is true

$$\mathbf{c}^*_{\mu,\delta} \xrightarrow[\mu,\delta\to 0]{} \mathbf{c}^{**} := \arg\min_{\mathbf{c}^*} (\mathbf{c}^*)^T \left(I + V^T V + \sum_{m=1}^{M} \Xi_m^T \mathbf{e}^K (\mathbf{e}^K)^T \Xi_m \right) \mathbf{c}^*. \qquad (6.32)$$

(the minimization is done over all the solutions $\mathbf{c}^*$ of the linear programming problem (6.12)).

The proof of this lemma is given in [17]

To find the minimum point $\mathbf{c}^*_\delta$ (6.28) of the function $\mathcal{P}_{\mu,\delta}(\mathbf{c})$ (6.25) when the parameters (v_{il}, π^l_{ij}) are unknown, we will use the stochastic approximation technique [4] which will permit us to define an iterative procedure

$$\mathbf{c}_{n+1} = \mathbf{c}_{n+1}(x_n, u_n, \boldsymbol{\eta}_n, x_{n+1}, \mathbf{c}_n)$$

generating the sequence $\{\mathbf{c}_n\}$ which converges in some probability sense to the solution $\mathbf{c}^{**}$ of the initial problem. This adaptive control algorithm [6] performs normalization, learning and action selection procedures. These procedures bear some resemblance to the optimization algorithms developed on the basis of learning automata (learning deals with the ability of systems to improve their responses based on past experience) [5, 7]. This adaptive control algorithm performs the following steps:

Step 1 (normalization procedure): using the available information

$$x_n = x(\alpha), u_n = u(\beta), \boldsymbol{\eta}_n, x_{n+1} = x(\gamma), \mathbf{c}_n \left(c_n^{il} > 0 \right)$$

to construct the following function

$$\xi_n := \mu_n \eta_n^0 + \sum_{j=1}^{K} \left(\sum_{l=1}^{N} c_n^{jl} - \sum_{i=1}^{K} \sum_{l=1}^{N} \left(\widehat{\pi}^l_{ij} \right)_n c_n^{il} \right) \left(\chi(x_n = x(j)) - \left(\widehat{\pi}^\beta_{\alpha j} \right)_n \right) \qquad (6.33)$$

$$+ \sum_{m=1}^{M} \left[(\mathbf{e}^K)^T \left(\widehat{\Xi}_m \right)_n \mathbf{c}_n \right]^+ \eta_n^m + \delta_n c_n^{\alpha\beta}, \qquad (6.34)$$

where $\left(\widehat{\pi}^l_{ij} \right)_n$ and $\left(\widehat{\Xi}_m \right)_n$ represent respectively the estimations at time n of the transition probability π^l_{ij} and the matrices Ξ_m. The operator $\chi(x(j) = x_{n+1})$ is defined as follows:

$$\chi(x(j) = x_{n+1}) = \begin{cases} 1 & \text{if } x(j) = x_{n+1} \\ 0 & \text{otherwise} \end{cases}.$$

Then normalize it (ξ_n) using the following affine transformation

$$\zeta_n := \frac{a_n \xi_n + b_n}{c_n^{\alpha\beta}}, \qquad (6.35)$$

where the numerical sequences $\{a_n\}$ and $\{b_n\}$ are given by

$$a_n := \left(2\frac{\left(\mu_n \sigma + 4 + \sum_{m=1}^{M}(\sigma^m)^2\right)}{\varepsilon_n} + \frac{N \cdot K}{N \cdot K - 1}\delta_n \right)^{-1}, \qquad (6.36)$$

$$b_n := a_n \left(\mu_n \sigma + 4 + \sum_{m=1}^{M}(\sigma^m)^2 \right).$$

The positive sequences $\{\delta_n\}, \{\varepsilon_n\}$ and $\{\mu_n\}$ will be specified below.

This normalization procedure is a kind of mapping or projection scheme for obtaining a new variable, namely ζ_n which belongs to some given interval. It has been mainly introduced by Najim and Poznyak [5, 7] in the development of optimization algorithms based on learning stochastic automata.

Step 2 (learning procedure): calculate the elements c_{n+1}^{il} using the following recursive algorithm

$$c_{n+1}^{il} = \begin{cases} c_n^{\alpha\beta} + \gamma_n(1 - c_n^{\alpha\beta} - \zeta_n) & i = \alpha \wedge l = \beta \\ c_n^{il} - \gamma_n\left(c_n^{il} - \dfrac{\zeta_n}{N \cdot K - 1}\right) & i \neq \alpha \vee l \neq \beta \end{cases}, \qquad (6.37)$$

$$\left(\hat{\pi}_{ij}^l\right)_{n+1} = \left(\hat{\pi}_{ij}^l\right)_n \qquad (6.38)$$

$$- \begin{cases} \left[\left(\hat{\pi}_{\alpha j}^l\right)_n - \chi(x_{n+1} = x(j))\right]\left(S_{\alpha j}^\beta\right)_{n+1}^{-1} & i = \alpha \wedge l = \beta \wedge j = \gamma \\ 0 & i \neq \alpha \vee l \neq \beta \vee j \neq \gamma \end{cases},$$
$$(6.39)$$

$$\left(S_{ij}^l\right)_{n+1} = \left(S_{ij}^l\right)_n + \chi\left(x_n = x(i) \wedge u_n = u(l) \wedge x_{n+1} = x(j)\right), \quad (6.40)$$

$$\left(\hat{\Xi}_m\right)_{n+1}^{il} = \left(\hat{\Xi}_m\right)_n^{il} - \begin{cases} \left[\left(\hat{\Xi}_m\right)_n^{\alpha\beta} - 1\right]\left(\sum_{j=1}^{K}\left(S_{\alpha j}^\beta\right)_{n+1}\right)^{-1} & i = \alpha \wedge l = \beta \\ 0 & i \neq \alpha \vee l \neq \beta \end{cases}$$
$$(6.41)$$

The deterministic sequence $\{\gamma_n\}$ will be specified below.

The recursion presented above are closely connected with the stochastic approximation techniques [4]

6.3. ADAPTIVE LEARNING ALGORITHM

Step 3 (new action selection): construct the stochastic matrix

$$d_{n+1}^{il} = c_{n+1}^{il} \left(\sum_{k=1}^{N} c_{n+1}^{ik} \right)^{-1} \quad (i=1,...,K, l=1,...,N) \qquad (6.42)$$

and according to

$$\Pr\{u_{n+1} = u(l) \mid x_{n+1} = x(\gamma) \wedge \mathcal{F}_n\} = d_{n+1}^{\gamma l}$$

generate randomly a new discrete random variable u_{n+1} as in learning stochastic automata implementation [5, 7], and get the new observation (realization) η_{n+1} which corresponds to the transition to state x_{n+1}.

Step 4 return to Step 1.

The schematic diagram of this adaptive control algorithm is given in figure 6.1. Its Matlab^{TM} mechanization is given in Appendix B.

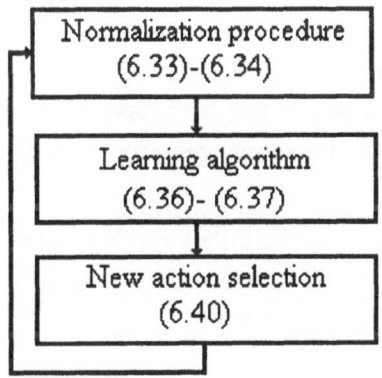

Figure 6.1: Scheme for the adaptive control algorithm.

An adaptive control algorithm for constrained Markov chains has been presented. It is based on learning automata and stochastic approximation techniques. We are now in the position to state some theoretical results.

The transformation (6.35) has many nice properties. The next lemma shows that the normalized function ζ_n belongs to the unit segment $(0,1)$ and that

$$\mathbf{c}_{n+1} \in S_{\varepsilon_{n+1}}^{KN} \text{ if } \mathbf{c}_n \in S_{\varepsilon_n}^{KN}.$$

Lemma 2 *If*

1. in the normalization procedure (6.35)

$$c_n^{\alpha\beta} \geq \varepsilon_n, \quad \delta_n \downarrow 0 \qquad (6.43)$$

then ζ_n belongs to the segment $[\zeta_n^-, \zeta_n^+]$, i.e.,

$$\zeta_n \in \left[\zeta_n^-, \zeta_n^+\right] \subset [0,1] \qquad (6.44)$$

where

$$\zeta_n^- = a_n \delta_n, \quad \zeta_n^+ = 1 - \frac{a_n \delta_n}{N \cdot K - 1}; \qquad (6.45)$$

2. in the algorithm (6.37) associated with the optimization of the regularized penalty function (6.25)

$$\zeta_n \in \left[\zeta_n^-, \zeta_n^+\right] \subset [0,1], \; \gamma_n \in [0,1], \; \mathbf{c}_n \in S_{\varepsilon_n}^{KN}$$

then

$$\mathbf{c}_{n+1} \in S_{\varepsilon_{n+1}}^{KN}. \qquad (6.46)$$

Proof. Claims (1) and (2) state that the normalized function ξ_n and the argument $\mathbf{c}_{n+1}$ belong to the interval $[\zeta_n^-, \zeta_n^+] \subset [0,1]$ and to the simplex $S_{\varepsilon_{n+1}}^{KN}$ respectively. To prove the first assertion (6.44) of this lemma, let us recall the properties of the sequences $\{\eta_n^i\}$ $(i = 0, 1, ..., M)$:

$$\sup_n \left|\eta_n^0\right| \overset{a.s.}{\leq} \sigma < \infty,$$

and

$$\sup_n \left|\eta_n^m\right| \overset{a.s.}{\leq} \sigma < \infty, (m = 1, ..., M).$$

Based on these properties, it follows:

$$\zeta_n = \frac{a_n}{c_n^{\alpha\beta}} \left\{ \left(\mu_n \eta_n^0 + \sum_{j=1}^{K} \left(\sum_{l=1}^{N} c_n^{jl} - \sum_{i=1}^{K}\sum_{l=1}^{N} \left(\widehat{\pi}_{ij}^l\right)_n c_n^{il} \right) \cdot \right. \right.$$

$$\left. \cdot \left(\chi\left(x_n = x(j)\right) - \left(\widehat{\pi}_{\alpha j}^\beta\right)_n \right) \right)$$

$$\left. + \sum_{m=1}^{M} \left[\left(\mathbf{e}^K\right)^T \left(\widehat{\Xi}_m\right)_n \mathbf{c}_n\right]^+ \eta_n^m + b_n \right\} + a_n \delta_n$$

$$\geq \frac{-a_n \left(\mu_n \sigma + 4 + \sum_{m=1}^{M} (\sigma^m)^2 \right) + b_n}{c_n^{\alpha\beta}} + a_n \delta_n = a_n \delta_n = \zeta_n^- > 0$$

6.3. ADAPTIVE LEARNING ALGORITHM

and

$$\zeta_n \leq \frac{a_n \left(\mu_n \sigma + 4 + \sum_{m=1}^{M} (\sigma^m)^2\right) + b_n}{\varepsilon_n} + a_n \delta_n$$

$$\leq a_n \left(\frac{2}{\varepsilon_n}\left(\mu_n \sigma + 4 + \sum_{m=1}^{M} (\sigma^m)^2\right) + \delta_n\right) = \zeta_n^+.$$

Notice that the procedure (6.37) corresponds to Bush-Mosteller reinforcement scheme [5, 7] with time varying correction factor γ_n, and simple algebraic calculations demonstrate that (6.44) holds and $\mathbf{c}_{n+1} \in S_0^{KN}$. Indeed, from (6.37) and the upper and lower bounds of ζ_n, it follows

$$c_{n+1}^{il} \geq (1 - \gamma_n)\, c_n^{il} + \gamma_n \min\left\{1 - \zeta_n^+; \frac{\zeta_n^-}{N \cdot K - 1}\right\}$$

$$\geq (1 - \gamma_n)\, c_n^{il} + \gamma_n \min\left\{1 - \zeta_n^+; \frac{\zeta_n^-}{N \cdot K - 1}\right\}.$$

To fulfill the following condition

$$\zeta_n^+ = 1 - \frac{\zeta_n^-}{N \cdot K - 1} < 1. \tag{6.47}$$

a_n has to be selected as in (6.36).

Taking into account this expression (6.47), the previous inequality can be rewritten as follows

$$c_{n+1}^{il} \geq (1 - \gamma_n)\, c_n^{il} + \gamma_n \frac{\zeta_n^-}{N \cdot K - 1} \geq \min\left\{c_n^{il}, \frac{\zeta_n^-}{N \cdot K - 1}\right\}$$

$$\geq \min\left\{\min\left\{c_{n-1}^{il}, \frac{\zeta_{n-1}^-}{N \cdot K - 1}\right\}, \frac{\zeta_n^-}{N \cdot K - 1}\right\}.$$

If ζ_n^- is monotonically decreasing $\zeta_n^- \downarrow 0$, from the last expression we obtain

$$c_{n+1}^{il} \geq \min\left\{\min\left\{c_{n-1}^{il}, \frac{\zeta_{n-1}^-}{N \cdot K - 1}\right\}, \frac{\zeta_n^-}{N \cdot K - 1}\right\}$$

$$= \min\left\{c_{n-1}^{il}, \frac{\zeta_n^-}{N \cdot K - 1}\right\}$$

$$\geq \cdots \geq \min\left\{c_1^{il}, \frac{\zeta_n^-}{N \cdot K - 1}\right\} = \frac{\zeta_n^-}{N \cdot K - 1} \geq \frac{\zeta_{n+1}^-}{N \cdot K - 1} \equiv \varepsilon_{n+1}.$$

From (6.37) it follows

$$\sum_{i=1}^{K}\sum_{l=1}^{N} c_{n+1}^{il} = \sum_{i=1}^{K}\sum_{l=1}^{N} c_{n}^{il} = 1.$$

To conclude, let us notice that in view of (6.36) ζ_n^- is monotically decreasing $\zeta_n^- \downarrow 0$. So, (6.46) is fulfilled. ∎

A remark is in order here. We have seen that the normalization procedure seems to be helpful in our analysis, and this will be reinforced as we proceed.

The properties related to this novel adaptive control algorithm are discussed in the next section.

6.4 Convergence analysis

In this section we establish the convergence properties of the adaptive control algorithm [6] described in the previous section. The proofs of the theorems and corollaries contained in this section are in many respects, similar to those presented in the previous chapters; however, some deep results for optimization and probability theories are also required. In any ways, these proofs can be skipped without loss of continuity. The analysis in this section was geared towards methods used in the analysis of the behaviour of learning automata and stochastic optimization techniques developed in connection with learning automata [5, 7].

To derive this adaptive control procedure, we have considered stationary randomized strategies. We obtained a policy sequence $\{d_n\}$ which according to (6.37) and (6.42) is essentially nonstationary. As a consequence, we need to prove the convergence of this sequence to the solution $\mathbf{c}^{**}$ (6.32) of the initial optimization problem.

The main tool used for the next analytical development is the Lyapunov theory which plays an important role in stability and convergence analysis. It represents a pillar of the stability theory. The main obstacle to using Lyapunov theory is finding a suitable Lyapunov function. Let us consider the following Lyapunov function

$$W_n := \left\| \mathbf{c}_n - \mathbf{c}_{\delta_n}^* \right\|^2. \qquad (6.48)$$

The term "Lyapunov function" has already been used in various contexts. This is always an attractive function which ensures a certain stabilization.

For the adaptive control algorithm presented in previous section, we now establish the following properties:

6.4. CONVERGENCE ANALYSIS

- convergence with probability 1;

- convergence in the mean square.

Theorem 1 *Consider an ergodic controlled Markov chain with any fixed distribution of the initial states, and assume that the loss function Φ_n is given by (6.1). If the control policy generated by the adaptive algorithm (6.37)-(6.42) with design parameters ε_n, δ_n, μ_n and γ_n satisfying the following conditions*

$$0 < \varepsilon_n, \delta_n, \mu_n \downarrow 0, \; \gamma_n \in (0,1),$$

$$\frac{\delta_n}{\mu_n} \xrightarrow[n \to \infty]{} 0, \; \frac{(\mu_n)^{3/2}}{\delta_n} \xrightarrow[n \to \infty]{} 0, \; \sum_{n=1}^{\infty} \gamma_n \varepsilon_n \delta_n = \infty$$

is used, then

1. *if*

$$\sum_{n=1}^{\infty} \theta_n < \infty$$

where

$$\theta_n := \gamma_n^2 + (\delta_n - \delta_{n+1})^2 (\gamma_n \varepsilon_n \delta_n)^{-1} + |\delta_n - \delta_{n+1}| \gamma_n + \gamma_n (n \varepsilon_n \delta_n)^{-1},$$

then, the control policy (6.37)-(6.42) converges with probability 1 to the optimal solution, i.e.,

$$W_n \xrightarrow[n \to \infty]{a.s.} 0;$$

2. *if*

$$\frac{\theta_n}{\gamma_n \varepsilon_n \delta_n} \xrightarrow[n \to \infty]{} 0$$

then, we obtain the convergence in the mean squares sense, i.e.,

$$E\{W_n\} \xrightarrow[n \to \infty]{a.s.} 0.$$

Proof. We will mainly use the Borel-Cantelli lemma [18-19], the strong law of large numbers for dependent sequences [20], and the Robbins-Siegmund theorem [21], as tools for the proof of this theorem.

For $\delta > 0$, the regularized penalty function $\mathcal{P}_{\mu,\delta}(\mathbf{c})$ (6.25), is strictly convex. Indeed, for any convex function and for any point $\mathbf{c} \in R^{NK}$ we have

$$\left(\mathbf{c} - \mathbf{c}_{\mu,\delta}^*\right)^T \left(\nabla_{\mathbf{c}} \mathcal{P}_{\mu,\delta}(\mathbf{c}) - \nabla_{\mathbf{c}} \mathcal{P}_{\mu,\delta}(\mathbf{c}_{\mu,\delta}^*)\right) \geq \left(\mathbf{c} - \mathbf{c}_{\mu,\delta}^*\right)^T \nabla_{\mathbf{c}}^2 \mathcal{P}_{\mu,\delta}(\mathbf{c}_{\mu,\delta}^*) \left(\mathbf{c} - \mathbf{c}_{\mu,\delta}^*\right).$$

Appealing to (6.25) one sees that the last inequality implies

$$\left(\mathbf{c} - \mathbf{c}^*_{\mu,\delta}\right)^T \nabla_\mathbf{c} \mathcal{P}_{\mu,\delta}(\mathbf{c}) \tag{6.49}$$

$$\geq \left(\mathbf{c} - \mathbf{c}^*_{\mu,\delta}\right)^T \left[V^T V + \sum_{m=1}^{M} \Xi_m^T \mathbf{e}^K \left(\mathbf{e}^K\right)^T \Xi_m \left[\tilde{V}_m(\mathbf{c}^*)\right]^+ + \delta I\right] \left(\mathbf{c} - \mathbf{c}^*_{\mu,\delta}\right)$$

$$\geq \delta \left\|\mathbf{c} - \mathbf{c}^*_{\mu,\delta}\right\|^2$$

where $\mathbf{c}^*_{\mu,\delta}$ is the minimum point of the regularized penalty function (6.25). Recall that

$$E\{Z \mid x_n = x(\alpha) \wedge \mathcal{F}_{n-1}\} = E\left\{Z \frac{\chi(x_n = x(\alpha))}{p(\alpha)} \mid \mathcal{F}_{n-1}\right\}.$$

and let us introduce the following notation

$$\varphi_n\left(\hat{\pi}^l_{ij}; \left(\hat{\Xi}_m\right)_n\right) := \mu_n \eta_n$$

$$+ \sum_{j=1}^{K} \left(\sum_{l=1}^{N} c_n^{jl} - \sum_{i=1}^{K} \sum_{l=1}^{N} \left(\hat{\pi}^l_{ij}\right)_n c_n^{il}\right) \left(\chi(x_n = x(j)) - \left(\hat{\pi}^\beta_{\alpha j}\right)_n\right)$$

$$+ \sum_{m=1}^{M} \left[\left(\mathbf{e}^K\right)^T \left(\hat{\Xi}_m\right)_n \mathbf{c}_n\right]^+ \eta_n^m + \delta_n c_n^{\alpha\beta}.$$

Then, from (6.37) it follows that the gradient with respect to $\mathbf{c}$ of the regularized penalty function (6.25) can be expressed as a function of the conditional mathematical expectation of ζ_n:

$$E\{\zeta_n \mathbf{e}(x_n \wedge u_n) \mid \mathcal{F}_{n-1}\} = E\left\{\frac{a_n \varphi_n\left(\hat{\pi}^l_{ij}; \left(\hat{\Xi}_m\right)_n\right) + b_n}{c_n^{\alpha\beta}} \mathbf{e}(x_n \wedge u_n) \mid \mathcal{F}_{n-1}\right\}$$

$$= E\left\{\frac{a_n \varphi_n\left(\pi^l_{ij}; \Xi_m\right) + b_n}{c_n^{\alpha\beta}} \mathbf{e}(x_n \wedge u_n) \mid \mathcal{F}_{n-1}\right\}$$

$$+ E\left\{\frac{a_n}{c_n^{\alpha\beta}} \left[\varphi_n\left(\hat{\pi}^l_{ij}; \left(\hat{\Xi}_m\right)_n\right) - \varphi_n\left(\pi^l_{ij}; \Xi_m\right)\right] \mathbf{e}(x_n \wedge u_n) \mid \mathcal{F}_{n-1}\right\}$$

$$\stackrel{a.s.}{=} a_n \frac{\partial}{\partial \mathbf{c}} \nabla_\mathbf{c} \mathcal{P}_{\mu_n,\delta_n}(\mathbf{c}_n) + b_n \mathbf{e}^{N \cdot K} + \Delta_n.$$

where

$$\Delta_n := E\left\{\frac{a_n}{c_n^{\alpha\beta}} \left[\varphi_n\left(\hat{\pi}^l_{ij}; \left(\hat{\Xi}_m\right)_n\right) - \varphi_n\left(\pi^l_{ij}; \Xi_m\right)\right] \mathbf{e}(x_n \wedge u_n) \mid \mathcal{F}_{n-1}\right\}$$

6.4. CONVERGENCE ANALYSIS

and $\mathbf{e}(x_n \wedge u_n)$ is a $N \cdot K$-tuple vector defined as follows

$$\mathbf{e}(x_n \wedge u_n) := \begin{bmatrix} \chi(x_n = x(1), u_n = u(1)) \\ \ldots \\ \chi(x_n = x(K), u_n = u(1)) \\ \ldots \\ \chi(x_n = x(1), u_n = u(l)) \\ \ldots \\ \chi(x_n = x(K), u_n = u(l)) \\ \ldots \\ \chi(x_n = x(K), u_n = u(N)) \end{bmatrix} \in R^{N \cdot K}. \quad (6.50)$$

From the assumption

$$\sum_{n=1}^{\infty} \gamma_n \varepsilon_n \delta_n = \infty$$

it follows

$$\sum_{n=1}^{\infty} c_n^{il} \geq \sum_{n=1}^{\infty} \varepsilon_n \geq \sum_{n=1}^{\infty} \gamma_n \varepsilon_n \delta_n = \infty.$$

This latter point

$$\sum_{n=1}^{\infty} c_n^{il} = \infty$$

is an important one. In fact, it allows us to use the Borel-Cantelli lemma [18].

In view of the Borel-Cantelli lemma [18], and the strong law of large numbers for dependent sequences [20], we derive

$$\|\Delta_n\| = O\left(\sum_{i,j,l} \left\|\left(\widehat{\pi}_{ij}^l\right)_n - \pi_{ij}^l\right\|\right) + O\left(\sum_{m=1}^{M} \left\|\left(\widehat{\Xi}_m\right)_n - \Xi_m\right\|\right) \stackrel{a.s.}{=} o_\omega\left(\frac{1}{\sqrt{n}}\right)$$

where $o_\omega\left(n^{-1/2}\right)$ is a random sequence tending to zero more quickly than $n^{-1/2}$.

Hence

$$E\{\zeta_n \mathbf{e}(x_n \wedge u_n) \mid \mathcal{F}_{n-1}\} \stackrel{a.s.}{=} a_n \frac{\partial}{\partial \mathbf{c}} \nabla_{\mathbf{c}} \mathcal{P}_{\mu_n, \delta_n}(\mathbf{c}_n) + b_n \mathbf{e}^{N \cdot K} + o_\omega\left(\frac{a_n}{\varepsilon_n \sqrt{n}}\right). \quad (6.51)$$

Rewriting (6.37) in a vector form, we obtain

$$\mathbf{c}_{n+1} = \mathbf{c}_n + \gamma_n \left(\mathbf{e}(x_n \wedge u_n) - \mathbf{c}_n + \zeta_n \frac{\mathbf{e}^{N \cdot K} - N \cdot K \mathbf{e}(x_n \wedge u_n)}{N \cdot K - 1}\right). \quad (6.52)$$

Substituting (6.52) into W_{n+1} (6.48) we derive

$$W_{n+1} \leq \left\| \mathbf{c}_n + \gamma_n \left(\mathbf{e}\left(x_n \wedge u_n\right) - \mathbf{c}_n + \zeta_n \frac{e^{N \cdot K} - N \cdot K \mathbf{e}\left(x_n \wedge u_n\right)}{N \cdot K - 1} \right) \right.$$

$$\left. - \mathbf{c}^*_{\delta_{n+1}} \right\|^2 = \left\| \left(\mathbf{c}_n - \mathbf{c}^*_{\delta_n}\right) \right.$$

$$\left. + \gamma_n \left(\mathbf{e}\left(x_n \wedge u_n\right) - \mathbf{c}_n + \zeta_n \frac{e^{N \cdot K} - N \cdot K \mathbf{e}\left(x_n \wedge u_n\right)}{N \cdot K - 1} \right) + \left(\mathbf{c}^*_{\delta_n} - \mathbf{c}^*_{\delta_{n+1}}\right) \right\|^2.$$

Calculating the square of the norms appearing in this inequality, and estimating the resulting terms using the inequality

$$\|\zeta_n\| \leq \|\zeta_n^+\| \leq 1,$$

allows us to rewrite the previous inequality as follows

$$W_{n+1} \leq W_n + \gamma_n^2 Const + \left\| \mathbf{c}^*_{\delta_n} - \mathbf{c}^*_{\delta_{n+1}} \right\|^2$$

$$+ 2 \left\| \mathbf{c}^*_{\delta_n} - \mathbf{c}^*_{\delta_{n+1}} \right\| \sqrt{W_n} + 2 \left\| \mathbf{c}^*_{\delta_n} - \mathbf{c}^*_{\delta_{n+1}} \right\| \gamma_n Const$$

$$+ 2\gamma_n \left(\mathbf{c}_n - \mathbf{c}^*_{\delta_n}\right)^T \left(\mathbf{e}\left(x_n \wedge u_n\right) - \mathbf{c}_n + \zeta_n \frac{e^{N \cdot K} - N \cdot K \mathbf{e}\left(x_n \wedge u_n\right)}{N \cdot K - 1} \right)$$

where $Const$ is a positive constant.

Combining the terms of the right hand side of this inequality and in view of (6.29), it follows

$$W_{n+1} \leq W_n + Const \cdot \mu_{1,n} \sqrt{W_n} + Const \cdot \mu_{2,n} + w_n \quad (6.53)$$

where

$$\mu_{1,n} := |\delta_n - \delta_{n+1}|,$$
$$\mu_{2,n} := \gamma_n^2 + (\delta_n - \delta_{n+1})^2 + |\delta_n - \delta_{n+1}| \gamma_n,$$
$$w_n := 2\gamma_n \left(\mathbf{c}_n - \mathbf{c}^*_{\delta_n}\right)^T \left(\mathbf{e}\left(x_n \wedge u_n\right) - \mathbf{c}_n + \zeta_n \frac{e^{N \cdot K} - N \cdot K \mathbf{e}\left(x_n \wedge u_n\right)}{N \cdot K - 1} \right).$$

Notice that

$$\left(\mathbf{c}_n - \mathbf{c}^*_{\delta_n}\right)^T \frac{e^{N \cdot K}}{N \cdot K - 1} = 0.$$

(6.51) leads to

$$E\{w_n \mid \mathcal{F}_{n-1}\} \stackrel{a.s.}{=} -2 \frac{N \cdot K}{N \cdot K - 1} \gamma_n \left(\mathbf{c}_n - \mathbf{c}^*_{\delta_n}\right)^T E\{\zeta_n \mathbf{e}\left(x_n \wedge u_n\right) \mid \mathcal{F}_{n-1}\}$$

6.4. CONVERGENCE ANALYSIS

$$\stackrel{a.s.}{=} -\frac{2 \cdot N \cdot K}{N \cdot K - 1}\gamma_n \left(\mathbf{c}_n - \mathbf{c}^*_{\delta_n}\right)^T \left(a_n \frac{\partial}{\partial \mathbf{c}}\nabla_\mathbf{c}\mathcal{P}_{\mu_n,\delta_n}(\mathbf{c}_n) + b_n \mathbf{e}^{N\cdot K} + o_\omega\left(\frac{a_n}{\varepsilon_n \sqrt{n}}\right)\right)$$

$$= -2\frac{N \cdot K}{N \cdot K - 1}\gamma_n a_n \left(\mathbf{c}_n - \mathbf{c}^*_{\delta_n}\right)^T \frac{\partial}{\partial \mathbf{c}}\nabla_\mathbf{c}\mathcal{P}_{\mu_n,\delta_n}(\mathbf{c}_n) + o_\omega\left(\frac{\gamma_n a_n \left(\mathbf{c}_n - \mathbf{c}^*_{\delta_n}\right)}{\varepsilon_n \sqrt{n}}\right).$$

Taking into account the assumptions of this theorem, and the strict convexity property (6.49) we deduce

$$E\{w_n \mid \mathcal{F}_{n-1}\} \stackrel{a.s.}{=} -2\frac{N \cdot K}{N \cdot K - 1}\gamma_n a_n \left(\mathbf{c}_n - \mathbf{c}^*_{\delta_n}\right)^T \frac{\partial}{\partial \mathbf{c}}\nabla_\mathbf{c}\mathcal{P}_{\mu_n,\delta_n}(\mathbf{c}_n)$$

$$+ o_\omega\left(\frac{\gamma_n a_n}{\varepsilon_n \sqrt{n}}\right)\sqrt{W_n}$$

$$\leq -\frac{N \cdot K}{N \cdot K - 1}\gamma_n a_n \delta_n W_n + o_\omega\left(\frac{\gamma_n a_n}{\varepsilon_n \sqrt{n}}\right)\sqrt{W_n}.$$

Calculating the conditional mathematical expectation of both sides of (6.53), and in view of the last inequality, we get

$$E\{W_{n+1} \mid \mathcal{F}_{n-1}\} \stackrel{a.s.}{\leq} \left(1 - \frac{N \cdot K}{N \cdot K - 1}\gamma_n a_n \delta_n\right) W_n + Const\left(\mu_{3,n}\sqrt{W_n} + \mu_{2,n}\right). \quad (6.54)$$

where

$$\mu_{3,n} := \mu_{1,n} + o_\omega\left(\frac{\gamma_n a_n}{\varepsilon_n \sqrt{n}}\right).$$

In view of

$$2\mu_{1,n}\sqrt{W_n} \leq \mu_{1,n}^2 \rho_n^{-1} + W_n \rho_n$$

(which is valid for any $\rho_n > 0$) for

$$\rho_n := \gamma_n a_n \delta_n$$

it follows

$$2\mu_{1,n}\sqrt{W_n} \leq \mu_{3,n}^2 (\gamma_n a_n \delta_n)^{-1} + W_n \gamma_n a_n \delta_n.$$

From this inequality and (6.54), and in view of the following estimation

$$\mu_{3,n}^2 (\gamma_n a_n \delta_n)^{-1} + \mu_{2,n} \leq Const \cdot \theta_n$$

we finally, obtain the following inequality

$$E\{W_{n+1} \mid \mathcal{F}_{n-1}\} \stackrel{a.s.}{\leq} \left(1 - \frac{1}{N \cdot K - 1}\gamma_n a_n \delta_n\right) W_n + Const \cdot \theta_n. \quad (6.55)$$

which is similar to the inequality involved in the Robbins-Siegmund theorem [21] (see Appendix A) for $x_n = W_n$, etc.

From (6.36) we deduce

$$a_n = O(\varepsilon_n).$$

From the assumptions of this theorem and in view of condition (6.31) and the Robbins-Siegmund theorem [21] (see Appendix A), the convergence with probability 1 follows.

The mean squares convergence follows from (6.55) after applying the operator of conditional mathematical expectation to both sides of this inequality and using lemma A5 [7] ∎

Theorem 1 shows that this adaptive learning control algorithm possess all the properties that one would desire, i.e., convergence with probability 1 as well as convergence in the mean square.

The next corollary states the conditions associated with the sequences $\{\mu_n\}$, $\{\varepsilon_n\}$, $\{\delta_n\}$, and $\{\gamma_n\}$.

Corollary 1 *If in theorem 1*

$$\varepsilon_n := \frac{\varepsilon_0}{1+n^\varepsilon} \left(\varepsilon_0 \in \left[0, (N \cdot K)^{-1}\right), \varepsilon \geq 0\right), \; \delta_n := \frac{\delta_0}{1+n^\delta \ln n} \; (\delta_0, \delta > 0),$$

$$\mu_n := \frac{\delta_0}{1+n^\mu \ln n} \; (\mu_0, \mu > 0,), \; \gamma_n := \frac{\gamma_0}{n^\gamma} \; (\gamma_0 \in (0,1), \; \gamma \geq 0),$$

with

$$\frac{2}{3}\delta < \mu \leq \delta, \; \gamma + \mu + \varepsilon + \delta \leq 1$$

then

1. *the convergence with probability 1 will take place if*

$$2\gamma > 1,$$

2. *the mean square convergence is guaranteed if*

$$\varepsilon + \delta < \frac{1}{2}.$$

It is easy to check up on these conditions by substituting the parameters given in this corollary in the assumptions of theorem 1.

Remark 4 *In the optimization problem related to the regularized penalty function $\mathcal{P}_{\mu_n,\delta_n}(\mathbf{c}_n)$ the parameters μ_n and δ_n must decrease less slowly than ε_n, i.e.,*

$$\mu \leq \varepsilon, \; \delta \leq \varepsilon.$$

6.4. CONVERGENCE ANALYSIS

The majority of methods for optimization are iterative. Even if it can be proved theoretically that this sequence will converge in the limit to the required point, a method will be practicable only if convergence occurs with some rapidity.

Having considered the convergence of the adaptive control algorithm presented in the previous section, we are now ready for our next topic, the estimation of the order of the convergence rate.

The next theorem states the convergence rate of the adaptive learning algorithm described above.

Theorem 2 *If the conditions of theorem 1 and corollary 1 are fulfilled, then*

$$W_n \stackrel{a.s.}{=} o_\omega \left(\frac{1}{n^\nu}\right)$$

where

$$0 < \nu < \min\{2\gamma - 1; 2\delta; 2\mu\} := \nu^*(\gamma, \mu, \delta)$$

and the positive parameters γ, δ, and μ satisfy the following constraints

$$\frac{2}{3}\delta < \mu \leq \delta,\ \gamma + \varepsilon + \delta \leq 1,\ 2\gamma > 1,\ \delta \leq \varepsilon,\ \mu \leq \varepsilon.$$

Proof. From (6.29), it follows:

$$W_n^* := \|\mathbf{c}_n - \mathbf{c}^{**}\|^2 = \|(\mathbf{c}_n - \mathbf{c}_n^*) + (\mathbf{c}_n^* - \mathbf{c}^{**})\|^2$$

$$\leq 2\|\mathbf{c}_n - \mathbf{c}_n^*\|^2 + 2\|\mathbf{c}_n^* - \mathbf{c}^{**}\|^2 \leq 2W_n + C\left(\delta_n^2 + \mu_n^2\right).$$

Multiplying both sides of the previous inequality by ν_n, we derive

$$\nu_n W_n^* \leq 2\nu_n W_n + \nu_n C\left(\delta_n^2 + \mu_n^2\right).$$

Selecting $\nu_n = n^\nu$ and in view of lemma 2 given in [22], and taking into account that

$$\frac{\nu_{n+1} - \nu_n}{\nu_n} = \frac{\nu + o(1)}{n}$$

we obtain

$$0 < \nu < \min\{2\gamma - 1; 2\delta; 2\mu\} := \nu^*(\gamma, \mu, \delta)$$

where the positive parameters γ, δ, ε and λ satisfy the following constraints

$$\frac{2}{3}\delta < \mu \leq \delta,\ \gamma + \varepsilon + \delta \leq 1,\ 2\gamma > 1,\ \delta \leq \varepsilon,\ \mu \leq \varepsilon.$$

∎

The following corollary provides the optimal convergence rate of the adaptive control algorithm (6.37)-(6.42).

Corollary 2 *The maximum convergence rate is achieved with the optimal parameters ε^*, δ^*, λ^* and γ^**

$$\varepsilon = \varepsilon^* = \mu = \mu^* = \delta = \delta^* = \frac{1}{6}, \ \gamma = \gamma^* = \frac{2}{3}$$

and is equal to

$$\nu^*(\gamma, \varepsilon, \delta) = 2\gamma^* - 1 = \nu^{**} = \frac{1}{3}.$$

Proof. The solution of the linear programming problem

$$\nu^*(\gamma, \mu, \delta) \to \max_{\gamma, \mu, \delta}$$

is given by

$$2\gamma - 1 = 2\delta = 2\mu = 2\delta, \ \gamma + \varepsilon + \delta = 1$$

or, in equivalent form,

$$\gamma = \frac{1}{2} + \delta = 1 - \delta - \varepsilon, \ \mu = \delta.$$

From these equalities it follows that

$$\varepsilon = \frac{1}{2} - 2\delta.$$

Taking into account that δ_n must decrease less slowly than ε_n (see Remark 4), we derive

$$\delta \leq \varepsilon$$

and, as a result, the smallest ε maximizing γ is equal to

$$\varepsilon = \delta.$$

Hence,

$$\delta = \frac{1}{6}, \ \gamma = \frac{1}{2} + \delta = \frac{2}{3}.$$

So, the optimal parameters are

$$\gamma = \gamma^* = \frac{2}{3}, \ \varepsilon = \varepsilon^* = \delta = \delta^* = \mu = \mu^* = \frac{1}{6}.$$

The maximum convergence rate is achieved with this choice of parameters and is equal to

$$\nu^*(\gamma, \varepsilon, \delta) = 2\gamma^* - 1 = \nu^{**} = \frac{1}{3}.$$

■ The theory of adaptive (self-learning) systems at the present time is able to solve many problems arising in practice. The algorithm developed and analyzed in this chapter, in very wide conditions of indeterminacy ensures the achievement of the stated control objective (learning goal). Several simulations results have been carried out. In order to illustrate the feasibility, the efficiency and the performance of the previous adaptive control algorithm, some simulations are presented in the last chapter. A Matlab mechanization of this self-learning control algorithm is given Appendix B.

6.5 Conclusions

We have described and analyzed an adaptive control algorithm for constrained finite Markov chains whose transition probabilities are unknown. This control algorithm is closely connected to stochastic approximation techniques. The control policy is designed to achieve the minimization of a loss function under a set of inequality constraints. The average values of the conditional mathematical expectations of this loss function and constraints are also assumed to be unknown. A regularized penalty function is introduced to derive an adaptive control algorithm. In this algorithm the transition probabilities of the Markov chain and the average values of the constraints are estimated at each time n. The control policy is adjusted using the Bush-Mosteller reinforcement scheme with time varying correction factor. The convergence properties (convergence with probability 1 as well as convergence in the mean squares) have been stated using Lyapunov approach and martingales theory. We establish that the optimal convergence rate is equal to $n^{-\frac{1}{3}+\delta}$ (δ is any small positive parameter).

6.6 References

1. O. H. Lerma, *Adaptive Markov Control Processes*, Springer-Verlag, London, 1989.

2. O. H. Lerma and J. B. Lasserre, *Discrete-time Markov Control Processes*, Springer-Verlag, London, 1996.

3. A. Arapostathis, V. S. Borkar, E. Fernandez-Gaucherand, M. K. Ghosh and S. I. Marcus, Discrete-time controlled Markov processes with average cost criterion: a survey, *SIAM Journal of Control and Optimization*, vol. 31, pp. 282-344, 1993.

4. M. Duflo, *Random Iterative Models*, Springer-Verlag, London, 1997.

5. A. S. Poznyak and K. Najim, *Learning Automata and Stochastic Optimization*, Springer-Verlag, London, 1997.

6. K. Najim and A. S. Poznyak, Penalty function and adaptive control of constrained finite Markov chains, *International Journal of Adaptive Control and Signal Processing*, vol. 12, pp. 545-565, 1998.

7. K. Najim, K. and A. S. Poznyak, *Learning Automata Theory and Applications*, Pergamon Press, London, 1994.

8. Ya. Z. Tsypkin, *Fundations of the Theory of Learning Systems*, Academic Press, New York, 1973.

9. Ya. Z. Tsypkin, *Adaptive and Learning in Automatic Systems*, Academic Press, New York, 1971.

10. A. Benveniste, M. Metivier and P. Priouret, *Stochastic Approximations and Adaptive Algorithms*, Springer-Verlag, Berlin, 1990.

11. H. Kushner and G. G. Yin, *Stochastic Approximation Algorithms*, Springer-Verlag, Berlin, 1997.

12. H. Walk, Stochastic iteration for a constrained optimization problem, *Commun. Statist.-Sequential Analysis*, vol. 2, pp. 369-385, 1983-84.

13. K. Najim, A. Rusnak, A. Mészaros and M. Fikar, Constrained long-range predictive control algorithm based on artificial neural networks, *International Journal of Systems Science*, vol. 28, pp. 1211-1226, 1997.

14. R. A. Howard, *Dynamic Programming and Markov Processes*, J. Wiley, New York, 1962.

15. W. I. Zangwill, Nonlinear programming via penalty functions, *Management Science*, vol. 13, pp. 344-358, 1967.

16. P. E. Gill, W. Murray and M. H. Wright, *Practical Optimization*, Academic Press, New York, 1981.

17. A. V. Nazin and A. S. Poznyak, *Adaptive Choice of Variants*, (in Russian) Nauka, Moscow, 1986.

18. J. L. Doob, *Stochastic Processes*, J. Wiley, New York, 1953.

19. R. B. Ash, *Real Analysis and Probability*, Academic Press, New York, 1972.

20. D. Hall and C. Heyde, *Martingales Limit Theory and its Applications*, Academic Press, New York, 1980.

21. H. Robbins and D. Siegmund, A Convergence theorem for nonnegative almost supermartingales and some applications, in *Optimizing Methods in Statistics*, ed. by J. S. Rustagi, Academic Press, New York, pp. 233-257, 1971.

22. A. S. Poznyak and K. Najim, Learning automata with continuous input and changing number of actions, *International Journal of Systems Science*, vol. 27, pp. 1467-1472, 1996.

Chapter 7

Nonregular Markov Chains

7.1 Introduction

In this chapter we shall be concerned with the control of a class of nonregular Markov chains including

- *ergodic or communicating* controlled Markov chains characterized by (see chapter 1):
$$L = 1, \ r_1 \geq 2, \ X^+(0) = \emptyset;$$

- controlled Markov chains of general type characterized by
$$L \geq 2, \ X^+(0) \neq \emptyset.$$

In what follows, we will show that the adaptive control problem of ergodic and general type Markov chains should be formulated in an absolutely different manner:

if in the ergodic case this problem can be formulated as an optimization problem, in the case of general type chains the adaptive control problem turns out to be equivalent to an inequality type problem.

The results presented in this chapter were stated in [1].
The next section deals with the adaptive control of ergodic Markov chains.

7.2 Ergodic Markov chains

The main differences between regular and ergodic Markov chains are:

1. for ergodic chains, the property (4.27) of chapter 4 is not fulfilled:
$$c_- = 0;$$

2. if we use the Projection Gradient Technique, within the intervals of the frozen control (when the corresponding Markov chain turns to be homogeneous), the inequality (4.26) (see chapter 4) is not valid. We have no exponential convergence of the state probability vector to its stationary point, i.e., the conditions of the Rozanov theorem [2] are not fulfilled (see chapter 1).

Recall that in the adaptive control algorithms presented in the previous chapters, the components of the randomized control strategy at each time n are calculated as

$$d_n^{il} = \frac{c_n^{il}}{\sum_{s=1}^{N} c_n^{is}}. \qquad (7.1)$$

For any n the denominator in (7.1) must be greater than zero, i.e.,

$$\sum_{s=1}^{N} c_n^{is} > 0. \qquad (7.2)$$

As in the nonregular case $c_- = 0$, we can not guarantee that this property (7.2) will be automatically fulfilled. Nevertheless, this property can be fulfilled by using the projection operator $\mathcal{P}^{\widehat{C}}\varepsilon_k\{\cdot\}$ [3-4] which ensures the projection onto the set $\widehat{C}_{\varepsilon_k}$ ((4.11) of chapter 4) and provides

$$c_n^{il} \geq \varepsilon_n > 0, \ i = \overline{1,K}; l = \overline{1,N}$$

for any n and leads to

$$\sum_{s=1}^{N} c_n^{is} \geq N\varepsilon_n > 0.$$

Take into account that the transition matrix $\Pi(d)$ of an ergodic Markov chain controlled by any nonsingular stationary randomized strategy $\{d\} \in \Sigma_s^+$ turns out to be irreducible. It follows that there exists a unique stationary probability distribution with components given by

$$p_i(d) = \sum_{s=1}^{N} c^{is} \ \left(i = \overline{1,K}\right).$$

In this situation we can also prove a similar result as in (4.26) of chapter 4. It concerns only the average values of the probability vector calculated within the period r which is equal to the number of cyclic subclasses of the given Markov chain. The next lemma states this result.

7.2. ERGODIC MARKOV CHAINS

Lemma 1 *For an ergodic Markov chain controlled by any non-singular stationary randomized strategy $\{d\} \in \Sigma_s^+$ satisfying*

$$d^{il} \geq \varepsilon > 0, \ i = \overline{1, K}; l = \overline{1, N}$$

the following estimate holds

$$\left| \frac{1}{r} \sum_{l=0}^{r-1} p_i(d, x, n+l) - \sum_{s=1}^{N} c^{is} \right|$$

$$\leq K^{\frac{5}{2}} \left[1 - C_1 \varepsilon^{r(K-r)} \right]^{\frac{n-r+1}{r(K-r)} - 1} \tag{7.3}$$

where C_1 is a positive constant, r is the number of cyclic subclasses of the considered homogeneous Markov chain with transition probability $\Pi(d)$ and, $p_i(d, x, n)$ is the probability of transition from the state $x \in X$ to the state $x = x(i)$ after n steps.

Proof. To start the proof, let us notice that

$$p_i(d, x(j), n) \ \left(i = \overline{1, K}; l = \overline{1, N} \right)$$

represents the elements of the matrix

$$[\Pi(d)]^n = [p_i(d, x(j), n)]_{i,j=1,\ldots,K}.$$

Let us introduce the stochastic matrix $A(d)$ (see (1.5) of chapter 1) satisfying the equality

$$A(d)\Pi(d) = A(d). \tag{7.4}$$

For ergodic homogeneous Markov chains, the equation

$$p(d) = \Pi^T(d)p(d),$$

which defines the stationary distribution, has a unique solution. So, we conclude that the matrix $A(d)$ (7.4) has identical rows equal to $p^T(d)$. Hence, to prove (7.4) it is sufficient to prove that for a large enough n the following inequality

$$\left\| \frac{1}{r} \sum_{l=0}^{r-1} [\Pi(d)]^{n+l} - A(d) \right\|$$

$$\leq K^{\frac{5}{2}} \left[1 - C_1 \varepsilon^{r(K-r)} \right]^{\frac{n-r+1}{r(K-r)} - 1} \tag{7.5}$$

is valid where $\|\cdot\|$ is the Euclidean norm.

Using the structure representation (1.15) of chapter 1 (in our case $l = L = 1$)

$$\Pi(d) = \begin{bmatrix} 0 & \Pi_{12}(d) & 0 & \cdots & 0 & \cdots & 0 \\ 0 & 0 & \Pi_{23}(d) & \cdots & 0 & \cdots & 0 \\ \cdot & \cdot & \cdot & \cdots & \cdot & \cdots & \cdot \\ \Pi_{r1}(d) & 0 & 0 & \cdots & 0 & \cdots & 0 \end{bmatrix},$$

we conclude that the rth-order of this matrix has a diagonal structure:

$$[\Pi(d)]^T = \begin{bmatrix} \tilde{\Pi}_1(d) & 0 & \cdots & 0 \\ 0 & \tilde{\Pi}_2(d) & \cdots & 0 \\ \cdot & \cdot & \cdot & \cdot \\ 0 & 0 & \cdots & \tilde{\Pi}_r(d) \end{bmatrix} \quad (7.6)$$

where $\tilde{\Pi}_l(d)$ ($l = 1, ..., r$) is a square stochastic matrix corresponding to the regular cyclic subclass of states with index l. Hence, the limit

$$A_0(d) := \lim_{k \to \infty} [\Pi(d)]^{kr}$$

exists.

Using this fact and the result from [5] that

$$\frac{1}{r} \sum_{l=0}^{r-1} [\Pi(d)]^l = I,$$

we conclude that

$$\lim_{k \to \infty} \frac{1}{r} \sum_{l=0}^{r-1} [\Pi(d)]^{kr+l} = A_0(d) \frac{1}{r} \sum_{l=0}^{r-1} [\Pi(d)]^l = A_0(d).$$

Let us denote the integer part of n/r by

$$k_n := int\,[n/r]$$

and the remainder $n - k_n r$ by

$$l_n := n - k_n r.$$

Based on these notations and taking into account relation (7.4), we derive

$$\frac{1}{r} \sum_{l=0}^{r-1} [\Pi(d)]^{n+l} - A(d)$$

$$= \left([\Pi(d)]^{k_n r} - A_0(d)\right) \left(\frac{1}{r} \sum_{l=0}^{r-1} [\Pi(d)]^l \Pi^{l_n}(d)\right).$$

7.2. ERGODIC MARKOV CHAINS

Using the fact that the norm of a stochastic matrix does not exceed $\sqrt{K}$, from the last inequality we obtain

$$\left\| \frac{1}{r} \sum_{l=0}^{r-1} [\Pi(d)]^{n+l} - A(d) \right\| \leq \sqrt{K} \left| [\Pi(d)]^{k_n r} - A_0(d) \right|.$$

Notice that the lth-diagonal block of the matrix $A_0(d)$ is equal to

$$[A_0(d)]_{ll} = \lim_{k \to \infty} \left[\tilde{\Pi}_l(d) \right]^k.$$

In view of the Rozanov theorem (see chapter 1), we get

$$\left\| \frac{1}{r} \sum_{l=0}^{r-1} [\Pi(d)]^{n+l} - A(d) \right\| < K^{\frac{5}{2}} \max_{l=\overline{1,r}} (1 - \rho_l(d))^{\frac{k_n}{\kappa_l} - 1} \quad (7.7)$$

where κ_l is a minimal positive number k corresponding to the situation when the matrix $\left[\tilde{\Pi}_l(d) \right]^k$ has no elements equal to zero and, $\rho_l(d)$ is the coefficient of ergodicity (Rozanov theorem, see chapter 1) corresponding to the lth-regular Markov chain with transition matrix $\tilde{\Pi}_l(d)$.

For the number κ_l we have the evident estimate:

$$\kappa_l \leq K - r.$$

We also have

$$k_n = \frac{n - l_n}{r} \geq \frac{n - (r-1)}{r}.$$

To estimate $\rho_l(d)$, let us use the formula given in Remark 2 of chapter 1

$$\rho_l(d) \geq \max_j \min_i \left(\left[\tilde{\Pi}_l(d) \right]^{\kappa_l} \right)_{ij}.$$

In the last inequality, the operator max and min are taken over all the states belonging to the lth-subchain. Taking into account the evident inequality

$$\pi_{ij}(d) = \sum_{l=1}^{N} \pi_{ij}^l d^{il} \geq \varepsilon \sum_{l=1}^{N} \pi_{ij}^l.$$

we derive

$$\left(\left[\tilde{\Pi}_l(d) \right]^{\kappa_l} \right)_{ij} = \left(\left[\tilde{\Pi}(d) \right]^{r\kappa_l} \right)_{ij}$$

$$\geq (N\varepsilon)^{r\kappa_l} \left(\left[\tilde{\Pi}(\overline{d}) \right]^{r\kappa_l} \right)_{ij} \quad (7.8)$$

where the strategy $\bar{d}$ corresponds to the selection of the control action with uniform probabilities, i.e.,

$$\bar{d} = \left[\bar{d}^{ij}\right]_{\substack{i=\overline{1,K} \\ j=\overline{1,N}}}, \quad \bar{d}^{ij} := \frac{1}{N}.$$

So we conclude that

$$\left(\left[\tilde{\Pi}(\bar{d})\right]^{r\kappa_l}\right)_{ij} > 0,$$

and from (7.7) and (7.8) the estimation (7.5) follows. The lemma is proved. ∎

Taking into account this lemma, we can prove the following result.

Theorem 1 *If for a controlled ergodic Markov chain, the following conditions*

$$\sum_{k=1}^{\infty}\left[(kg_k)^{-1} + \left(\frac{\Delta n_k}{n_k}\right)^2\right] < \infty \tag{7.9}$$

$$\varlimsup_{k\to\infty} \frac{\Delta n_k \sqrt{kg_k}}{\sum_{t=1}^{k} \varepsilon_t^2 \Delta n_t} < \infty \tag{7.10}$$

$$n_k \varepsilon_k^{\frac{K^2}{4}} \underset{k\to\infty}{\to} \infty, \quad \varepsilon_k \Delta n_k \underset{k\to\infty}{\to} \infty, \quad n_k^{-1}\varepsilon_k^{\frac{K^2}{4}} |\ln \varepsilon_k| \underset{k\to\infty}{\to} \infty \tag{7.11}$$

are fulfilled, then, for any initial state distribution and, for any $i = 1, ..., K$ and $l = 1, ..., N$ we have

$$\lim_{n\to\infty} s_{n+1}^{il} \mid \sum_{t=1}^{t(n)} \varepsilon_t^2 \Delta n_t > 0 \tag{7.12}$$

and

$$\left(\hat{\pi}_{ij}^l\right)_n - \pi_{ij}^l \overset{a.s.}{=} o\left(\left(\sum_{t=1}^{t(n)} \varepsilon_t^2 \Delta n_t\right)^{-\frac{\rho}{2}}\right), \tag{7.13}$$

where

$$\rho \in [0, 1), \quad j = \overline{1, K}$$

Proof. The proof of this theorem is absolutely similar to the proof of theorem 1 of chapter 4: the difference lies only in the calculations of the terms θ_t^i. In view of the previous theorem, for large enough T and C, we have:

$$\theta_t^i = \sum_{s=1}^{N} c_t^{is} + \frac{1}{\Delta n_t} \sum_{\tau=n_t}^{n_{t+1}-1} \left[E\left\{ \chi\left(x_\tau = x(i) \mid \hat{\mathcal{F}}_t\right)\right\} - \sum_{s=1}^{N} c_t^{is}\right]$$

7.2. ERGODIC MARKOV CHAINS

$$\geq N\varepsilon_t - \frac{r}{\Delta n_t} \sum_{k=0}^{[\Delta n_t/r]} \left| \frac{1}{r} \sum_{l=0}^{r-1} \left[p_i(d_{n_t}, x_{n_t}, n_t + rk + l) - \sum_{s=1}^{N} c_t^{is} \right] \right|$$

$$+ O\left(\frac{1}{\Delta n_t}\right) \geq N\varepsilon_t - \frac{r}{\Delta n_t} \sum_{k=0}^{\infty} \left[1 - C\varepsilon_t^{r(K-r)}\right]^{\frac{n_t + rk + 1 - r}{r(K-r)}} + O\left(\frac{1}{\Delta n_t}\right)$$

$$= \varepsilon_t \left[N + O\left((\Delta n_t)^{-1} \varepsilon_t^{-1-r/(K-r)} e^{-c_{n_t \varepsilon_t^{r(K-r)}}}\right) + O\left((\Delta n_t \varepsilon_t)^{-1}\right) \right] \geq \frac{N}{2} \varepsilon_t.$$

So, the following inequality

$$\overline{S}_{n+1}^{il} \leq \frac{N}{2n} \sum_{t=T}^{t(n)} \varepsilon_t^2 \Delta n_t$$

holds with probability 1. Following the proof of theorem 1 of chapter 4, we obtain (7.12) and (7.13). The theorem is proved. ■

This has an immediate corollary.

Corollary 1 *If the sequences $\{\varepsilon_k\}$ and $\{n_k\}$ satisfy*

$$\varepsilon_k \sim k^{-\theta}, \quad n_k \sim k^{-\kappa}, \quad \kappa > \max\left\{1 + \theta, \theta \frac{K^2}{4}\right\}, \quad 0 \leq \theta \leq \frac{1}{4}$$

then for any $i = \overline{1, K}; l = \overline{1, N}$ the following inequality:

$$\lim_{n \to \infty} \frac{S_{n+1}^{il}}{n^{1 - \frac{2\theta}{\kappa}}} > 0. \tag{7.14}$$

holds with probability 1.

Proof. The proof is similar to the proof of the corollary of theorem 1 of chapter 4. ■

Based on these estimates, we can state some results concerning the convergence of the Projection Gradient Scheme (4.6)-(4.11) of chapter 4.

Theorem 2 *If the sequences $\{\gamma_k\}$, $\{\varepsilon_k\}$ and $\{n_k\}$ in (4.6)-(4.11) of chapter 4 are selected as follows*

$$0 > \varepsilon_k \xrightarrow[k \to \infty]{} \infty, \quad \gamma_k > 0, \quad \varepsilon_k^{1 + \frac{K^2}{4}} \Delta n_k \xrightarrow[k \to \infty]{} \infty,$$

$$\lim_{n \to \infty} \gamma_k^{-1} \left[n_k^{-1} \Delta n_k + \Delta n_k \left(\sum_{t=1}^{k} \varepsilon_t^2 \Delta n_t \right)^{-1} \right] = 0, \quad \gamma_{k+1}^{-1} \Delta n_{k+1} \geq \gamma_k^{-1} \Delta n_k,$$

$$\sum_{k=1}^{\infty}\left[\gamma_k\varepsilon_k^{-2}\left(\Delta n_k\right)n_k^{-1}+\left(\frac{\Delta n_k}{n_k}\right)^2\right]<\infty$$

and there exist a positive constant $\delta > 0$ and, a positive sequence $\{h_k\}$ such that

$$\sum_{k=1}^{\infty}h_k^2<\infty,\ \varlimsup_{n\to\infty}\left[h_k\varepsilon_k^{-\delta}+\Delta n_k\left(h_k\sum_{t=1}^{k}\varepsilon_t^2\Delta n_t\right)^{-1}\right]<\infty$$

then for any initial conditions

$$c_1 \in \widehat{\mathbf{C}}_{\varepsilon_1},\ x_1 \in X$$

the loss function Φ_n converges to its minimal value Φ_* with probability 1, i.e.,

$$\Phi_n \underset{n\to\infty}{\overset{a.s.}{\to}} \Phi_*.$$

Proof. For any point $\tilde{c} \in \mathbf{C}$, let us consider the affine transformation given by

$$\widetilde{\tilde{c}}_t^{il} := a_t \tilde{c}^{il} + b_t^i,\ i = \overline{1,K}; l = \overline{1,N}$$

where the sequences $\{a_t\}$ and $\{b_t\}$ are defined by

$$a_t := 1 - N\varepsilon_t\left(\min_i \overline{b}^i\right)^{-1},\ b_t^i := 2\overline{b}^i\varepsilon_t\left(\min_i \overline{b}^i\right)^{-1}.$$

The vector $\overline{b} = \left(\overline{b}^1, ..., \overline{b}^K\right)^T$ satisfies the following system of equations

$$\overline{b}^j = \frac{1}{N}\sum_{i=1}^{K}\overline{b}^i\sum_{l=1}^{N}\pi_{ij}^l,\ \sum_{j=1}^{K}\overline{b}^j = 1.$$

Notice that the solution of this system of equations exists and is unique and positive, i.e.,

$$\min_i \overline{b}^i > 0,$$

because it corresponds to the solution of the system

$$p(d) = \Pi(d)p(d),\ \mathbf{e}^T p(d) = 1$$

with $d^{il} = 1/N\ \left(i = \overline{1,K}; l = \overline{1,N}\right)$. Notice also that

$$\widetilde{\tilde{c}}_t \in \mathbf{C}_{2\varepsilon_t}\ \text{and}\ \tilde{c} := \widehat{\pi}_{\varepsilon_t}\left\{\widetilde{\tilde{c}}_t\right\},\ t = 1, 2, ...$$

Under the assumptions of the previous theorem and taking into account that $\mathbf{C}_{2\varepsilon_t} \subset \mathbf{C}_{\varepsilon_t}$, it follows:

$$\left|\tilde{c} - \widetilde{\tilde{c}}_t\right| \leq const\ \|\widehat{\pi}_{n_t} - \pi\|$$

7.2. ERGODIC MARKOV CHAINS

$$= o\left(\left(\sum_{s=1}^{t} \varepsilon_s^2 \Delta n_s\right)^{-\frac{\rho}{2}}\right) \quad \forall \rho \in [0,1)$$

and

$$|\tilde{c}_{t+1} - \tilde{c}_t| \leq const \left\|\widehat{\pi}_{n_{t+1}} - \widehat{\pi}_{n_t}\right\| O^*\left(\Delta n_t \left(\left(\sum_{s=1}^{t} \varepsilon_s^2 \Delta n_t\right)^{-1}\right)\right).$$

By following the lines of the proof of theorem 2 of chapter 4, we have to prove again relation (4.38) of chapter 4. In our case, condition (4.39) is not fulfilled. In view of lemma 3 (see Appendix A), we have to prove the convergence, with probability 1, of the following series:

$$\sum_{\tau=1}^{\infty} \tau^{-2} \sum_{i,l} E\left\{\left(c_{t(\tau)}^{il} - \tilde{c}_{t(\tau)}^{il}\right)^2 \frac{\eta_\tau^2 \chi\left(x_\tau = x(i), u_\tau = u(l)\right)}{\left(c_{t(\tau)}^{il}\right)^2} \mid \mathcal{F}_\tau\right\}$$

$$\overset{a.s.}{\leq} const \sum_{\tau=1}^{\infty} \tau^{-2} \sum_{i,l} \left(c_{t(\tau)}^{il} \sum_{l=1}^{N} c_{t(\tau)}^{il}\right)^{-1} \leq const \sum_{\tau=1}^{\infty} \tau^{-2} \varepsilon_{t(\tau)}^{-2}$$

$$\leq const \sum_{t=1}^{\infty} \varepsilon_t^{-2} \left(\frac{1}{n_t - 1} - \frac{1}{n_{t+1} - 1}\right) < \infty.$$

The last series converges because:

$$\sum_{k=1}^{\infty} \gamma_k \varepsilon_k^{-2} \Delta n_k n_k^{-1} < \infty$$

and

$$n_{k+1} \gamma_k \underset{k \to \infty}{\to} \infty.$$

Based on this result, we derive

$$\frac{1}{n} \sum_{\tau=1}^{n} E\left\{\theta_\tau \mid \mathcal{F}_\tau\right\} \overset{a.s.}{=} \frac{1}{n} \sum_{\tau=1}^{n} \sum_{i,l} v_{ij} \chi\left(x_\tau = x(i)\right) d_{n'(\tau)}^{ij}$$

$$- \sum_{i,l} v_{ij} \tilde{c}^{ij} - r_{1n} + O\left(\frac{1}{n} \sum_{k=1}^{t(n)} \Delta n_k \varepsilon_k\right).$$

The elements r_{1n} will be decomposed into two terms:

$$r_{1n} := r'_{1n} + r''_{1n}$$

where

$$r'_{1n} := \sum_{i,l} v_{ij} \frac{1}{n} \sum_{\tau=1}^{n} \left(\tilde{c}^{il}_{t(\tau)} - \bar{\tilde{c}}^{il}_{t(\tau)} \right) \chi \left(x_\tau = x(i) \right) \left(\sum_{s=1}^{N} c^{is}_{t(\tau)} \right)^{-1}$$

$$r''_{1n} := \sum_{i,l} v_{ij} \frac{1}{n} \sum_{\tau=1}^{n} \bar{\tilde{c}}^{il}_{t(\tau)} \left[\chi \left(x_\tau = x(i) \right) \left(\sum_{s=1}^{N} c^{is}_{t(\tau)} \right)^{-1} - 1 \right].$$

1) First, let us prove that

$$r_{1n_k} \xrightarrow[k \to \infty]{a.s.} 0.$$

In view of this new estimate, we will prove that the right hand side of (4.40) of chapter 4 tends to zero.

a) To prove that

$$r'_{1n_k} \xrightarrow[k \to \infty]{a.s.} 0$$

it is sufficient to state that for some $\rho \in (0,1)$ we have

$$\frac{1}{n_k} \sum_{t=1}^{k} \frac{\Delta n_t}{\varepsilon_t} \left(\sum_{s=1}^{t} \varepsilon_s^2 \Delta n_s \right)^{-\frac{\rho}{2}} \xrightarrow[k \to \infty]{} 0.$$

Based on Toeplitz lemma (lemma 8 of Appendix A), we derive

$$\varepsilon_t^{-\frac{2}{\rho}} \left(\sum_{s=1}^{t} \varepsilon_s^2 \Delta n_s \right)^{-1} = O \left(\varepsilon_t^{-\frac{2}{\rho}} \Delta n_t^{-1} h_t \right)$$

$$= O \left(h_t \varepsilon_t^{-\frac{2}{\rho} + 1 + \frac{K^2}{4}} \Delta n_t^{-1} \varepsilon_t^{-1 - \frac{K^2}{4}} \right)$$

$$= \left(\Delta n_t \varepsilon_t^{1 + \frac{K^2}{4}} \right)^{-1} O \left(h_t \varepsilon_t^{-\delta} \right) \xrightarrow[t \to \infty]{} 0$$

for

$$\delta := 2 \left(\rho^{-1} - 1 \right).$$

b) Let us now prove that

$$r''_{1n_k} \xrightarrow[k \to \infty]{a.s.} 0.$$

Again, according to lemma 3 (see Appendix A), we obtain

$$\frac{1}{n_{k+1} - 1} \sum_{\tau=1}^{n_{k+1}-1} \bar{\tilde{c}}^{il}_{t(\tau)} \chi \left(x_\tau = x(i) \right) \left(\sum_{s=1}^{N} c^{is}_{t(\tau)} \right)^{-1}$$

7.2. ERGODIC MARKOV CHAINS

$$= \frac{1}{n_{k+1}-1} \sum_{t=1}^{k} \left(\sum_{s=1}^{N} c_{t(\tau)}^{is}\right)^{-1} \widetilde{c}_t^{il} \sum_{\tau=n_t}^{n_{t+1}-1} \chi(x_\tau = x(i))$$

$$\stackrel{a.s.}{=} \frac{1}{n_{k+1}-1} \sum_{t=1}^{k} \left(\sum_{s=1}^{N} c_{t(\tau)}^{is}\right)^{-1} \widetilde{c}_t^{il} \sum_{\tau=n_t}^{n_{t+1}-1} P(x_\tau = x(i) \mid \mathcal{F}_{n_t}) + o(1).$$

This equality is valid because of the convergence of the following series

$$\sum_{\tau}^{\infty} \tau^{-2} E \left\{ \left[\widetilde{c}_{t(\tau)}^{il} \left(\sum_{s=1}^{N} c_{t(\tau)}^{is}\right)^{-1} \chi(x_\tau = x(i)) \right. \right.$$

$$\left. \left. - P\left(x_\tau = x(i) \mid \mathcal{F}_{n'(\tau)}\right) \right]^2 \mid \mathcal{F}_{n'(\tau)} \right\}$$

$$\leq \sum_{k}^{\infty} \sum_{\tau=n_k}^{n_{k+1}-1} \tau^{-2} E \left\{ \left[\chi(x_\tau = x(i)) \left(\sum_{s=1}^{N} c_{t(\tau)}^{is}\right)^{-1} \right. \right.$$

$$\left. \left. - P\left(x_\tau = x(i) \mid \mathcal{F}_{n'(\tau)}\right) \right]^2 \mid \mathcal{F}_{n'(\tau)} \right\}$$

$$\stackrel{a.s.}{\leq} C(\omega) \sum_{k}^{\infty} \varepsilon_k^{-2} \sum_{\tau=n_k}^{n_{k+1}-1} \tau^{-2} \leq C(\omega) \sum_{k}^{\infty} \varepsilon_k^{-2} \left(\frac{1}{n_k-1} - \frac{1}{n_{k+1}-1}\right) < \infty.$$

In view of the Toeplitz lemma (lemma 8 of Appendix A) and lemma 1, we get

$$r''_{1n_k} \leq \frac{C_1}{n_{k+1}-1} \sum_{t=1}^{k} \varepsilon_k^{-1} \sum_{i=1}^{K} \sum_{\tau=0}^{[\Delta n_t/r]} \left(\left| \sum_{l=0}^{r-1} P\{x_{n_t+l+\tau r} \right. \right.$$

$$= x(i) \mid \mathcal{F}_{n(\tau)} \} - \sum_{s=1}^{N} c_t^{is} \bigg| + r \bigg) + o(1)$$

$$\stackrel{a.s.}{\leq} \frac{C_1}{n_{k+1}-1} \sum_{t=1}^{k} \varepsilon_k^{-1} \left(1 + \sum_{\tau=0}^{[\Delta n_t/r]} \left[1 - C\varepsilon_t^{r(K-r)}\right]^{\frac{n_t-r+1+\tau r}{r(K-r)}}\right)$$

$$\leq \frac{C_1}{n_{k+1}-1} \sum_{t=1}^{k} \varepsilon_k^{-1-r(K-r)} + o(1) \stackrel{a.s.}{\underset{k\to\infty}{\to}} 0.$$

Combining all the previous estimates, we conclude that

$$r_{1n_k} \stackrel{a.s.}{\underset{k\to\infty}{\to}} 0.$$

2) The sequence $\{r_{2n_k}\}$ tends to zero absolutely under the same conditions as in theorem 2 of chapter 4

3) Let us now consider the term r_{3nk}. Based on the previous estimates, we obtain

$$r_{3nk} \leq \frac{K_2}{n_{k+1}-1} \sum_{\tau=1}^{n_{k+1}-1} \theta'_\tau + \frac{K_2 C(\omega)}{n_{k+1}-1} \sum_{t=1}^{k} \theta''_\tau \qquad (7.15)$$

where

$$\theta'_\tau := \gamma_{t(\tau)} \eta_\tau^2 \left(c_{t(\tau)}^{il}\right)^{-2},$$

and

$$\theta''_\tau := (\Delta n_t)^2 \left(\gamma_t \sum_{s=1s}^{t2} \Delta n_s\right)^{-1}.$$

Notice that

$$\frac{1}{n} \sum_{\tau=1}^{n} \theta'_\tau$$

because of the convergence of the following series (see lemma 2 of Appendix A):

$$\sum_{\tau=1}^{\infty} \tau^{-1} E\left\{\theta'_\tau \mid \mathcal{F}_{n'(\tau)}\right\} \leq C(\omega) \sum_{\tau=1}^{\infty} \tau^{-1} \gamma_{t(\tau)} \varepsilon_{t(\tau)}^{-2} \leq$$

$$\leq C(\omega) \sum_{k=1}^{\infty} \gamma_k \varepsilon_k^{-2} \sum_{\tau=n_k}^{n_{k+1}-1} \tau^{-1} \leq C(\omega) \sum_{k=1}^{\infty} \gamma_k \varepsilon_k^{-2} (\Delta n_k)^{-1} n_k^{-1} < \infty.$$

The second term in (7.15) tends to zero by the Toeplitz lemma (lemma 8 of Appendix A). So we proved that

$$r_{3n_k} \xrightarrow[k\to\infty]{a.s.} 0.$$

4) For the analysis of the behaviour of the sequence $\{\Phi_n\}$ we have to prove its asymptotic equivalence with the sequence $\{\overline{\Phi}_n\}$ (see (4.49)). It follows from lemma 3 (see Appendix A) because of the convergence of the following series

$$\sum_{\tau=1}^{\infty} \tau^{-1} E\left\{\eta_\tau^2 \mid \mathcal{F}_{n'(\tau)}\right\} \stackrel{a.s.}{\leq} const \sum_{\tau=1}^{\infty} \tau^{-1} \varepsilon_{n'(\tau)}^{-2} \leq$$

$$\leq const \sum_{k=1}^{\infty} \varepsilon_k^{-2} \frac{\Delta n_k}{n_k n_{k+1}} < \infty.$$

To finish the proof of this theorem, we have only to notice that in the beginning we considered any point $\tilde{c} \in \mathbf{C}$. Hence, we have

$$\overline{\lim_{n\to\infty}} \Phi_n = \overline{\lim_{n\to\infty}} \Phi_{n_k} \leq \min_{\tilde{c}\in\mathbf{C}} \sum_{i,j} v_{ij} \tilde{c}^{ij} = \Phi^*.$$

The theorem is proved. ■

7.2. ERGODIC MARKOV CHAINS

Corollary 2 *Let us consider the subclass of the parameter sequences $\{\gamma_k\}$, $\{\varepsilon_k\}$ and $\{n_k\}$ involved in the adaptive algorithm (4.6)-(4.11) of chapter 4 given by*

$$\gamma_k = \gamma k^{-\nu}, \ \varepsilon_k = \varepsilon k^{-\theta}, \ n_k = [k^\kappa]. \qquad (7.16)$$

Then, the conditions of the previous theorem will be fulfilled if the constants θ, ν and κ satisfy

$$0 < 2\theta < \nu < 1 - 2\theta, \ \kappa > 1 + \theta\left(1 + \frac{K^2}{4}.\right) \qquad (7.17)$$

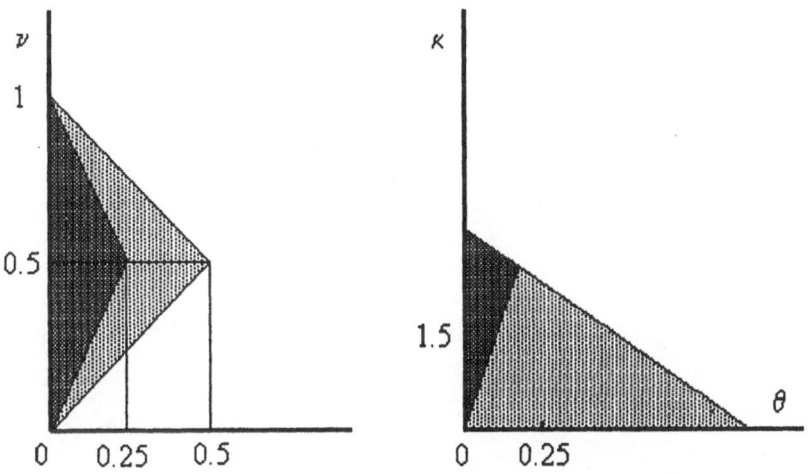

Figure 7.1: Convergence domain.

Figure 7.1 represents the areas associated with the parameters θ, ν and κ in the planes (θ, ν) and (θ, κ), which guarantee the convergence of the adaptive control algorithm (4.6)-(4.11) of chapter 4 for two different situations:

- the darkest shaded areas correspond to *Regular* Markov Chains;

- the lighter shaded areas correspond to *Ergodic* Markov Chains.

As it seen from this figure, the area associated with the design parameters for Ergodic Markov Chains is smaller than the area associated with the design parameters of Regular Markov Chains.

The following theorem gives the estimation of the order of the corresponding adaptation process within the subclass of the parameters given by (7.16) and (7.17).

Theorem 3 *Within the parameter subclass (7.16) and (7.17) for any initial conditions*
$$c_1 \in \mathbf{C}_{\varepsilon_1}, \quad x_1 \in X$$
and for any $\delta > 0$, we have
$$\Phi_n - \Phi^* \leq o\left(n^{\delta-\varphi_1}\right) + O^*\left(n^{-\varphi_2}\right), \qquad (7.18)$$
where
$$\varphi_1 = \kappa^{-1} \min\left\{\frac{1}{2} - \theta, \nu - 2\theta\right\}, \qquad (7.19)$$
and
$$\varphi_2 = \kappa^{-1} \min\left\{\kappa - 1 - \theta\left(1 + \frac{K^2}{4}\right), 1 - \nu - 2\theta, \theta\right\}, \qquad (7.20)$$
i.e., the order of the adaptation rate is equal to

- $o\left(n^{\delta-\varphi_1}\right)$, *if $\varphi_1 \leq \varphi_2$ or equal to*

- $o\left(n^{\delta-\varphi_2}\right)$, *if $\varphi_1 > \varphi_2$.*

Proof. Following the scheme of the proof of theorem 2 of chapter 4, we obtain:
$$\overline{\Phi}_{n_{k+1}-1} - \tilde{V}(\tilde{c}) \leq \tilde{r}_{1n_k} + r_{2n_k} + r_{3n_k} + r_{4n_k} + O^*(\varepsilon_k) \qquad (7.21)$$
where
$$\tilde{r}_{1n_k} := \left(r'_{1n_k} + r''_{2n_k}\right)\big|_{n=n_{k+1}-1},$$
the sequences $\{r'_{1n_k}\}$ and $\{r''_{2n_k}\}$ are defined in the previous theorem and the other sequences $\{r_{2n_k}\}$, $\{r_{3n_k}\}$ and $\{r_{4n_k}\}$ are the same as in theorem 2 of chapter 4. Using the estimates derived in this chapter, we get

$$r_{4n_k} \stackrel{a.s.}{=} o\left(n_k^{\delta - \frac{\kappa-2\theta}{2\kappa}}\right),$$

$$r'_{1n_k} \stackrel{a.s.}{=} O^*\left(n^{\delta - \frac{\kappa-2\theta}{2\kappa}}\right),$$

$$r''_{n_k} \stackrel{a.s.}{\leq} o\left(n_k^{\delta - \frac{1/2-\theta}{\kappa}}\right) + O^*\left(n_k^{\frac{1-\kappa+\theta(1+K^2/4)}{\kappa}}\right),$$

$$r_{1n_k} = r'_{1n_k} + r''_{1n_k},$$

$$|r_{2n_k}| \stackrel{a.s.}{\leq} O^*\left(n_k^{-\frac{1-\nu}{\kappa}}\right),$$

7.2. ERGODIC MARKOV CHAINS

$$r_{3n_k} \overset{a.s.}{\leq} o\left(n_k^{\delta - \frac{\nu-2\theta}{\kappa}}\right) + O^*\left(n_k^{\frac{2\theta+\nu-1}{\kappa}}\right).$$

These estimates hold for any $\delta > 0$. Taking again into account that all the previous results have been derived for any $\tilde{c} \in \mathbf{C}$, we conclude:

$$\Phi_n - \Phi_* = \overline{\Phi}_n - \Phi_* + o\left(n^{\delta - \frac{1-2\theta}{2}}\right) =$$

$$= \overline{\Phi}_{n'(n)} - \Phi_* + O(n^{-\frac{1}{\kappa}}) + o\left(n^{\delta - \frac{1-2\theta}{2}}\right),$$

from which (7.18) follows. The theorem is proved. ∎

We have now the following convergence result (optimal order of convergence), which is in many ways analogous to the results presented in the previous chapters.

Corollary 3 *The best order of adaptation rate within the subclass of parameters (7.16) and (7.17) is equal to $o\left(n^{\delta - \varphi^*}\right)$ with*

$$\varphi^* = (8 + \frac{K^2}{4})^{-1},$$

i.e.,

$$\Phi_n - \Phi_* \overset{a.s.}{\leq} o\left(n^{\delta - \varphi^*}\right) \quad \forall \delta > 0.$$

This optimal order of convergence rate is achieved for

$$\nu = \nu^* := \frac{1}{2}, \quad \theta = \theta^* := \frac{1}{6}, \quad \kappa = \kappa^* := \frac{4}{3} + \frac{K^2}{24}.$$

Let us notice that if the number K of states of a given Finite Markov Chain increases, then the adaptation rate decreases (see Figure 7.2).

The maximum possible adaptation rate φ^* is achieved for a simple Markov chain containing only two states $K = 2 \left(\kappa^* = \frac{3}{2}\right)$ and is equal to

$$\varphi^* = \frac{1}{9}.$$

Remark 1 *Let us recall that for Regular Markov Chains the optimal order of the adaptation rate does not depend on the number K of states (see chapter 4).*

The remainder of this chapter is dedicated to the self-adjusting (adaptive) control of general type Markov chains.

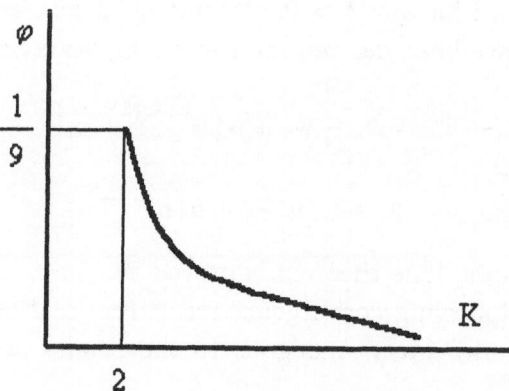

Figure 7.2: Evolution of the optimal convergence rate as a function of the number of states.

7.3 General type Markov chains

The previous chapters of this book dealt with regular and ergodic controlled Markov chains that allow us to use the technique of Markov Process Theory for the analysis of adaptive control algorithms. In the case of General Type Markov Chains we now deal with, they are called non-Markov processes because the behaviour of a controlled Markov chain in the future depends on the history of this process including possible random transitions from the class $X^+(0)$ of non-return states to one of the ergodic subclass $X^+(l)$ $(l = 1, ..., L)$. In this situation, we are not able to formulate the adaptive control problem as an optimization one. We have to formulate this problem as an inequality problem dealing with a performance index which includes the operation of maximization over all the ergodic subclasses. So, using the results of lemma 1 of chapter 2, for General Type Markov Chains, we can formulate the following **Adaptive Control Problem**:

Construct a randomized strategy $\{d_n\} \in \sum$ generating an adaptive control policy $\{u_n\}$ $(u_n \in U)$ to achieve the following objective

$$\varlimsup_{n \to \infty} \Phi_n \overset{a.s.}{\leq} \Phi^* \qquad (7.22)$$

with probability 1, where

$$\Phi^* := \min_{(d) \in \sum_s} \max_{k=\overline{1,L}} \min_{p^{(k)} \in Q^{(k)}(d)} \sum_{x(i) \in X^+(l)} \sum_{l=1}^{N} v_{il} d^{il} p_i^{(k)}, \qquad (7.23)$$

7.3. GENERAL TYPE MARKOV CHAINS

the ergodic subclasses $X^+(l)$ are defined in chapter 1 and, the sets $Q^{(k)}(d)$ $(k=1,...,L)$ containing K_k states are defined by

$$Q^{(k)}(d) := \left\{ p^{(k)} \in R^{K_k} \mid p^{(k)} \geq 0 \, (i=1,...,K_k), \sum_{i=1}^{K_k} p_i^{(k)} = 1, \right.$$

$$\left. p^{(k)} = \left[\Pi^{(k)}(d)\right]^T p^{(k)} \right\}. \tag{7.24}$$

It is clear that this problem does not have a unique solution. To solve it, we have to impose more restrictive conditions on the parameters of the Projection Gradient Algorithm to guarantee the success of the corresponding adaptation process.

Theorem 4 *If under the assumptions of the previous theorem, we assume in addition that*

$$\lim_{n \to \infty} \left(\sum_{t=1}^{k} \Delta n_t \varepsilon_t^K \right)^{-1} \ln k = 0, \tag{7.25}$$

then for any initial conditions $c_1 \in \mathbf{C}_{\varepsilon_1}$, $x_1 \in X$ of any controlled Markov Chain (not obligatory ergodic), the objective (7.22)-(7.23) is achieved i.e.,

$$\overline{\lim_{n \to \infty}} \, \Phi_n \overset{a.s.}{\leq} \Phi^*. \tag{7.26}$$

Proof. To prove this theorem, let us first prove that after a finite (may be random) number of transitions, any Controlled Finite Markov chain will evolve into one of the ergodic subclasses $X^+(k)$ $(k=1,...,L)$ and, will remain there. Let us denote by K_0 the number of the non-return states constituting the set $X^+(0)$ and by $\Pi^{(0)}(d)$ the corresponding transition matrix within the subclass $X^+(0)$, defined for any stationary non-singular strategy $\{d\} \in \Sigma_s^+$.

To simplify the study, let us assume that the states of the subclass $X^+(0)$ are numbered as follows: $1, 2, ..., K_0$.

There exists a row (numbered α) such that

$$\sum_{j=1}^{K} \sum_{l=1}^{N} \pi_{\alpha j}^l d^{\alpha l} < 1$$

for any $d \in int\mathbf{D}$. It follows that the matrix $\Pi^{(0)}(d)$ is non-stochastic one (this fact has been already mentioned in chapter 1).

1) Let us now demonstrate that under the condition (7.25) of this theorem, the following inequality

$$\sum_{k=1}^{\infty} P\left\{ x_{n_k} \in X^+(0) \right\} < \infty. \tag{7.27}$$

is valid.

Within the interval $\overline{n_k, n_{k+1} - 1}$ the randomized control strategy d_n is "frozen" (remain constant); then, it follows

$$P\left\{x_{n_{k+1}} = x(j) \mid \mathcal{F}_{n_k}, x_{n_k} = x(i)\right\} = \left[\left(\Pi^{(0)}(d_k)\right)^{\Delta n_k}\right]_{ij}.$$

Let us introduce the following notation

$$m_k := [\Delta n_k / K_0], \quad \overline{i = 1, K_0}.$$

For any $i = \overline{1, K_0}$ we obtain:

$$\sum_{j=1}^{K_0} P\left\{x_{n_{k+1}} = x(j) \mid \mathcal{F}_{n_k}, x_{n_k} = x(i)\right\}$$

$$= \sum_{j=1}^{K_0} \sum_{j_1=1}^{K_0} \cdots \sum_{j_{m_k}=1}^{K_0} \left[\left(\Pi^{(0)}(d_k)\right)^{K_0}\right]_{ij_1} \cdots$$

$$\cdot \left[\left(\Pi^{(0)}(d_k)\right)^{K_0}\right]_{j_{m_k-1}, j_{m_k}} \left[\left(\Pi^{(0)}(d_k)\right)^{K_0}\right]_{j_{m_k}, j}$$

$$\leq \sum_{j_1=1}^{K_0} \left[\left(\Pi^{(0)}(d_k)\right)^{K_0}\right]_{ij_1} \cdot \sum_{j_2=1}^{K_0} \left[\left(\Pi^{(0)}(d_k)\right)^{K_0}\right]_{j_1 j_2} \cdots$$

$$\cdot \sum_{j_{m_k}=1}^{K_0} \left[\left(\Pi^{(0)}(d_k)\right)^{K_0}\right]_{j_{m_k-1}, j_{m_k}}$$

$$\leq \left(\max_{i=\overline{1,K_0}} \max_{d \in \mathbf{D}_{\varepsilon_k}} \sum_{j=1}^{K_0} \left[\left(\Pi^{(0)}(d_k)\right)^{K_0}\right]_{ij}\right)^{m_k}$$

where

$$\mathbf{D}_\varepsilon := \left\{\|d^{il}\| \mid d^{il} > \varepsilon, \sum_{l=1}^N d^{il} = 1 \ \left(i = \overline{1, K}; l = \overline{1, N}\right)\right\}.$$

$X^+(0)$ is the set of non-return states. It can not contain subsets of communicating states. Hence, for any $\varepsilon > 0$

$$\max_{d \in \mathbf{D}_\varepsilon} \sum_{j=1}^{K_0} \left[\left(\Pi^{(0)}(d_k)\right)^{K_0}\right]_{ij} < 1.$$

7.3. GENERAL TYPE MARKOV CHAINS

This estimation can be made more precisely, if we take into account the linear dependence of the matrix $\Pi^{(0)}(d)$ on d and, the property of the set $\mathbf{D}_\varepsilon$:

$$\max_{d\in \mathbf{D}_\varepsilon} \sum_{j=1}^{K_0} \left[\left(\Pi^{(0)}(d_k)\right)^{K_0}\right]_{ij} < 1 - b\varepsilon^{K_0}. \tag{7.28}$$

which is valid for any $i = \overline{1, K_0}, \varepsilon > 0$ and some $b > 0$.

From (7.28) we derive:

$$\sum_{j=1}^{K_0} P\left\{x_{n_{k+1}} = x(j) \mid \mathcal{F}_{n_k}, x_{n_k} = x(i)\right\} \leq \left(1 - b\varepsilon^{K_0}\right)^{m_k},$$

and, as a result, we obtain

$$P\left\{x_{n_{k+1}} \in X^+(0)\right\} = E\left\{\sum_{j=1}^{K_0} P\left[x_{n_{k+1}} = x(j) \mid \mathcal{F}_{n_k}\right]\right\}$$

$$= E\left\{\sum_{j=1}^{K_0} \chi\left(x_{n_k} = x(i)\right) \sum_{j=1}^{K_0} P\left[x_{n_{k+1}} = x(j) \mid \mathcal{F}_{n_k}, x_{n_k} = x(i)\right]\right\}$$

$$\leq \left(1 - b\varepsilon^{K_0}\right)^{m_k} P\left\{x_{n_{k+1}} \in X^+(0)\right\}.$$

From this inequality, we directly get (7.27).

2) Based on inequality (7.27) and, using the Borel-Cantelli lemma [6], we conclude that the process $\{x_n\}$ will stay in the set $X^+(0)$ only a finite (may be random) numbers of steps, and then evolves in to one of the ergodic (communicating) components $X^+(l)$. This process will never leave this component in the future:

$$\omega \in \Omega^+(l)$$

(the trajectory sets $\Omega^+(l)$ are defined in chapter 1). Hence, starting from this instant, theorem 2 can be applied and, we can formulate the optimization problem especially for the subclass $X^+(l)$ of states:

$$\varlimsup_{n\to\infty} \Phi_n = \Phi_*(l) := \inf_{\{d\}\in \sum_s^+} \sum_{x(i)\in X^+(l)} \sum_{k=1}^N v_{ik} d_{ik} p_i(d).$$

Combining these results for any $l = \overline{0, L}$, we get (7.26):

$$\varlimsup_{n\to\infty} \Phi_n \leq \max_{l=\overline{0,L}} \Phi_*(l) := \Phi^*.$$

The theorem is proved. ∎

The process $\{x_n\}$ remains in one of the ergodic components, so we conclude that
$$\Phi_*(l) \leq \Phi^*, l = \overline{0, L}$$
and, hence, the conditions of theorem 2 are fulfilled, which leads to the following result.

Theorem 5 *Under the assumptions of the previous theorem, the control objective (7.22) is achieved.*

Proof. This theorem is an evident consequence of the previous discussion. ■

Corollary 4 *The optimal order of the adaptation rate is equal to $o\left(n^{\delta-\varphi^*}\right)$ $\forall \delta > 0$ with*
$$\varphi^* = \frac{4}{K^2 + 32}$$
and, it is achieved for the following parameter values:
$$\nu = \nu^* = \frac{1}{2}, \ \theta = \theta^* = \frac{1}{6}, \ \kappa = \kappa^* \frac{K^2 + 32}{24}.$$

Proof. Taking into account the additional assumption (7.25), the proof is similar to the proof of theorem 3. The corollary is proved. ■

7.4 Conclusions

Let us notice that in the partial case when $K = 1$, any Markov chain leads to a simple static system, namely: a *Learning Automaton* [1, 3], and the adaptation algorithm of chapter 4, leads to the Stochastic Approximation Algorithm studied in [4] and for $n_k = k$, $k = 1, 2, \ldots$.

This chapter was concerned with the adaptive control of a class of nonregular ergodic Markov chains, including ergodic and general type Markov chains. The formulation of the adaptive control problem for this class of Markov chains is different from the formulation of unconstrained and constrained adaptive control problems stated in the previous chapters.

7.5 References

1. A. V. Nażin and A. S. Poznyak, *Adaptive Choice of Variants*, (in Russian) Nauka, Moscow, 1986.

7.5. REFERENCES

2. Yu. A. Rozanov, *Random Processes*, (in Russian) Nauka, Moscow, 1973.

3. K. Najim and A. S. Poznyak, *Learning Automata Theory and Applications*, Pergamon Press, London, 1994.

4. A. S. Poznyak and K. Najim, *Learning Automata and Stochastic Optimization*, Pergamon Press, London, 1997.

5. J. G. Kemeny and J. L. Snell, *Finite Markov Chains*, Springer-Verlag, Berlin, 1976.

6. J. L. Doob, *Stochastic Processes*, J. Wiley, New York, 1953.

Chapter 8

Practical Aspects

8.1 Introduction

Numerical simulation is an efficient tool which can be used independently for any theoretical developments or in connection with fundamental research. In fact,

i) a model representing a given system can be simulated to illustrate by, for example, graphics or tables, the behaviour of the concerned process;

ii) simulations can help the researcher in the development of theoretical results and are able to induce and involve new theoretical analysis or research.

In other words, there exists a feedback between theory and simulation. In the framework of optimal dual control, many problems can not be solved analytically; only in special simple or special cases is it possible to calculate an optimal control law. It is therefore interesting to study the effects of making different approximations (suboptimal control laws). In this situation, simulation represents a valuable tool to get a feeling for the properties of suboptimal control strategies.

It is interesting to note that the area of computer control and computer implementation is becoming increasingly important. The ever present microprocessor is not only allowing new applications but also is generating new areas for theoretical research.

This last chapter is devoted chiefly to the numerical implementation of the self-learning (adaptive) control algorithms developed on the basis of Lagrange multipliers and penalty function approaches. The behaviour (convergence and convergence rate) of these algorithms has been analyzed in the previous chapters. For the purpose of investigating the behaviour and the performance of the algorithms dealing with the adaptive control of both unconstrained and constrained Markov chains, several simulations have been carried out;

however, only a fraction of the most representative results are presented in what follows. The second purpose of this chapter is to help the reader to better understand and assimilate the contents of the previous algorithmic and analytical developments.

In other words, chapter 8 provides a comprehensive performance evaluation of the capabilities of the adaptive control algorithms developed in the previous chapters on the basis of Lagrange multipliers and penalty function approaches.

Two problems are simulated here to show that near-optimal performance can be attained by the adaptive schemes described thus far. For each example we present a set of simulation results dealing with the Lagrange multipliers and the penalty function approaches for both unconstrained and constrained cases.

8.2 Description of controlled Markov chain

Let us consider a finite controlled Markov chain with four states ($K = 4$) and three control actions ($N = 3$). The associated transition probabilities are:

$$\pi_{ij}^1 = \begin{bmatrix} 0.7 & 0.1 & 0.1 & 0.1 \\ 0 & 0.8 & 0.1 & 0.1 \\ 0 & 0 & 0.9 & 0.1 \\ 0.8 & 0.1 & 0.1 & 0 \end{bmatrix}$$

$$\pi_{ij}^2 = \begin{bmatrix} 0.1 & 0.6 & 0.1 & 0.2 \\ 0.8 & 0 & 0.1 & 0.1 \\ 0 & 0.8 & 0.2 & 0 \\ 0.1 & 0.7 & 0.1 & 0.1 \end{bmatrix}$$

$$\pi_{ij}^3 = \begin{bmatrix} 0 & 0 & 0.9 & 0.1 \\ 0 & 0.1 & 0.8 & 0.1 \\ 0.7 & 0.1 & 0.1 & 0.1 \\ 0 & 0.1 & 0.9 & 0 \end{bmatrix}$$

8.2.1 Equivalent Linear Programming Problem

We have used the Matlab^{TM} Optimization Toolbox (see Appendix B) to solve the following Linear Programming Problem:

$$\tilde{V}(c) := \sum_{i=1}^{K} \sum_{l=1}^{N} v_{il}^0 c^{il} \to \min_{c \in \mathbf{C}}$$

8.2. DESCRIPTION OF CONTROLLED MARKOV CHAIN

where the set $\mathbf{C}$ is given by

$$\mathbf{C} = \left\{ c \mid c = \left[c^{il}\right],\ c^{il} \geq 0,\ \sum_{i=1}^{K}\sum_{l=1}^{N} c^{il} = 1 \right.$$

$$\left. \sum_{l=1}^{N} c^{jl} = \sum_{i=1}^{K}\sum_{l=1}^{N} \pi_{ij}^{l} c^{il}\ (i,j=1,...,K;\ l=1,...,N) \right\}$$

subject to:

$$\widetilde{V}_m(c) := \sum_{i=1}^{K}\sum_{l=1}^{N} v_{il}^{m} c^{il} \leq 0\ (m=1,...,M)$$

We shall be concerned with the simplest case corresponding to only one constraint ($M = 1$).

To solve this Linear Programming Problem with Matlab, we have to reformulate it in the following vector form:

$$f^T x \to \min_x$$

subject to

$$Ax \leq b$$

where

$$x = \left[c^{1,1}, c^{1,2}, ..., c^{1,N}, c^{2,1}, ..., c^{K,N}\right]^T$$

$$f = \left[v_{1,1}^0, ..., v_{1,N}^0, v_{2,1}^0, ..., v_{K,N}^0\right]^T,$$

and the matrix A and the vector b are respectively given by

$$A = \begin{bmatrix} 1 & 1 & 1 & 1 & ... & ... & ... & ... & 1 \\ \pi_{1,1}^1 - 1 & \pi_{1,1}^2 - 1 & ... & \pi_{1,1}^N - 1 & \pi_{2,1}^1 & ... & \pi_{2,1}^N & ... & \pi_{K,1}^N \\ \pi_{1,2}^1 & \pi_{1,2}^2 & ... & \pi_{1,2}^N & \pi_{2,2}^1 - 1 & ... & \pi_{2,2}^N - 1 & ... & \pi_{K,2}^N \\ \vdots & \vdots & \vdots & \vdots & \vdots & \vdots & \vdots & \vdots & \vdots \\ v_{1,1}^1 & v_{1,2}^1 & ... & v_{1,N}^1 & v_{2,1}^1 & ... & v_{2,N}^1 & ... & v_{K,N}^1 \\ v_{1,1}^2 & v_{1,2}^2 & ... & v_{1,N}^2 & v_{2,1}^2 & ... & v_{2,N}^2 & ... & v_{K,N}^2 \\ \vdots & \vdots & \vdots & \vdots & \vdots & \vdots & \vdots & \vdots & \vdots \\ v_{1,1}^M & v_{1,2}^M & ... & v_{1,N}^M & v_{2,1}^M & ... & v_{2,N}^M & ... & v_{K,N}^M \\ -1 & 0 & ... & 0 & 0 & ... & 0 & ... & 0 \\ 0 & -1 & ... & 0 & 0 & ... & 0 & ... & 0 \\ \vdots & \vdots & \vdots & \vdots & \vdots & \vdots & \vdots & \vdots & \vdots \\ 0 & 0 & ... & 0 & 0 & ... & 0 & ... & -1 \end{bmatrix},$$

and

$$b = [1, 0, 0, ..., 0]^T \in R^{K+M+K \cdot N+1}.$$

In view of this formulation, the value of c can be computed using the following command:

$$c = lp(f, A, b, 0, 1, c_0, 1 + K)$$

where c_0 corresponds to the initial condition (see Appendix B, lpmc.m program).

- **Example 1.** *(Preference for the state numbered 1) In this example, we select v^0 in such a way that for any initial probability, the Markov process tends to the state numbered 1. To achieve this objective, v^0 is selected as follows:*

$$v^0 = \begin{bmatrix} 0 & 10 & 10 \\ 10 & 0 & 10 \\ 10 & 10 & 0 \\ 0 & 10 & 10 \end{bmatrix}.$$

The solution of this unconstrained problem ($v_{il}^m = 0$) using the Lagrange multipliers and the penalty function approaches is given in the following section.

8.3 The unconstrained case (example 1)

This section provides a comprehensive performance evaluation of the capabilities of the adaptive control algorithms developed on the basis of both Lagrange multipliers and penalty function approaches. The detailed specification of the adaptive control algorithms used in the trials described in the sequel may be considered as involving two distinct elements:

i) Specification of the design parameters (basic algorithmic parameters) that determine the performance of a given algorithm during the adaptation (learning) operation, such as the correction factor, the upper and lower bound of some parameters, etc.

ii) Specification of the initial value of some parameter estimates and the initial probability vector.

Before presenting the simulation results, let us recall how to select an action among a set of prespecified actions. The selection of an action is done as follows: Let us consider N actions $u(i)$ ($i = 1, ..., N$), and a probability distribution $p(i)$ ($i = 1, ..., N$). To each action $u(i)$, a probability $p(i)$ is associated. A practical method for choosing an action according to the probability distribution is to generate a uniformly distributed random variable $z \in [0, 1]$. The jth action is then chosen (see figure 8.1) such that j is equal

8.3. THE UNCONSTRAINED CASE (EXAMPLE 1)

to the least value of k, satisfying the following constraint:

$$\sum_{k=1}^{j} p(k) \geq z.$$

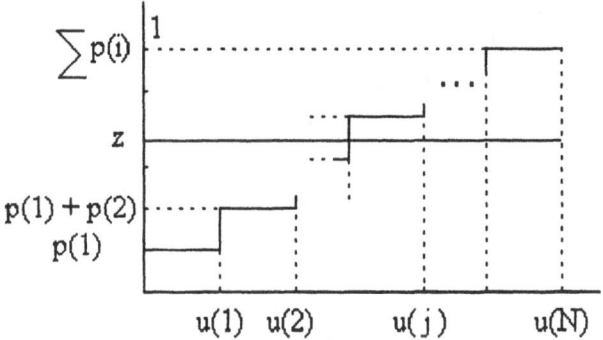

Figure 8.1: Action selection.

The graph in figure shows clearly the procedure dealing with the choice of actions out of a prescribed set to optimize responses from a random environment.

8.3.1 Lagrange multipliers approach

Using the lp Matlab command, we obtain:

$$c = \begin{bmatrix} 0.7182 & 0 & 0 \\ 0 & 0.090 & 0 \\ 0 & 0 & 0.1 \\ 0.09 & 0 & 0 \end{bmatrix},$$

and

$$\tilde{V}_0(c) = 0.$$

The adaptive control algorithm developed in chapter 2 has been implemented to solve the problem stated above. A convenient choice of the design parameters associated with this self-learning control algorithm, and one we shall make here, is that

$$\delta_0 = 0.3, \lambda_0^+ = 0.3, \gamma_0 = 0.006.$$

The value
$$p_0 = [0.25\ 0.25\ 0.25\ 0.25]^T$$
was chosen. The Matlab mechanization of the adaptive control algorithm based on Lagrange multipliers approach is given in Appendix B. The obtained results are:
$$c_L = \begin{bmatrix} 0.7316 & 0 & 0 \\ 0 & 0.1124 & 0 \\ 0 & 0 & 0.0880 \\ 0.0679 & 0 & 0 \end{bmatrix},$$

and
$$\tilde{V}_{0,L}(c) = 0.1191.$$

We have introduced the index L to characterize the results induced by the Lagrange multipliers approach. Simulation runs over $81,000$ samples (iterations) are reported here. The evolution of the loss function Φ_n is shown in figure 8.2.

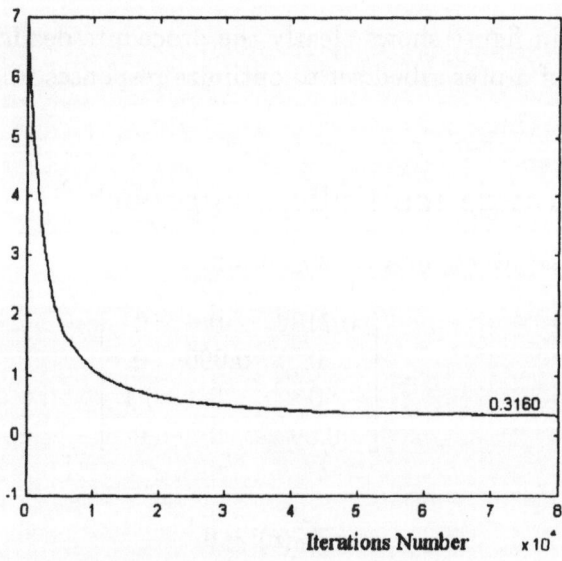

Figure 8.2: Evolution of the loss function Φ_n.

In this figure on the abscissa axis we plotted the time (iterations number), and on the ordinate axis the loss function Φ_n. The loss function converges

8.3. THE UNCONSTRAINED CASE (EXAMPLE 1)

to a value close to 0.3160. Taking into account the following relations:

$$p_{n+1}(j) = \sum_{i=1}^{K} \pi_n^{ij} p_n(i), \tag{8.1}$$

and

$$\pi_n^{ij} = \sum_{l=1}^{K} \pi_{ij}^{l} d_n^{il}, \tag{8.2}$$

We can see the effect of the control action in figure 8.3 where the probability vector p_n against the iterations number is plotted.

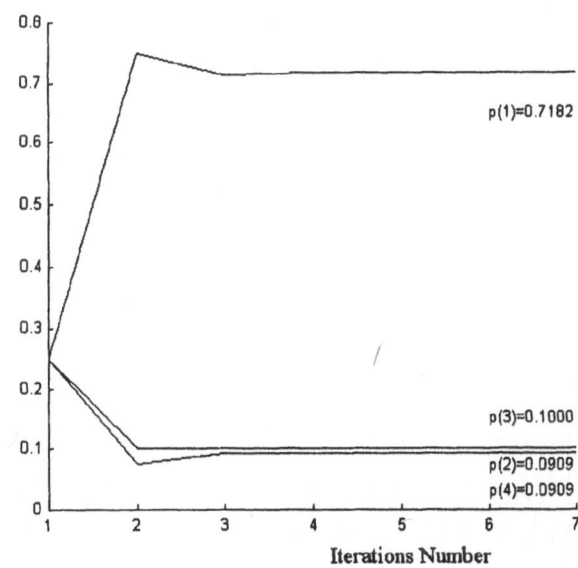

Figure 8.3: Time evolution of the state probability vector p_n.

This figure indicates, how the components of the probability p_n evolve with the iterations number. The limiting value of p_n depends on the values of v^0 and c_L. The system goes to the state numbered 1 ($p_n(1)$ tends to 0.7182). Typical convergence behaviour of the components of the matrix c_n is shown in figures 8.4-8.15.

The components of the matrix c_n converge after (approximately) 10^4 iterations. As is seen from figure 8.4 the component c_n^{11} tends to a value close to 0.7. The components c_n^{12}, c_n^{13}, c_n^{21}, c_n^{22}, c_n^{23}, c_n^{31}, c_n^{32}, c_n^{33}, c_n^{41}, c_n^{42} and c_n^{43} tend respectively to 0.0, 0.0, 0.0, 0.13, 0.0, 0.0, 0.02, 0.1, 0.08, 0.0 and 0.0.

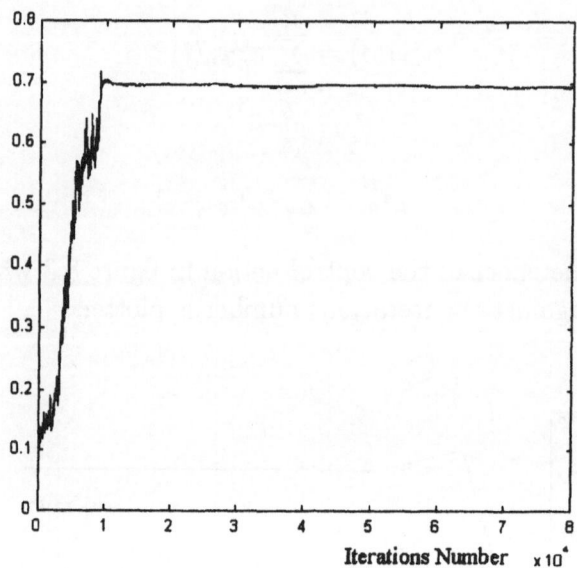

Figure 8.4: Evolution of c_n^{11}.

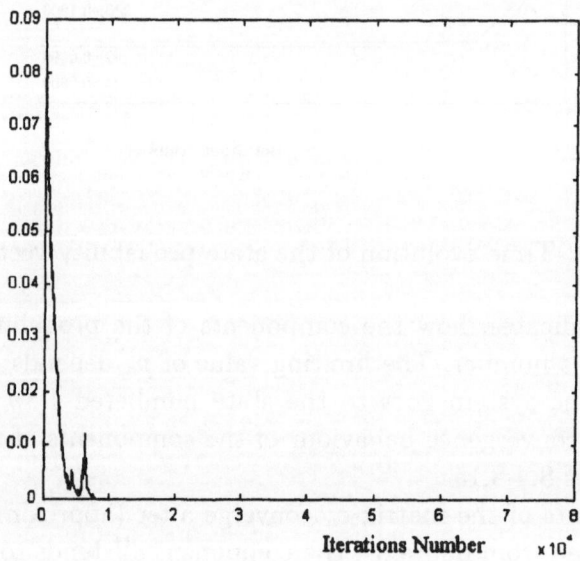

Figure 8.5: Evolution of c_n^{12}.

8.3. THE UNCONSTRAINED CASE (EXAMPLE 1)

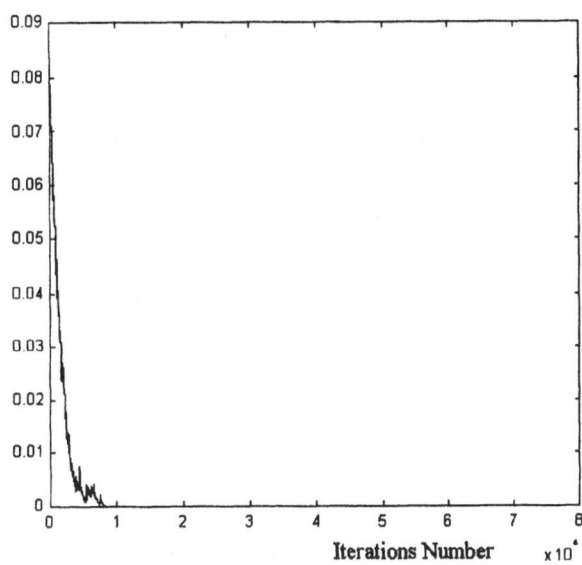

Figure 8.6: Evolution of c_n^{13}.

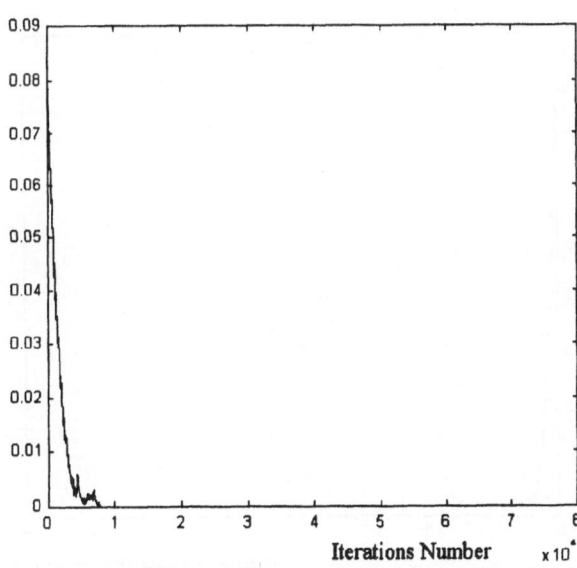

Figure 8.7: Evolution of c_n^{21}.

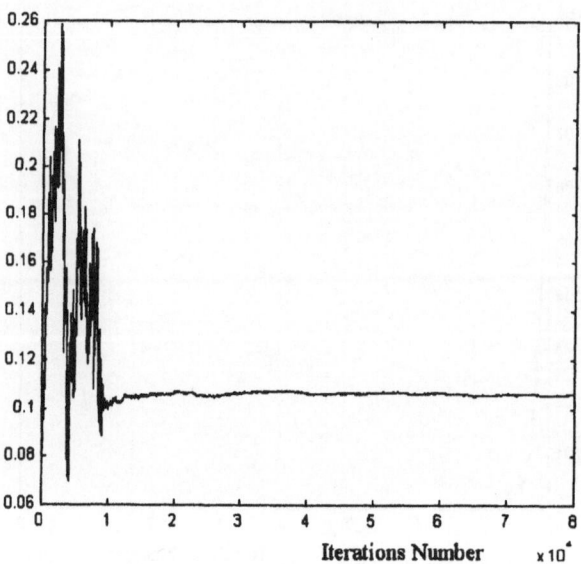

Figure 8.8: Evolution of c_n^{22}.

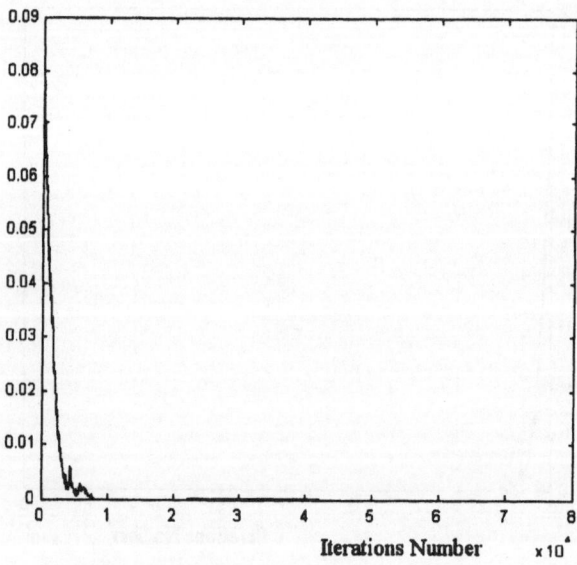

Figure 8.9: Evolution of c_n^{23}.

8.3. THE UNCONSTRAINED CASE (EXAMPLE 1)

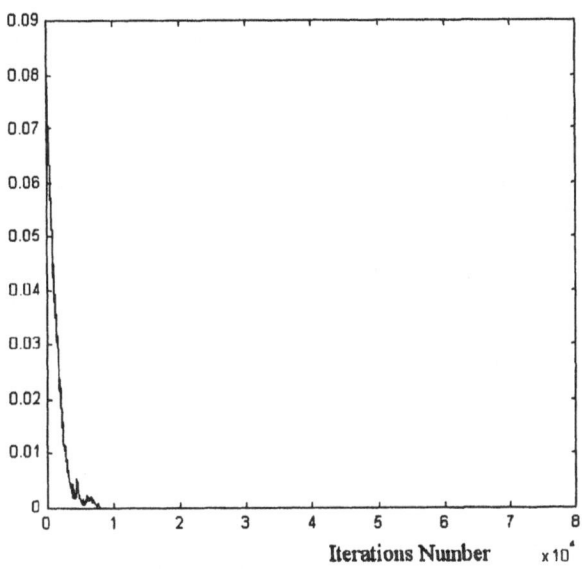

Figure 8.10: Evolution of c_n^{31}.

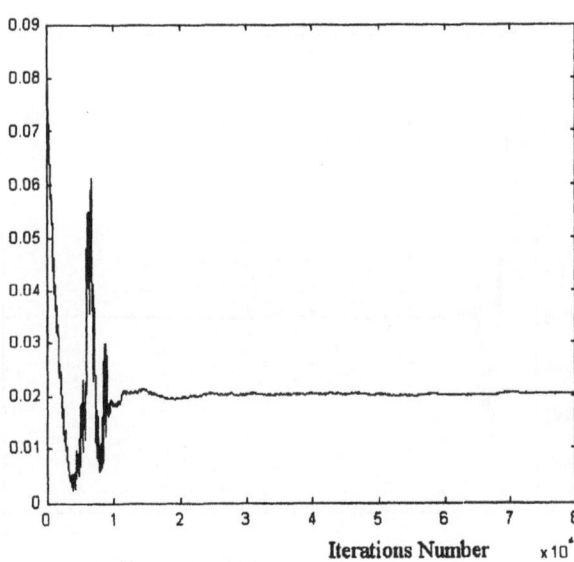

Figure 8.11: Evolution of c_n^{32}.

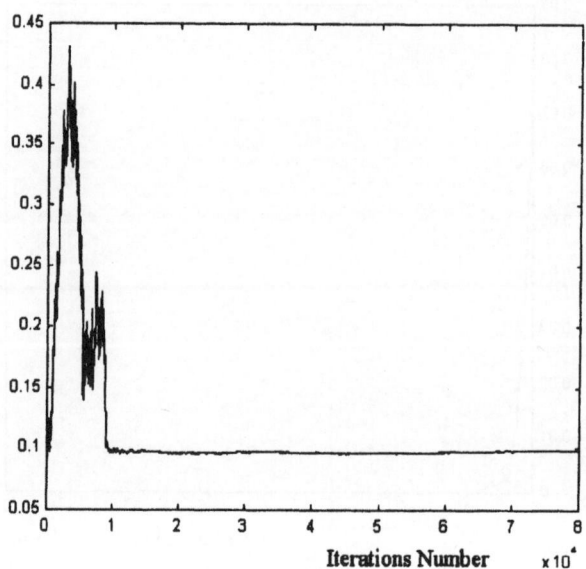

Figure 8.12: Evolution of c_n^{33}.

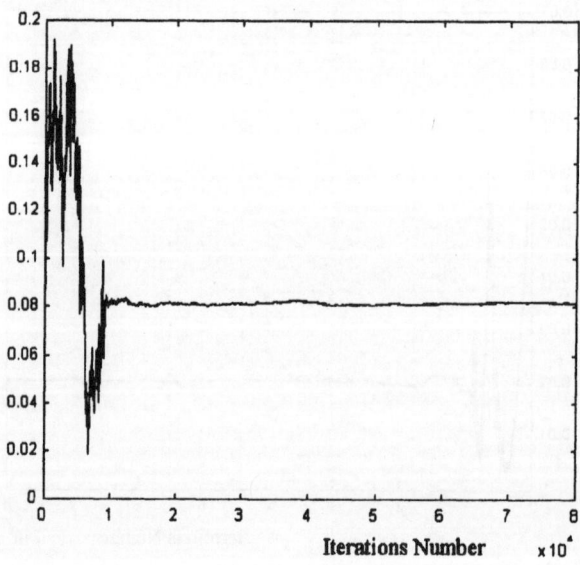

Figure 8.13: Evolution of c_n^{41}.

8.3. THE UNCONSTRAINED CASE (EXAMPLE 1)

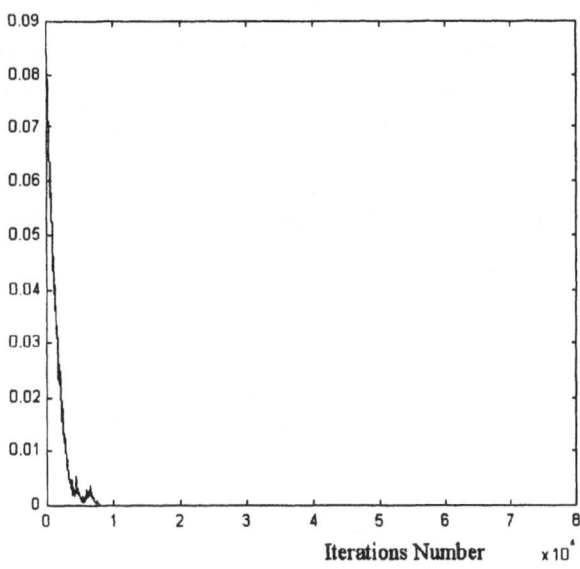

Figure 8.14: Evolution of c_n^{42}.

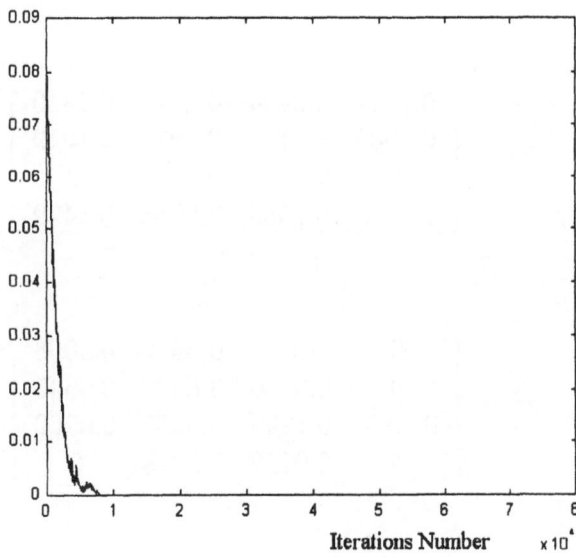

Figure 8.15: Evolution of c_n^{43}.

8.3.2 Penalty function approach

Let us now look at some simulations to examine the behaviour of the adaptive control algorithm described in chapter 3. The parameters associated with this adaptive control algorithm were chosen as follows: $\delta_0 = 0.5$, $\mu_0 = 4$, $\gamma_0 = 0.005$.

The algorithm based on the penalty function approach leads to the following results:

$$c_P = \begin{bmatrix} 0.6941 & 0 & 0 \\ 0 & 0.1061 & 0.0001 \\ 0 & 0.0203 & 0.0978 \\ 0.0816 & 0 & 0 \end{bmatrix}, \tilde{V}_{0,P}(c) = 0.3161,$$

and

$$\hat{\pi}^1_{ij} = \begin{bmatrix} 0.6999 & 0.0994 & 0.1026 & 0.0982 \\ 0 & 0.8000 & 0.1200 & 0.0800 \\ 0 & 0 & 0.9280 & 0.0720 \\ 0.7988 & 0.0983 & 0.1029 & 0 \end{bmatrix}$$

$$\hat{\pi}^2_{ij} = \begin{bmatrix} 0.1121 & 0.5984 & 0.1121 & 0.1810 \\ 0.8005 & 0 & 0.0975 & 0.1019 \\ 0 & 0.7976 & 0.2024 & 0 \\ 0.0420 & 0.7563 & 0.0588 & 0.1429 \end{bmatrix}$$

$$\hat{\pi}^3_{ij} = \begin{bmatrix} 0 & 0 & 0.8992 & 0.1008 \\ 0 & 0.0806 & 0.8548 & 0.0645 \\ 0.6974 & 0.1099 & 0.0965 & 0.0962 \\ 0 & 0.0982 & 0.9018 & 0 \end{bmatrix}$$

We can observe that the probabilities are well estimated. These simulations confirm that the estimator is consistent. The following graphs display results over $81,000$ iterations. The variation in the loss function Φ_n is illustrated in figure 8.16.

8.3. THE UNCONSTRAINED CASE (EXAMPLE 1)

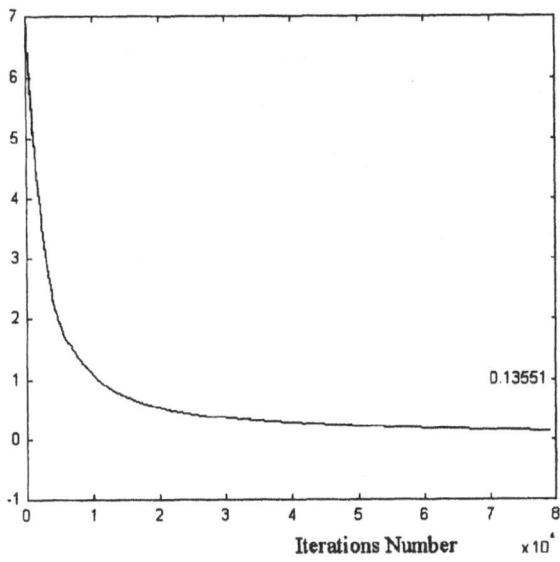

Figure 8.16: Evolution of the loss function Φ_n.

The decay of the loss function is clearly revealed in this figure. The evolution of the components of the state probability vector p_n are depicted in figure 8.17.

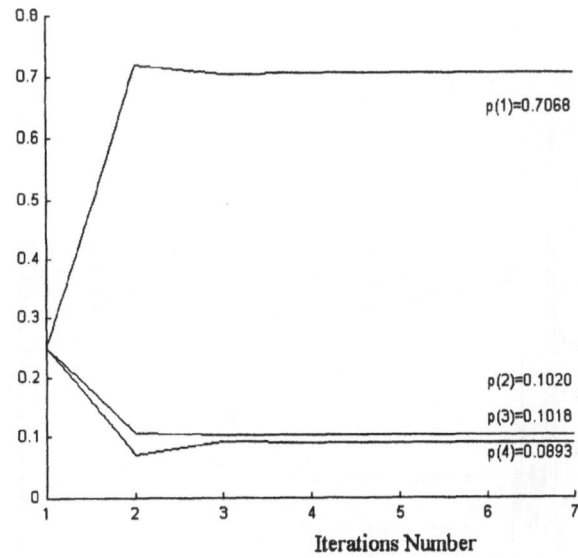

Figure 8.17: Evolution of the state probability vector p_n.

Taking into consideration relations (8.1) and (8.2), we observe again that the state numbered 1 constitutes the termination state ($p(1)$ converges to

0.7068). Learning curves corresponding to the components of the matrix c_n are drawn in figures 8.18-8.29. These figures record the evolution of the components c_n^{ij} for some values of the couple ij.

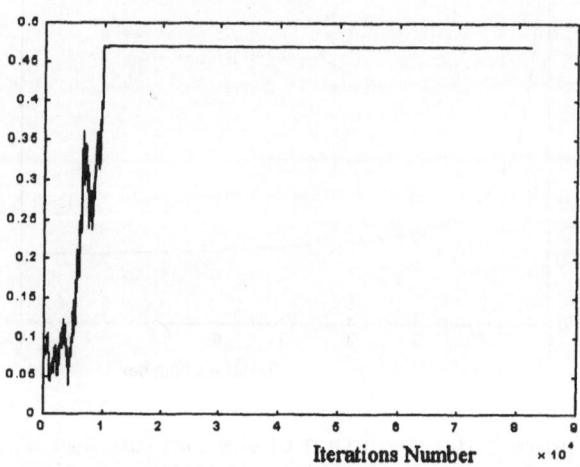

Figure 8.18: Evolution of c_n^{11}.

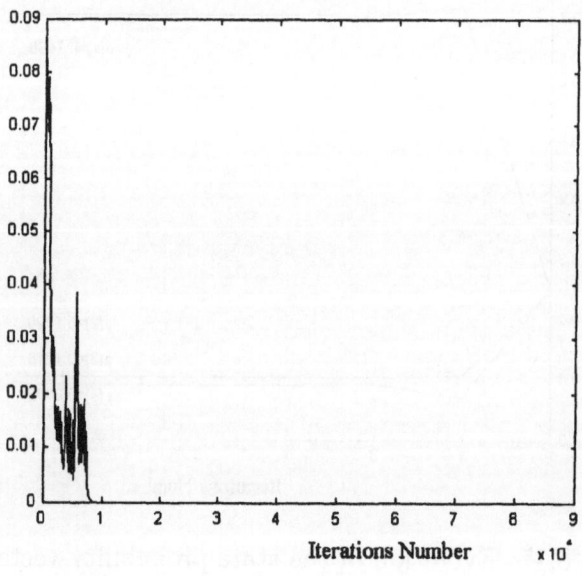

Figure 8.19: Evolution of c_n^{12}.

8.3. THE UNCONSTRAINED CASE (EXAMPLE 1)

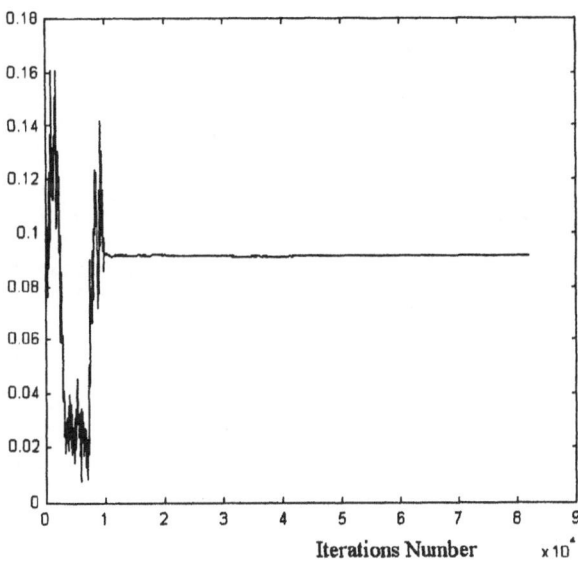

Figure 8.20: Evolution of c_n^{13}.

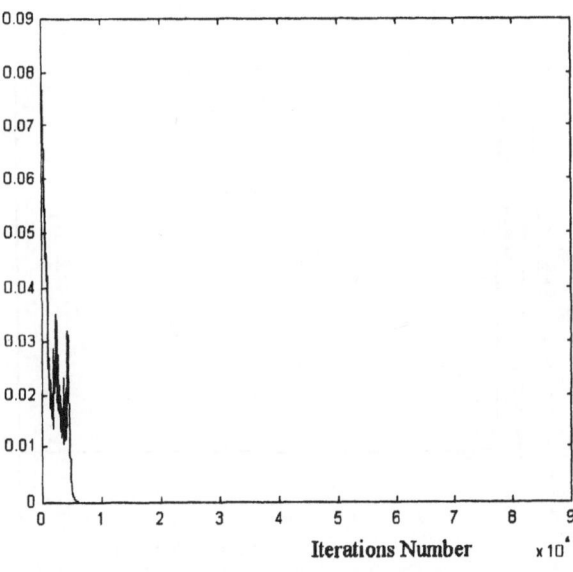

Figure 8.21: Evolution of c_n^{21}.

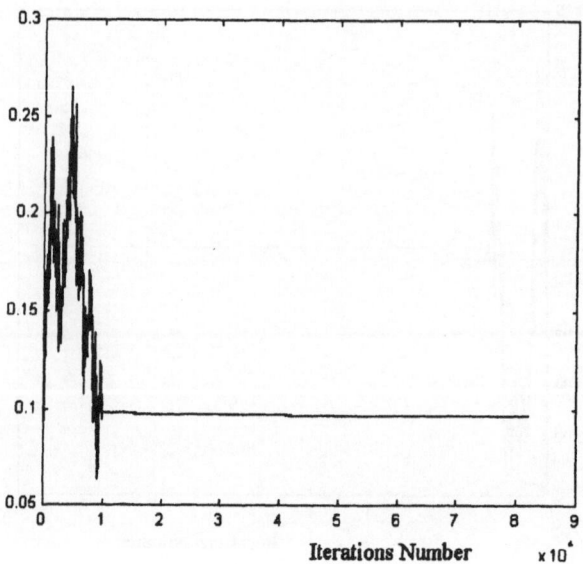

Figure 8.22: Evolution of c_n^{22}.

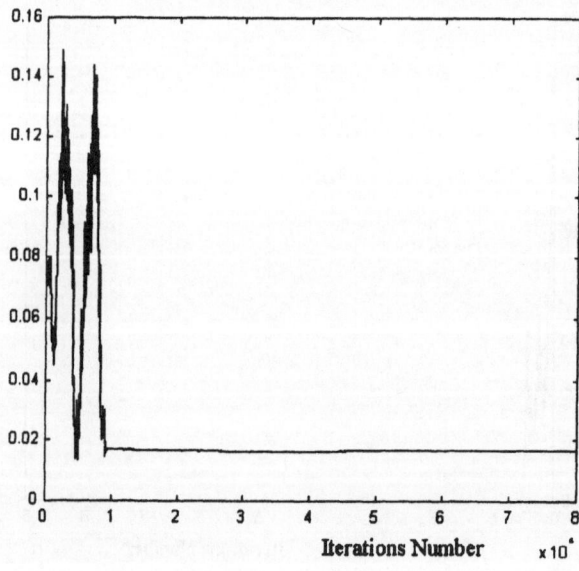

Figure 8.23: Evolution of c_n^{23}.

8.3. THE UNCONSTRAINED CASE (EXAMPLE 1)

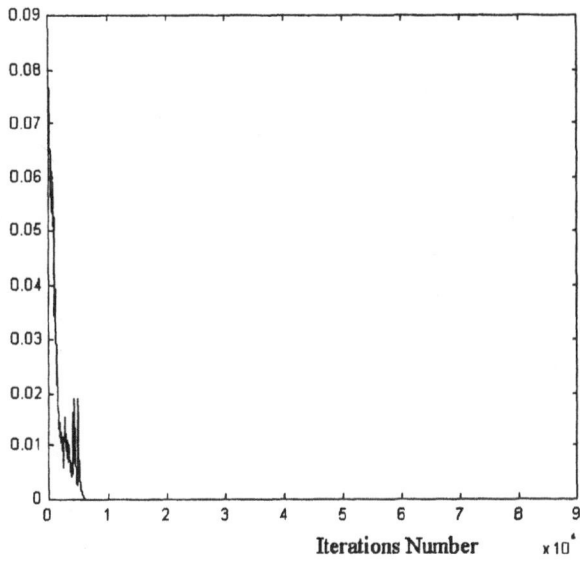

Figure 8.24: Evolution of c_n^{31}.

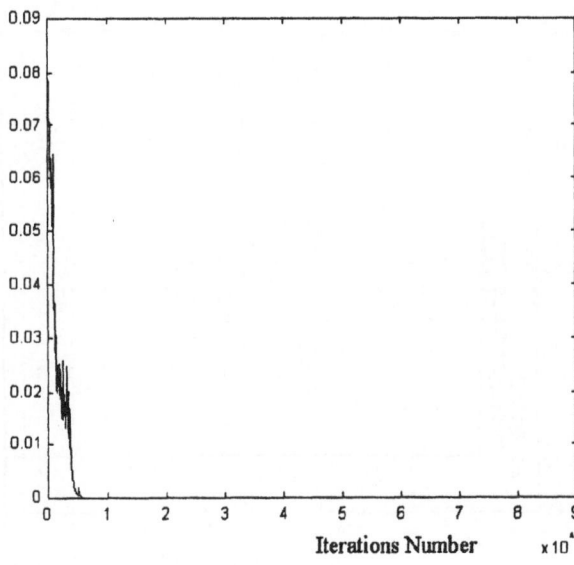

Figure 8.25: Evolution of c_n^{32}.

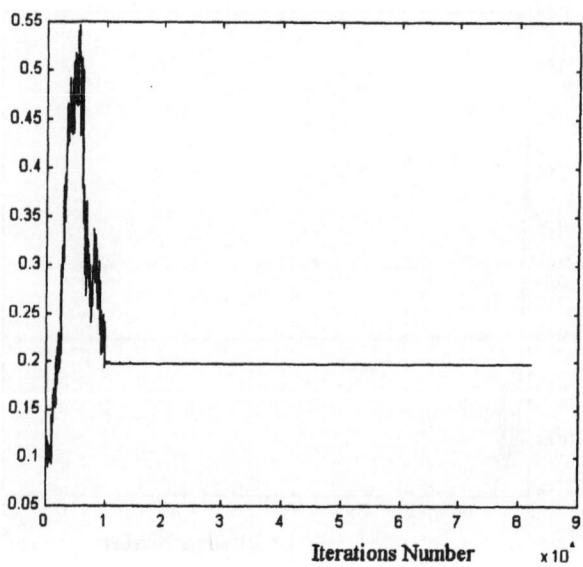

Figure 8.26: Evolution of c_n^{33}.

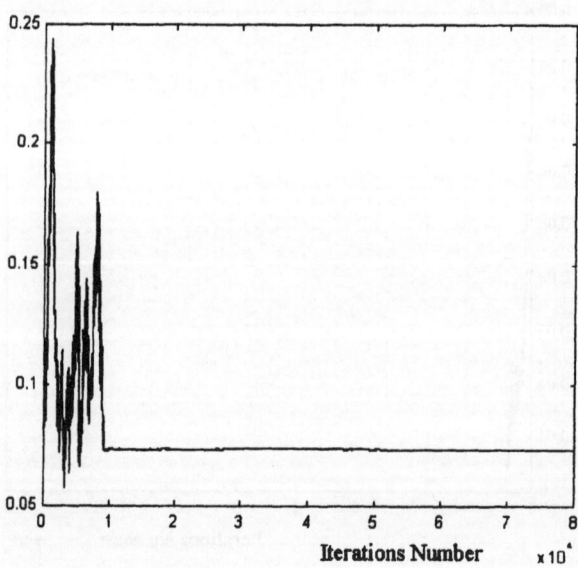

Figure 8.27: Evolution of c_n^{41}.

8.3. THE UNCONSTRAINED CASE (EXAMPLE 1)

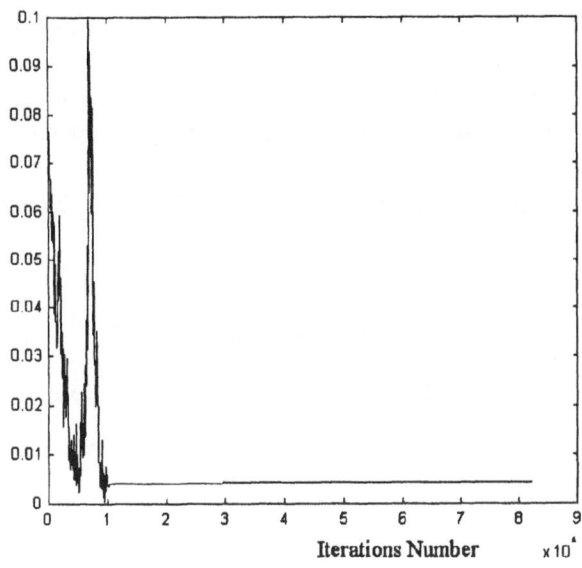

Figure 8.28: Evolution of c_n^{42}.

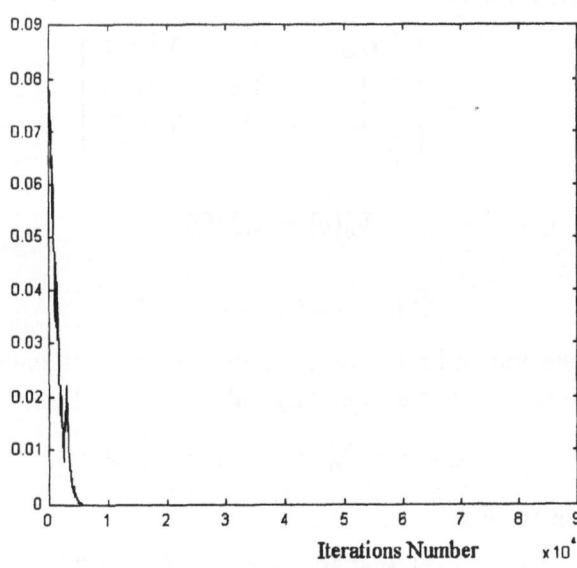

Figure 8.29: Evolution of c_n^{43}.

8.4 The constrained case (example 1)

In this section we present a set of simulation results in order to verify that the properties stated analytically also hold in practice. We examine the behaviour of the adaptive control algorithms developed for constrained controlled Markov chains on the basis of Lagrange multipliers and penalty function approaches.

8.4.1 Lagrange multipliers approach

We shall consider example 1 with a supplementary constraint defined by

$$v^1 = \begin{bmatrix} 10 & 10 & -10 \\ 10 & -10 & 10 \\ 10 & 10 & -10 \\ -10 & 10 & 10 \end{bmatrix}.$$

We consider the same transition probabilities as in the previous example. To take into consideration this constraints, we have to add some lines to the matrix A. Using the Matlab command

$$c = lp(f, A, b, 0, 1, c_0, 1 + K)$$

where c_0 corresponds to the initial condition, and has been given in the previous section, we derive:

$$c = \begin{bmatrix} 0.5 & 0 & 0.1277 \\ 0 & 0.0793 & 0 \\ 0 & 0 & 0.2021 \\ 0.0909 & 0 & 0 \end{bmatrix}$$

$$\tilde{V}_0(c) = 1.2766$$

and

$$\tilde{V}_1(c) = -8.8818 \cdot 10^{-16}.$$

The Lagrange multipliers approach has been implemented over 81,000 samples (iterations) with the following values of the design parameters

$$\delta_0 = 0.5, \lambda_0^+ = 0.3, \gamma_0 = 0.006.$$

The following results:

$$c_L = \begin{bmatrix} 0.4756 & 0 & 0.1379 \\ 0 & 0.0785 & 0 \\ 0 & 0 & 0.2244 \\ 0.0836 & 0 & 0 \end{bmatrix}$$

8.4. THE CONSTRAINED CASE (EXAMPLE 1)

$$\tilde{V}_{0,L}(c) = 1.4966$$

and

$$\tilde{V}_{1,L}(c) = -0.6229,$$

have been obtained.

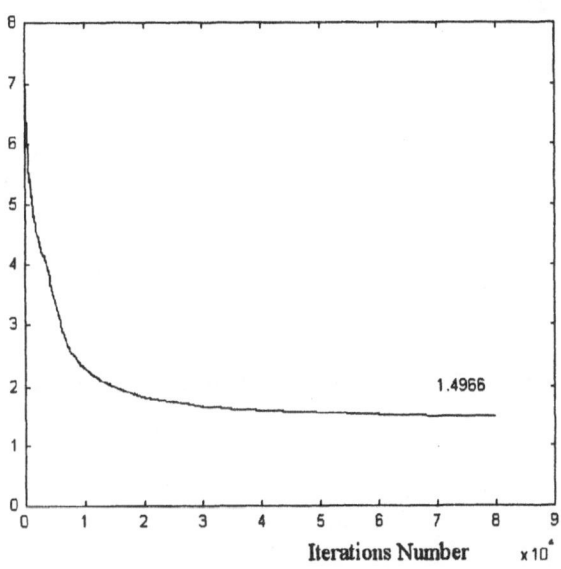

Figure 8.30: Evolution of the loss function Φ_n^0.

Figure 8.30 indicates how the loss function Φ_n^0 evolves with stage number (iterations number). This loss function decreases exponentially (practically) and converges to a value close to 1.4966. In figure 8.31, the loss function Φ_n^1 against the iteration number is plotted. After approximately 10^4 iterations, this loss function decreases to a final value close to -0.6229. The initial probability vector p_0 was selected as in the previous simulations. In figure 8.32, we can see the effect of the control actions on the controlled system. By inspection of this figure, we see that the probability $p(1)$ converges to 0.6194. Figures 8.33-8.44 plot the evolution of the corresponding components of the matrix c_n.

The simulations verified that the properties stated analytically also hold in practice. The conclusion that can be drawn from this example is that adaptive control algorithms are efficient tools for handling indeterminacy.

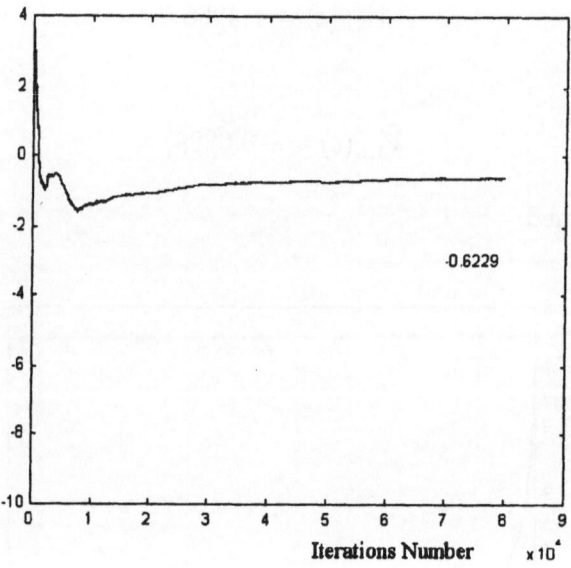

Figure 8.31: Evolution of the loss function Φ_n^1.

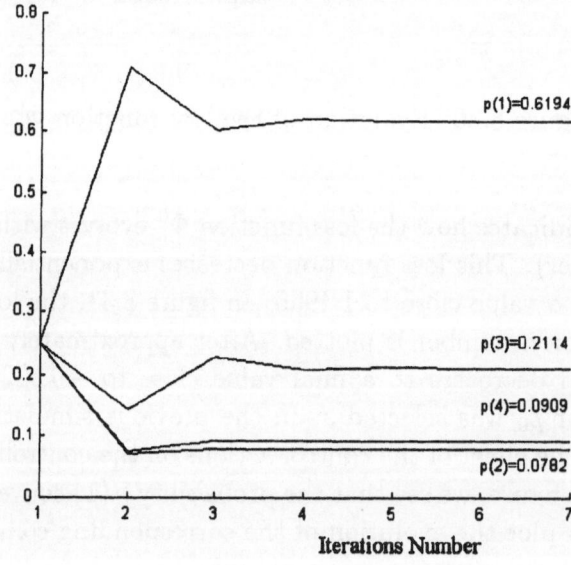

Figure 8.32: Time evolution of the state probability vector p_n.

8.4. THE CONSTRAINED CASE (EXAMPLE 1)

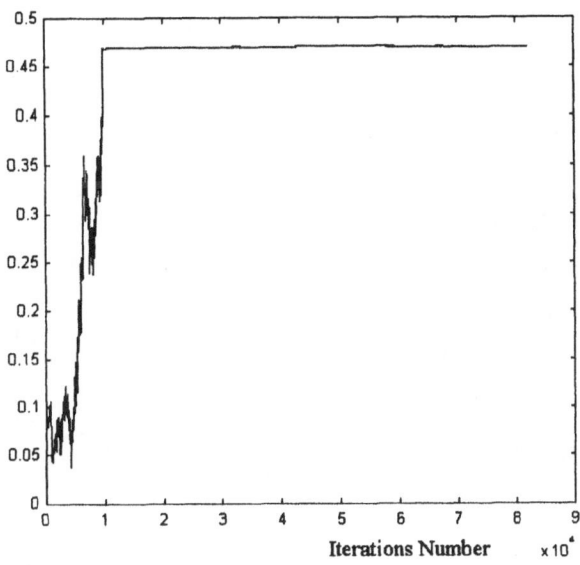

Figure 8.33: Evolution of c_n^{11}.

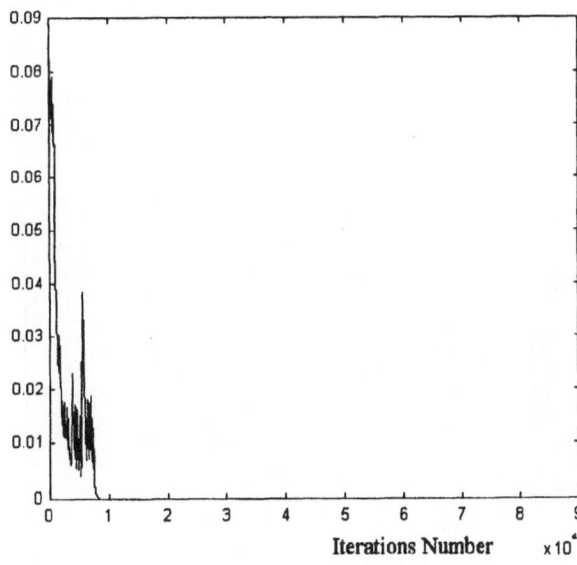

Figure 8.34: Evolution of c_n^{12}.

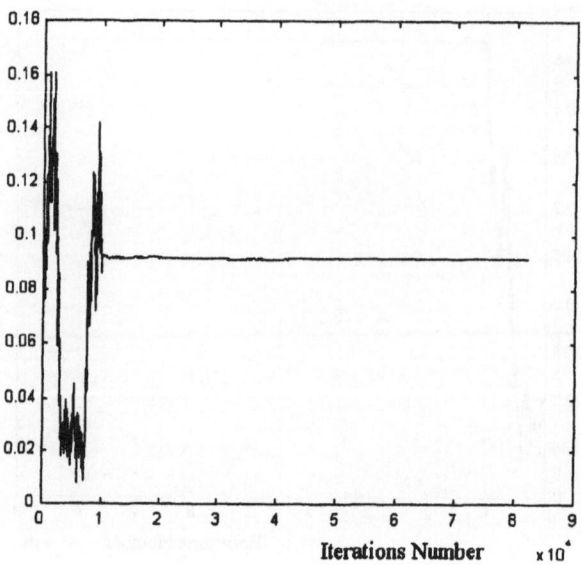

Figure 8.35: Evolution of c_n^{13}.

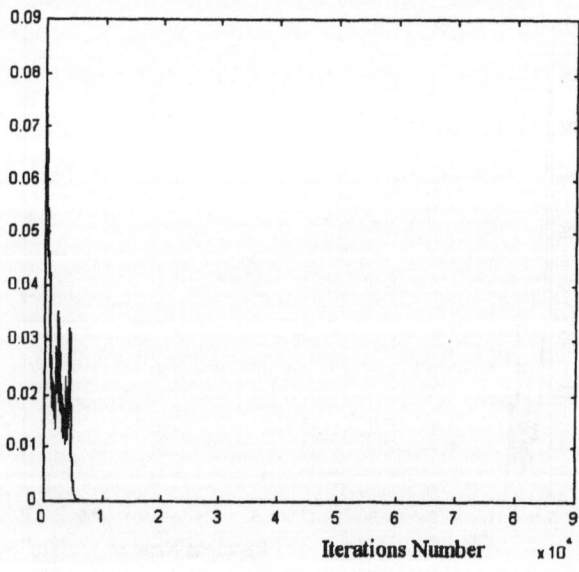

Figure 8.36: Evolution of c_n^{21}.

8.4. THE CONSTRAINED CASE (EXAMPLE 1)

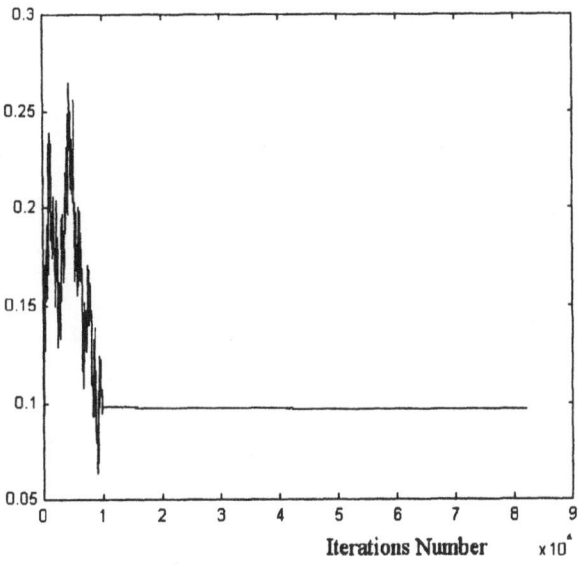

Figure 8.37: Evolution of c_n^{22}.

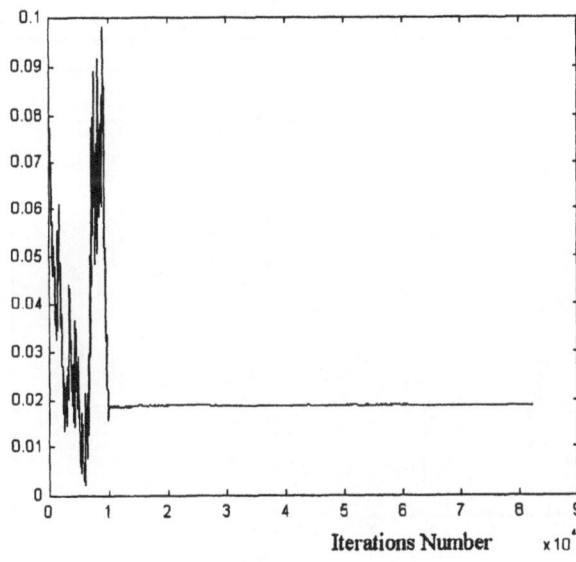

Figure 8.38: Evolution of c_n^{23}.

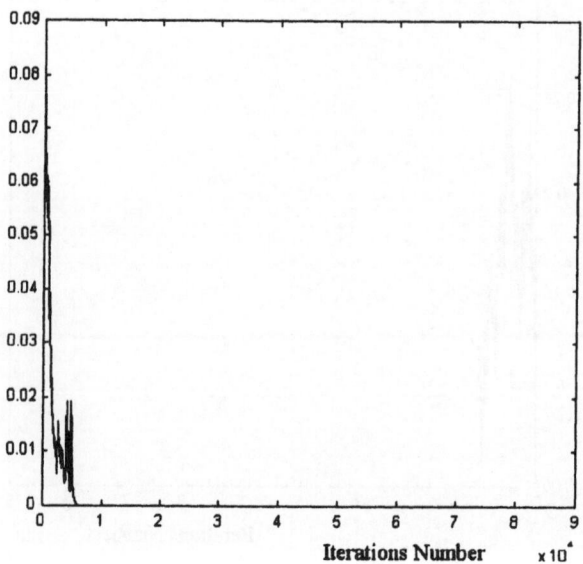

Figure 8.39: Evolution of c_n^{31}.

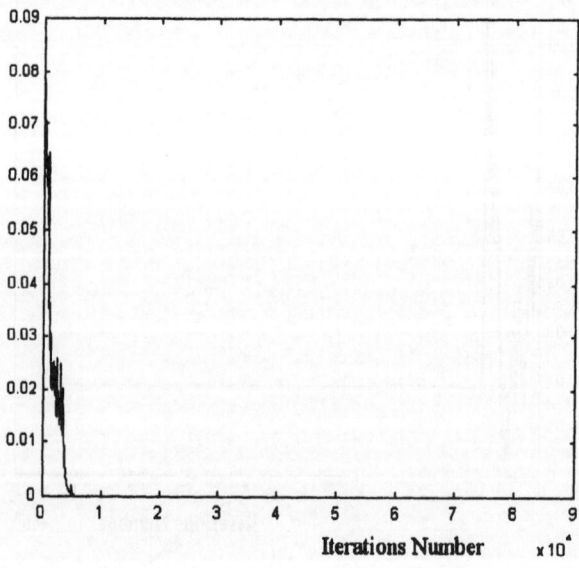

Figure 8.40: Evolution of c_n^{32}.

8.4. THE CONSTRAINED CASE (EXAMPLE 1)

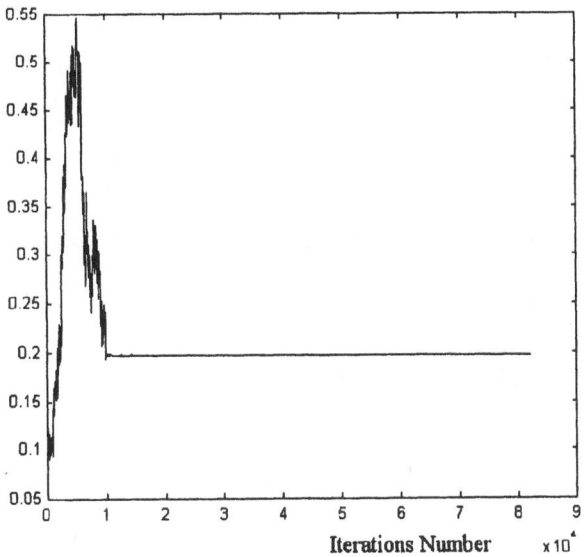

Figure 8.41: Evolution of c_n^{33}.

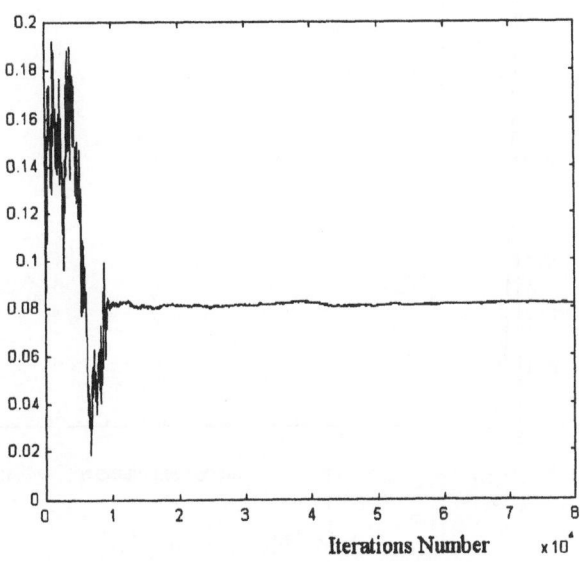

Figure 8.42: Evolution of c_n^{41}.

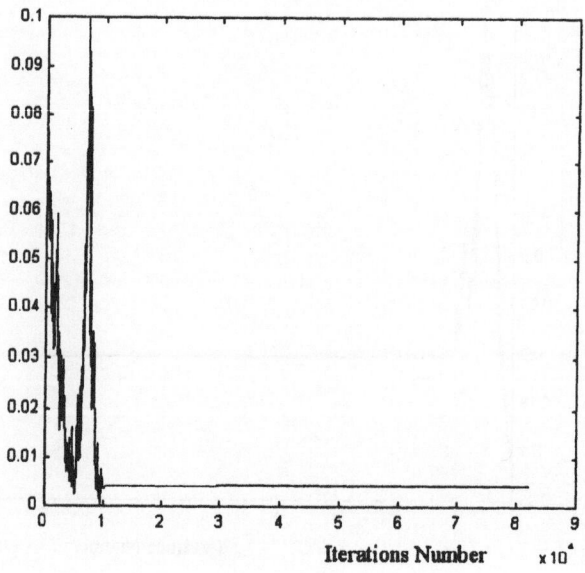

Figure 8.43: Evolution of c_n^{42}.

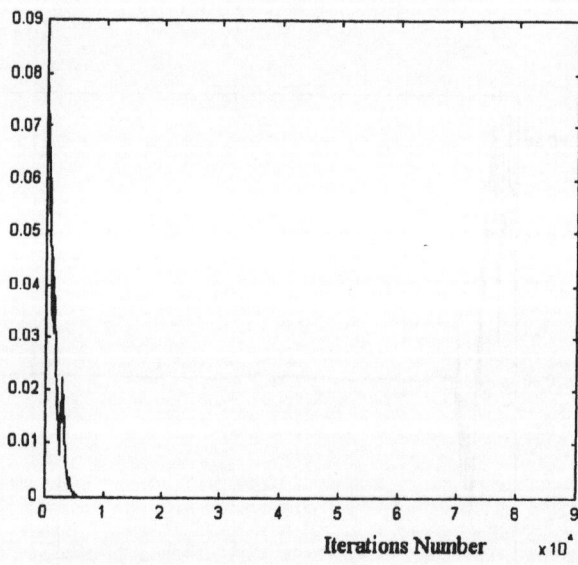

Figure 8.44: Evolution of c_n^{43}.

Simulation results concerning the penalty function approach are presented in the next subsection.

8.4.2 Penalty function approach

Let us now look at some simulations to examine the behaviour of the control scheme based on the penalty function approach. The design parameters were selected as follows:

$$\delta_0 = 2, \mu_0 = 2.8, \gamma_0 = 0.005.$$

A set of 80,000 iterations has been considered. The following results

$$c_P = \begin{bmatrix} 0.4756 & 0 & 0.1379 \\ 0 & 0.0785 & 0 \\ 0 & 0 & 0.2244 \\ 0.0836 & 0 & 0 \end{bmatrix}$$

$$\tilde{V}_{0,P}(c) = 1.4966$$

and

$$\tilde{V}_{1,P}(c) = -0.6229,$$

have been obtained. Recall that the index P corresponds to the penalty function approach. The estimation of the probabilities are given in the following:

$$\hat{\pi}_{ij}^1 = \begin{bmatrix} 0.6974 & 0.1021 & 0.1020 & 0.0985 \\ 0 & 0.7918 & 0.0936 & 0.1146 \\ 0 & 0 & 0.9122 & 0.0878 \\ 0.7955 & 0.158 & 0.0987 & 0 \end{bmatrix}$$

$$\hat{\pi}_{ij}^2 = \begin{bmatrix} 0.1163 & 0.5924 & 0.1023 & 0.1889 \\ 0.7988 & 0 & 0.1011 & 0.1001 \\ 0 & 0.8116 & 0.1884 & 0 \\ 0.1130 & 0.7006 & 0.0791 & 0.1073 \end{bmatrix}$$

$$\hat{\pi}_{ij}^3 = \begin{bmatrix} 0 & 0 & 0.8965 & 0.11035 \\ 0 & 0.0938 & 0.8069 & 0.0993 \\ 0.6927 & 0.1033 & 0.0990 & 0.1050 \\ 0 & 0.0946 & 0.9054 & 0 \end{bmatrix}$$

The evolutions of Φ_n^0 and Φ_n^1 are shown in figures 8.45-8.46.

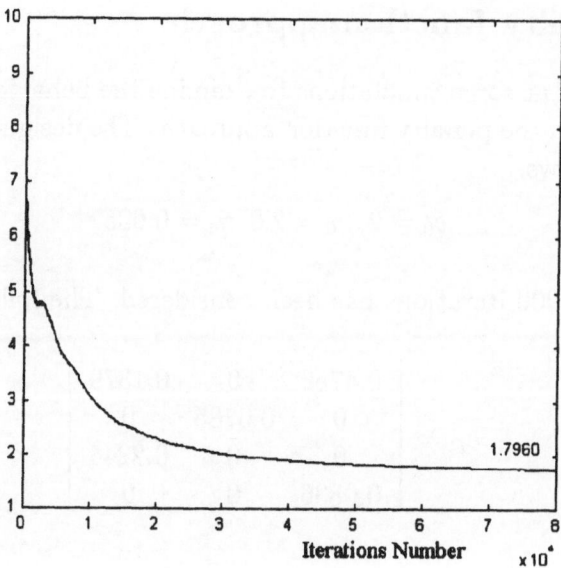

Figure 8.45: Evolution of the loss function Φ_n^0.

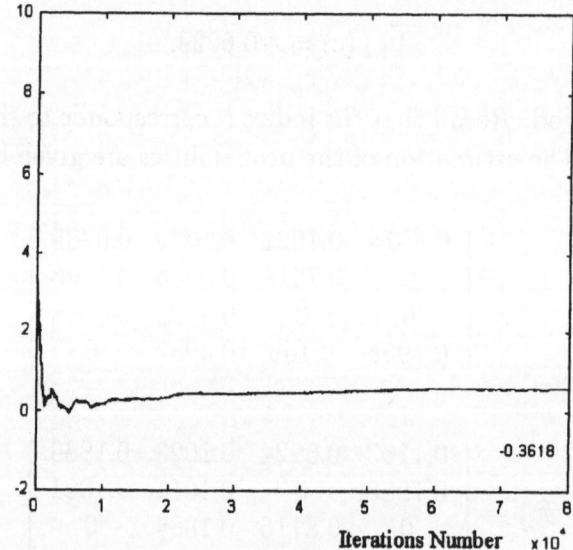

Figure 8.46: Evolution of the loss function Φ_n^1.

Using again the same initial condition

$$p_0 = [0.25\ 0.25\ 0.25\ 0.25]^T$$

for the probability vector, figure 8.47 shows the effect of the control actions on the controlled system.

8.4. THE CONSTRAINED CASE (EXAMPLE 1)

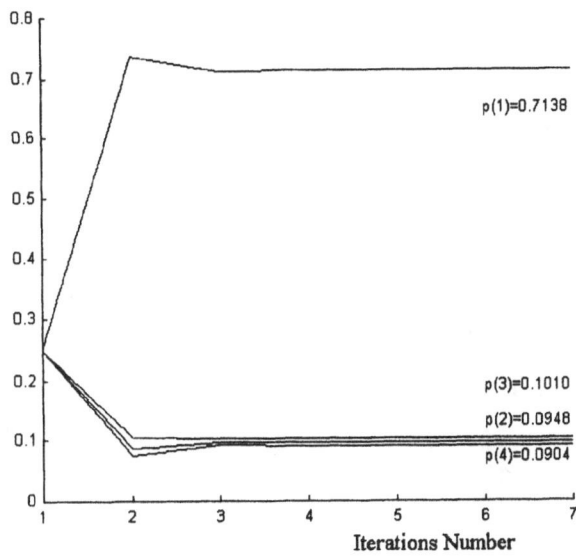

Figure 8.47: Evolution of the state probability vector p_n.

The probability $p(1)$ converges to a value close to 0.7138. The transient behaviour of the adaptive control algorithm is relatively short. This behaviour was expected from the theoretical results stated in the previous chapters. In figures 8.48-8.59 the components of the matrix c_n are depicted. After a relatively short learning period, the components of the matrix c_n converge to some constant values.

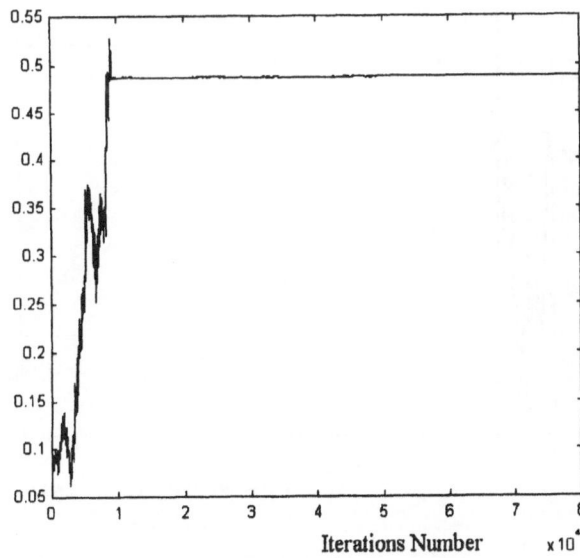

Figure 8.48: Evolution of c_n^{11}.

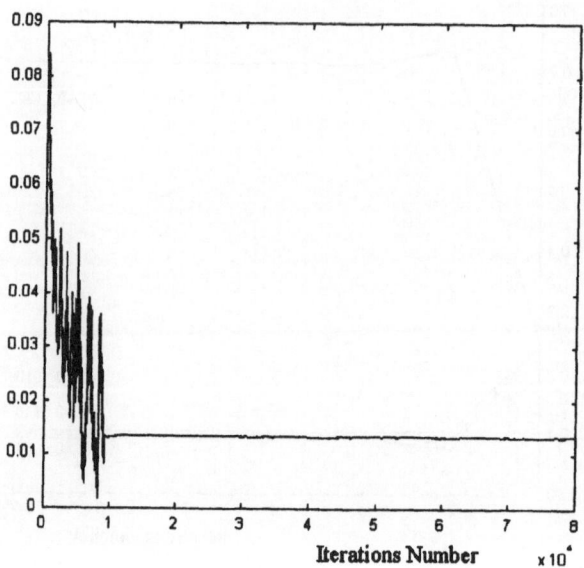

Figure 8.49: Evolution of c_n^{12}.

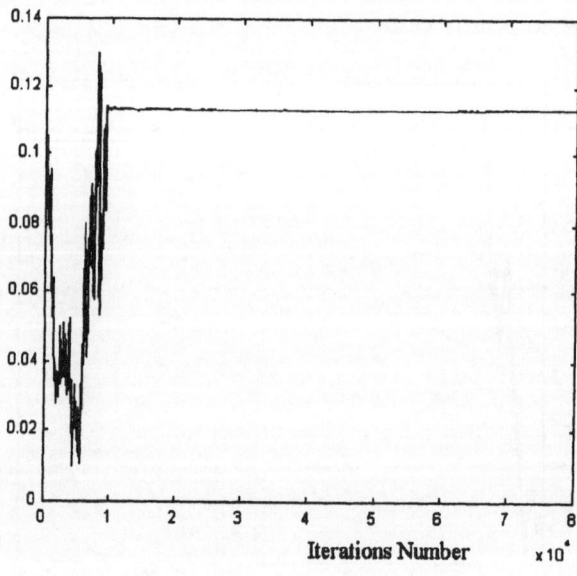

Figure 8.50: Evolution of c_n^{13}.

8.4. THE CONSTRAINED CASE (EXAMPLE 1)

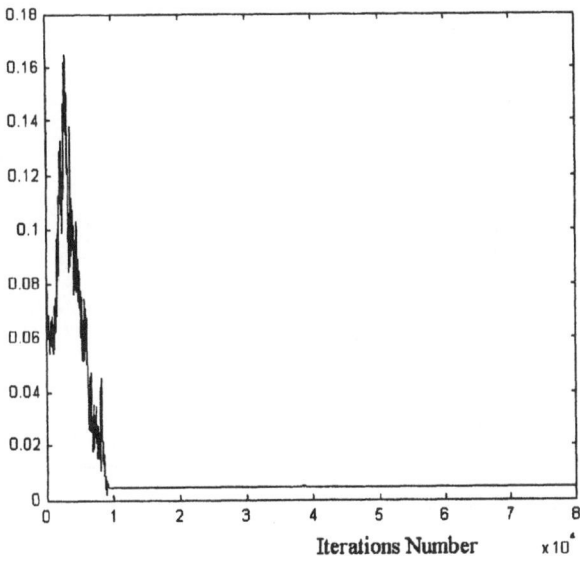

Figure 8.51: Evolution of c_n^{21}.

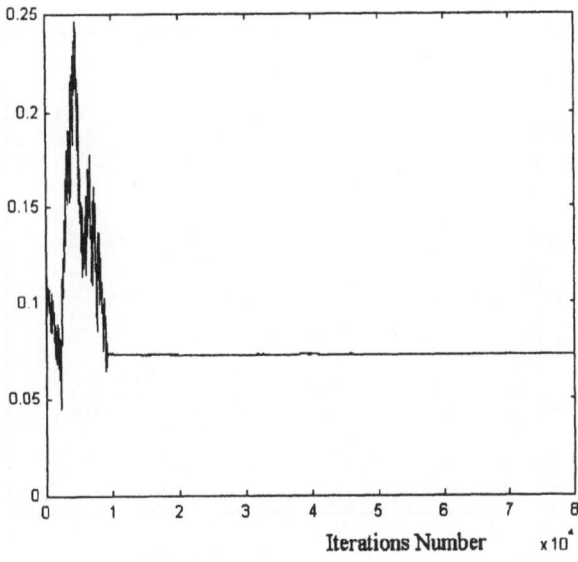

Figure 8.52: Evolution of c_n^{22}.

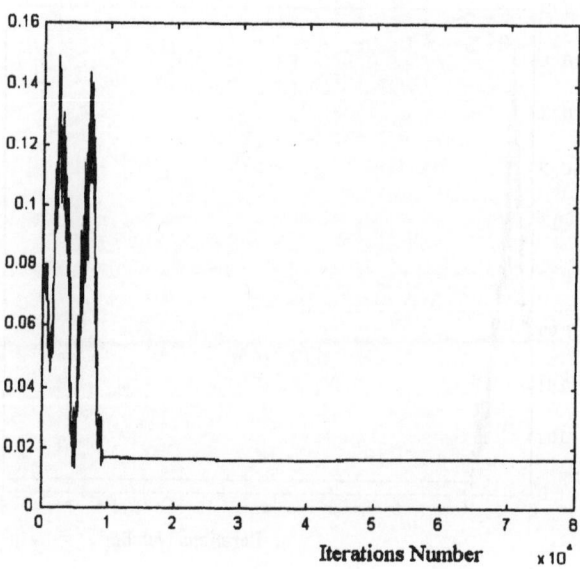

Figure 8.53: Evolution of c_n^{23}.

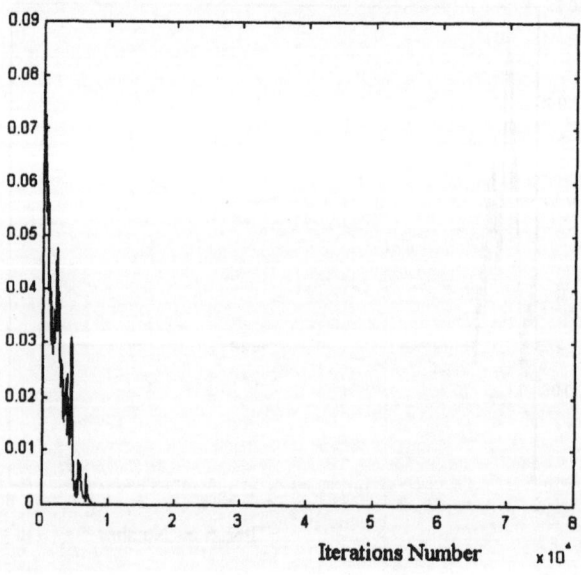

Figure 8.54: Evolution of c_n^{31}.

8.4. THE CONSTRAINED CASE (EXAMPLE 1)

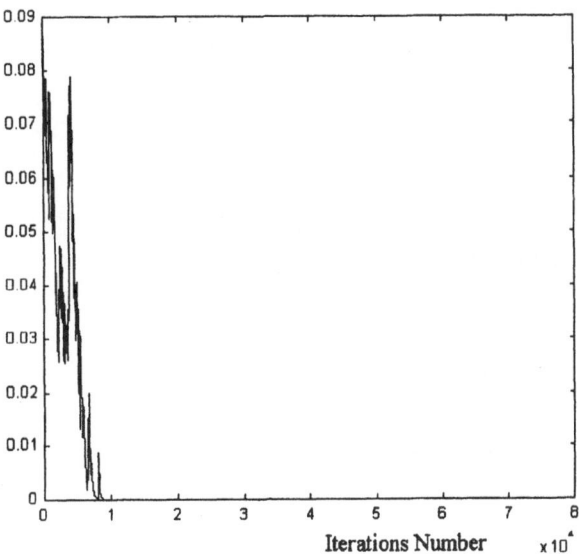

Figure 8.55: Evolution of c_n^{32}.

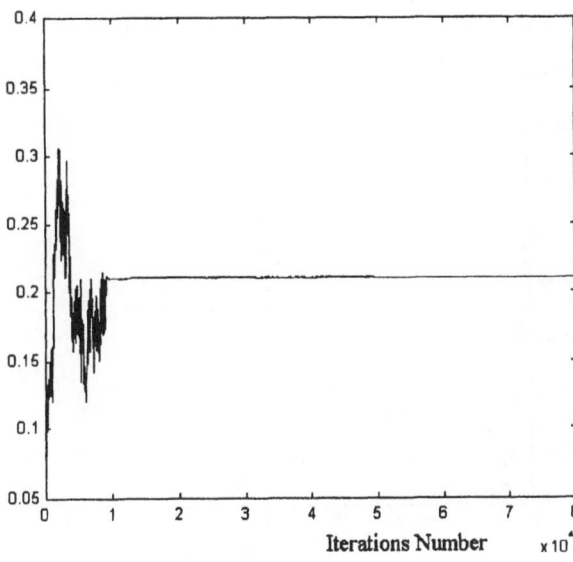

Figure 8.56: Evolution of c_n^{33}.

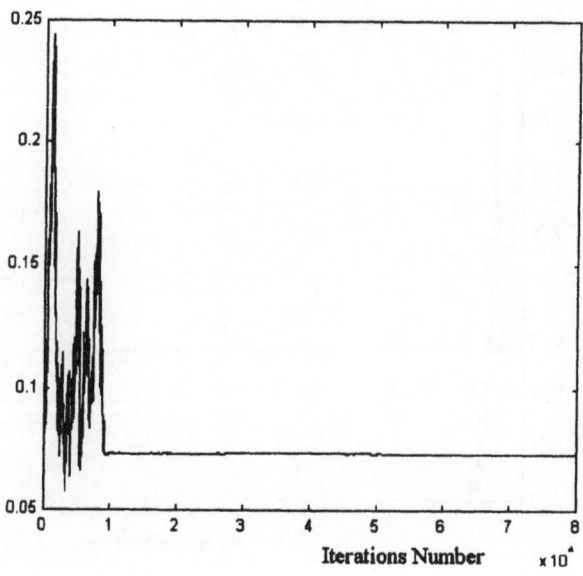

Figure 8.57: Evolution of c_n^{41}.

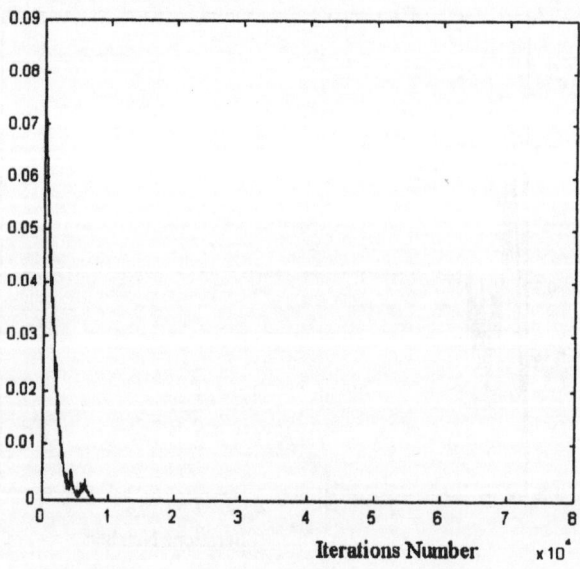

Figure 8.58: Evolution of c_n^{42}.

8.4. THE CONSTRAINED CASE (EXAMPLE 1)

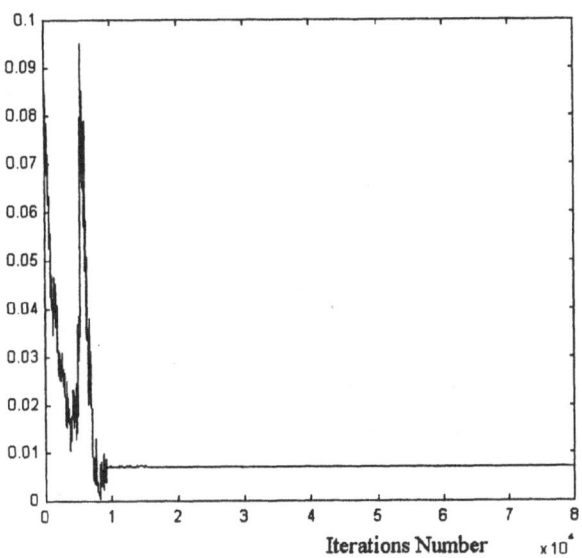

Figure 8.59: Evolution of c_n^{43}.

The nature of convergence of this adaptive control algorithm is clearly reflected by these figures. It is also clear from these figures that the behaviour of this adaptive control algorithm corresponds very closely to that observed for the self-learning algorithm based on the Lagrange multipliers approach. The implementation of these adaptive control algorithms show that they are computationally efficient, not sensitive to the round errors, and require relatively little memory. We can also notice that:

1. the desired control objective is achieved without the use of an extra input or dividing the control horizon (time) into a successive cycles (each cycle consists of a forcing or experimenting phase followed by the certainty equivalence control phase in which the unknown parameters are replaced by their estimates);

2. a stochastic approximation procedure is used for estimation purposes instead of the maximum likelihood approach which is commonly used.

All the simulation results presented here were carried out using a PC.

The next section deals with the second example presented in this chapter. In this example, the matrix v^0 is selected in such a way that for any initial probability, the Markov process tends to the state numbered 2. Both constrained and unconstrained cases will be considered.

- **Example 2.** *(Preference for the state numbered 2)* In this example, we use the same transition probabilities as in example 1. The matrix v^0 is selected in such a way that for any initial probability, the Markov process tends to the state numbered 2.

$$v^0 = \begin{bmatrix} 10 & 0 & 10 \\ 0 & 10 & 10 \\ 10 & 0 & 10 \\ 10 & 0 & 10 \end{bmatrix}.$$

As before, we use the same Matlab Optimization Toolbox to solve the linear programming problem related to this example.

$$c = lp(f, A, 0, 1, c_0, 1 + K)$$

where the initial condition c_0 is equal to

$$c_0 = \begin{bmatrix} 0 & 0.0100 & 0 \\ 0.7890 & 0 & 0 \\ 0 & 0.1111 & 0 \\ 0 & 0.0899 & 0 \end{bmatrix}$$

The corresponding losses are equal to

$$\tilde{V}(c) = 0.$$

The solution of this unconstrained problem using, respectively, the Lagrange multipliers and the penalty function approaches is given in the following section.

8.5 The unconstrained case (example 2)

This section presents simple numerical simulations, from which we can verify the viability of the design and analysis given in the previous chapters.

8.5.1 Lagrange multipliers approach

The application of the Lagrange multipliers approach is described here. There are few parameters (design parameters) that must be specified a priori. The control performance is achieved with the following choice of parameters:

$$\delta_0 = 0.5, \lambda_0^+ = 0.3, \gamma_0 = 0.005.$$

8.5. THE UNCONSTRAINED CASE (EXAMPLE 2)

Observe that it is preferable to assign a small value to the correction factor γ_0. The simulations have been carried out over 81,000 samples (iterations). The numerical implementation of the self-learning scheme based on the Lagrange multipliers approach leads to the following results:

$$c_L = \begin{bmatrix} 0 & 0.0239 & 0 \\ 0.7591 & 0 & 0 \\ 0 & 0.1157 & 0 \\ 0 & 0.1013 & 0 \end{bmatrix}$$

The corresponding losses are equal to

$$\tilde{V}_{0,L}(c) = 0.1036.$$

The evolution of the loss function Φ_n is depicted in figure 8.60. This figure shows a typical well behaved run using this self-learning algorithm. From figure 8.61, the final probabilities (after the learning period) are seen to be $p(1) = 0.0100$, $p(2) = 0.7890$, $p(3) = 0.1111$, $p(4) = 0.0899$. We can also see the effect of the value of the matrix v^0 on the control actions.

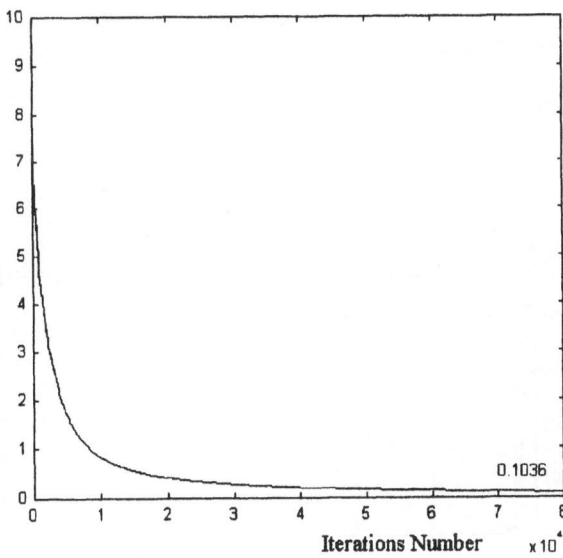

Figure 8.60: Evolution of the loss function Φ_n.

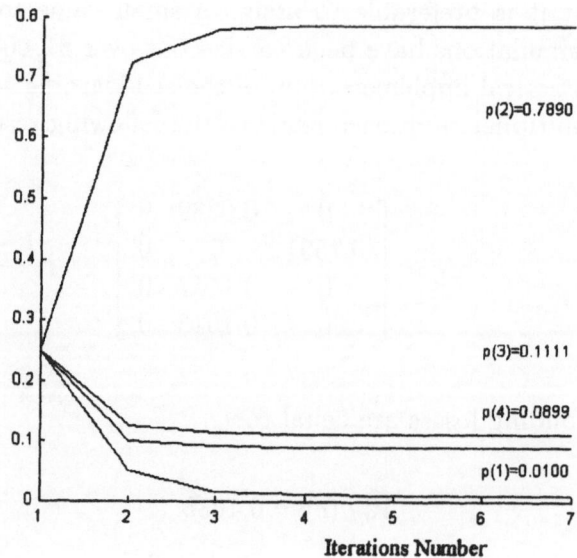

Figure 8.61: Time evolution of the state probability vector p_n.

Figures 8.62-8.73 indicate how the components of the matrix c_n evolve with the iterations number. The convergence occurs after a simulation horizon less than 10^4 samples (iterations number).

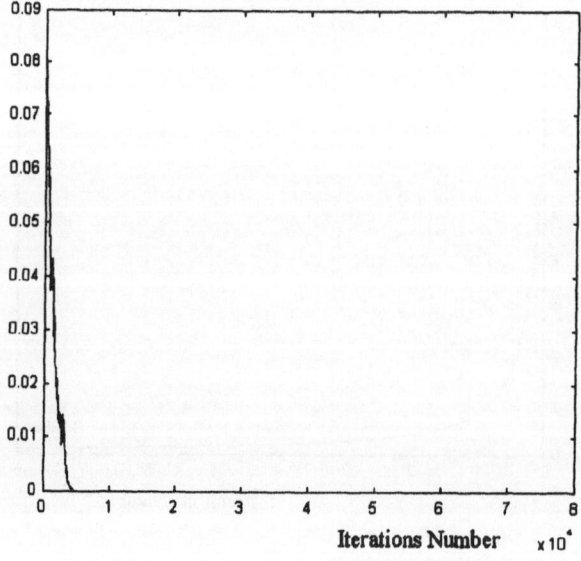

Figure 8.62: Evolution of c_n^{11}.

8.5. THE UNCONSTRAINED CASE (EXAMPLE 2)

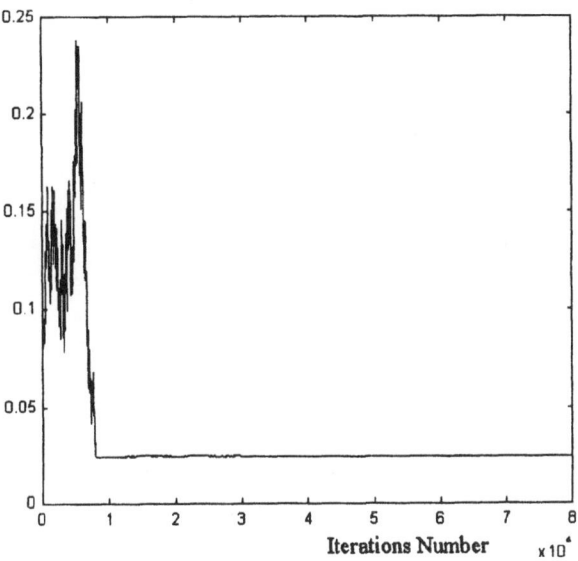

Figure 8.63: Evolution of c_n^{12}.

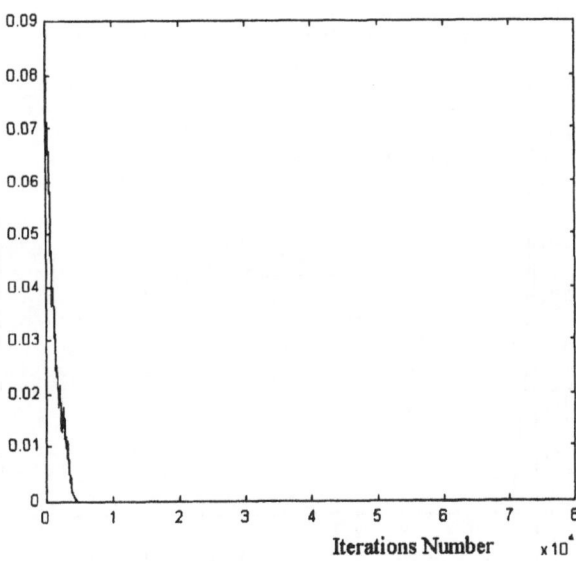

Figure 8.64: Evolution of c_n^{13}.

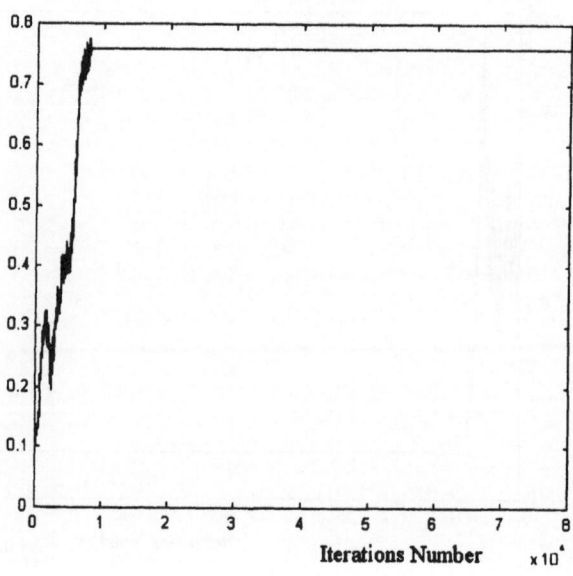

Figure 8.65: Evolution of c_n^{21}.

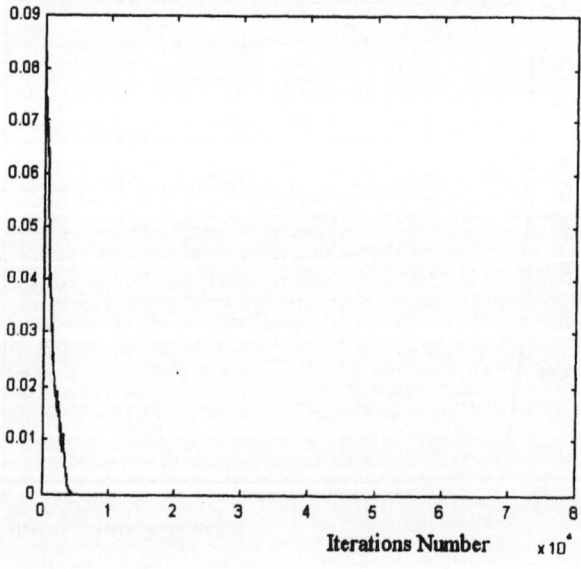

Figure 8.66: Evolution of c_n^{22}.

8.5. THE UNCONSTRAINED CASE (EXAMPLE 2)

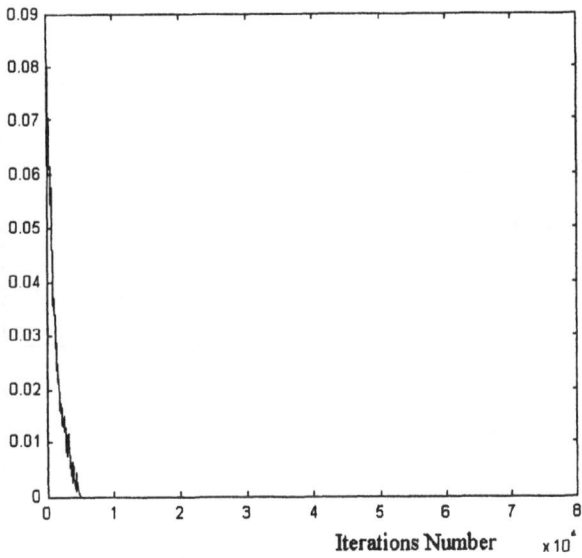

Figure 8.67: Evolution of c_n^{23}.

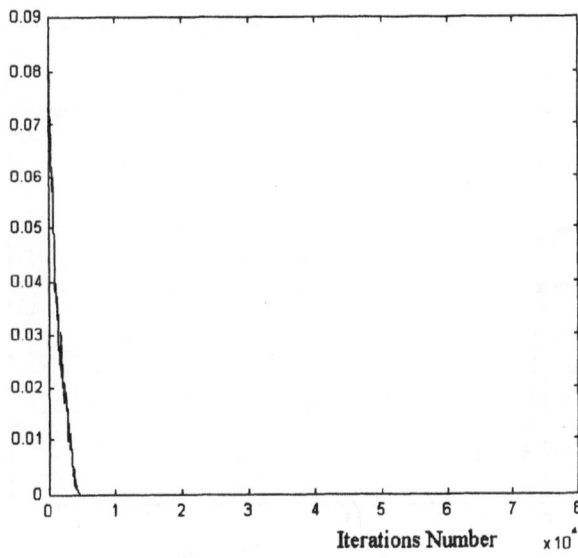

Figure 8.68: Evolution of c_n^{31}.

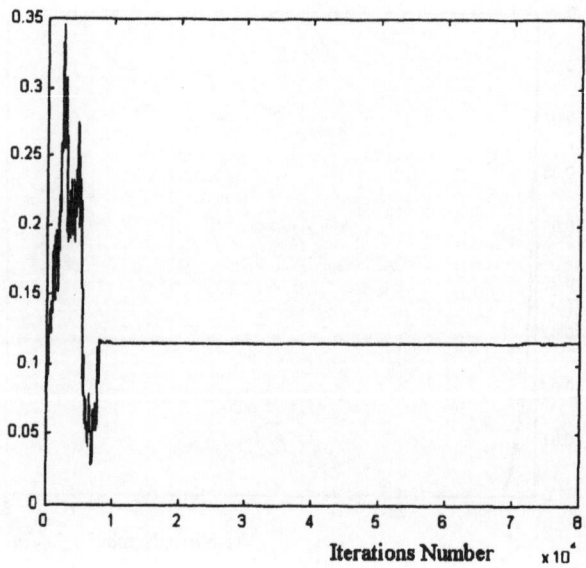

Figure 8.69: Evolution of c_n^{32}.

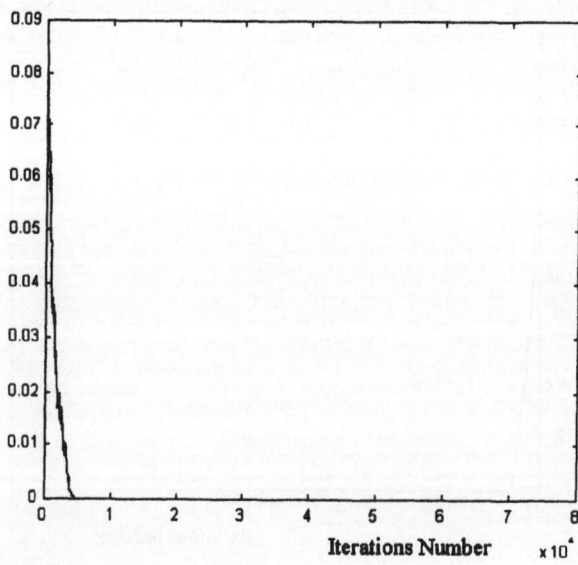

Figure 8.70: Evolution of c_n^{33}.

8.5. THE UNCONSTRAINED CASE (EXAMPLE 2)

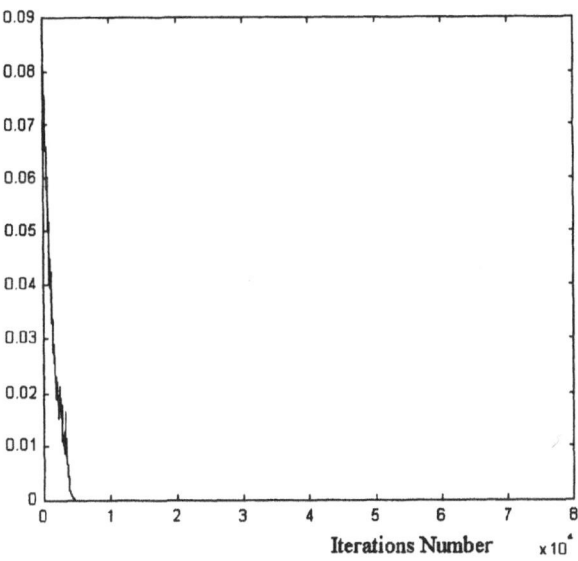

Figure 8.71: Evolution of c_n^{41}.

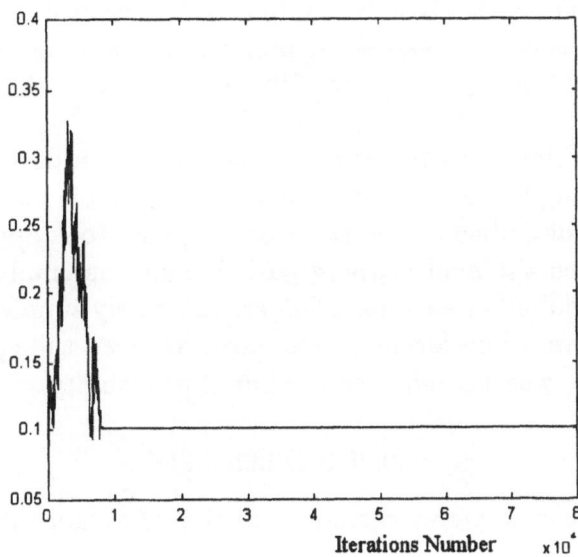

Figure 8.72: Evolution of c_n^{42}.

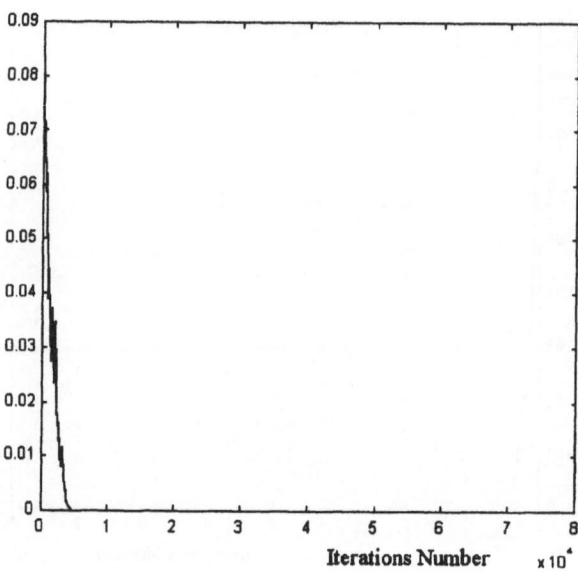

Figure 8.73: Evolution of c_n^{43}.

These simulation results show that the adaptive algorithm converges and achieves the desired goal. This behaviour was expected from the theoretical results stated in the previous chapters.

To provide a greater ease of implementation in the trial (experiment) the design parameters were set to a fixed values. It is conceivable that even better performance levels can be realized by permitting the design parameters also to be updated, although at the cost of increased computational and memory requirements.

After a set of iterations the adaptive control system develops enough experience in making better selection among the available control actions. In the experiments described above, the learning phase took approximately 10^4 iterations to reach the final learning goal. It must be emphasized that the computations required at each iteration are extremely simple. We must also note that this level of performance was achieved even though the recursive control procedure was started with the initial probability vector set at arbitrary values:

$$p_0 = [0.25\ 0.25\ 025\ 0.25]^T.$$

In view of this example we can conclude that the algorithm works well as an on-line controller.

The next subsection deals with the penalty function approach.

8.5. THE UNCONSTRAINED CASE (EXAMPLE 2)

8.5.2 Penalty function approach

The previous problem (example 2) was also solved using the penalty function approach. The adaptive control algorithm was designed with the following parameters:

$$\delta_0 = 0.5, \mu_0 = 4, \gamma_0 = 0.006.$$

A set of 81,000 iterations have been carried out. The implementation of the adaptive control algorithm outlined in the previous chapters was run on a PC. The obtained simulation results are:

$$c_P = \begin{bmatrix} 0 & 0 & 0 \\ 0.7887 & 0 & 0 \\ 0 & 0.1033 & 0 \\ 0 & 0.0978 & 0 \end{bmatrix}$$

and the corresponding losses are equal to:

$$\tilde{V}_{0,P}(c) = 0.1020.$$

The estimation of the probabilities are:

$$\widehat{\pi}_{ij}^1 = \begin{bmatrix} 0.7500 & 0.0862 & 0.1121 & 0.0517 \\ 0 & 0.7995 & 0.0999 & 0.1006 \\ 0 & 0 & 0.8846 & 0.1154 \\ 0.7374 & 0.1515 & 0.1111 & 0 \end{bmatrix}$$

$$\widehat{\pi}_{ij}^2 = \begin{bmatrix} 0.0934 & 0.6051 & 0.0895 & 0.2121 \\ 0.7568 & 0 & 0.1261 & 0.1171 \\ 0 & 0.7939 & 0.2061 & 0 \\ 0.1093 & 0.6979 & 0.0943 & 0.0986 \end{bmatrix}$$

$$\widehat{\pi}_{ij}^3 = \begin{bmatrix} 0 & 0 & 0.8878 & 0.1122 \\ 0 & 0.1290 & 0.7527 & 0.1183 \\ 0.6735 & 0.0816 & 0.1531 & 0.0918 \\ 0 & 0.0825 & 0.9175 & 0 \end{bmatrix}$$

Figure 8.74 depicts the variation of the loss function Φ_n. By inspection of this figure we see that the loss function decreases very quickly (exponentially) to its final value which is close to 0.1020.

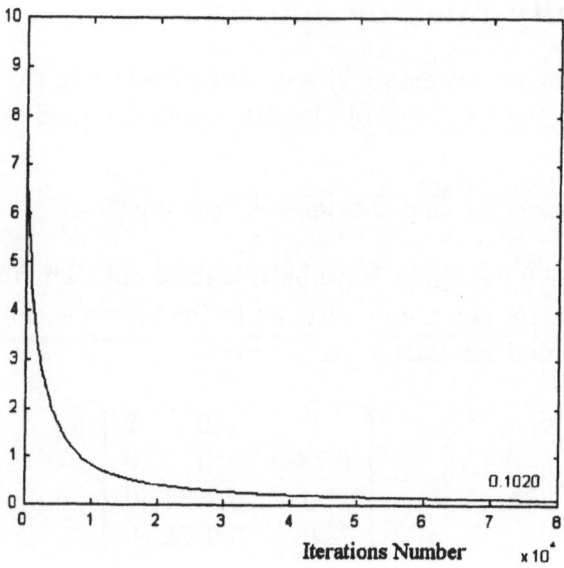

Figure 8.74: Evolution of the loss function Φ_n.

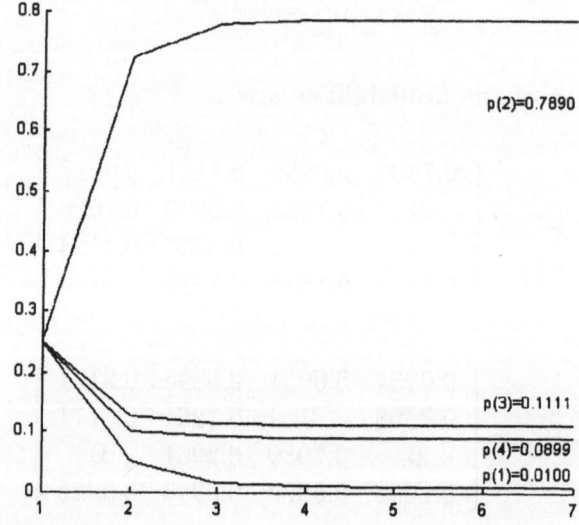

Figure 8.75: Time evolution of the state probability vector p_n.

The performance of the algorithm is illustrated in figure 8.75 which represents the evolution of the components of the probability vector versus the iterations number. Figures 8.76-8.87 show plots for components of the matrix c_n. Some of these components converge to zero. For most components the learning phase is less than $5 \cdot 10^3$ samples (iterations).

8.5. THE UNCONSTRAINED CASE (EXAMPLE 2)

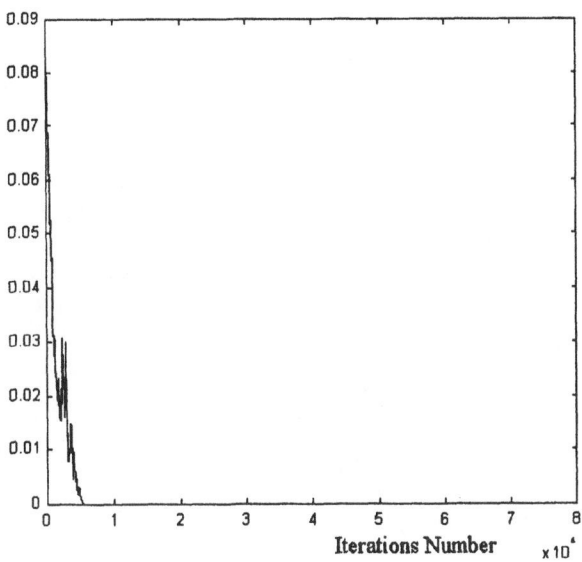

Figure 8.76: Evolution of c_n^{11}.

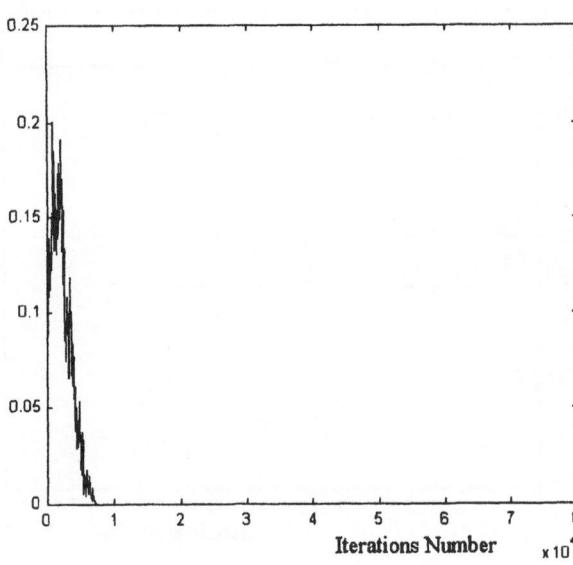

Figure 8.77: Evolution of c_n^{12}.

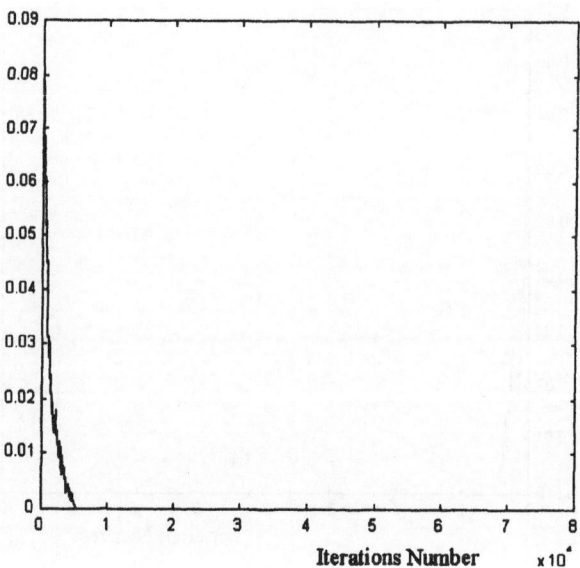

Figure 8.78: Evolution of c_n^{13}.

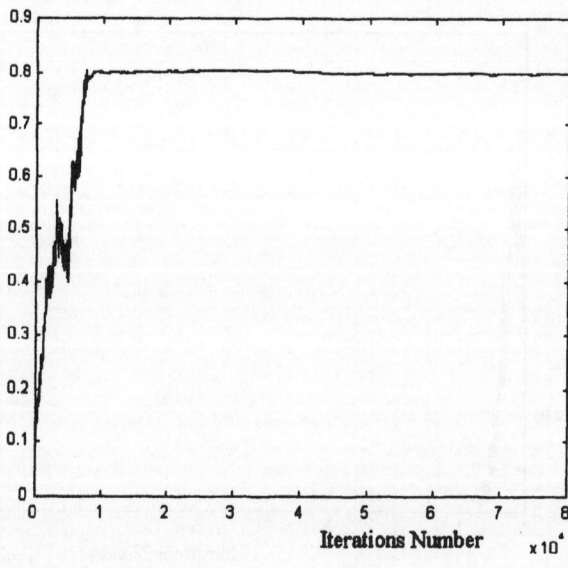

Figure 8.79: Evolution of c_n^{21}.

8.5. THE UNCONSTRAINED CASE (EXAMPLE 2)

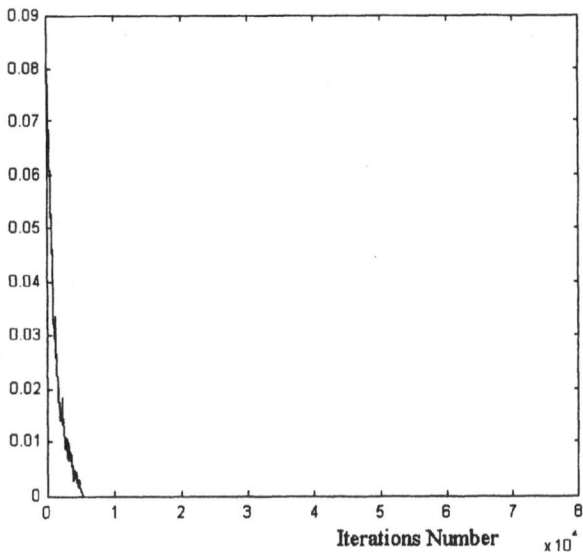

Figure 8.80: Evolution of c_n^{22}.

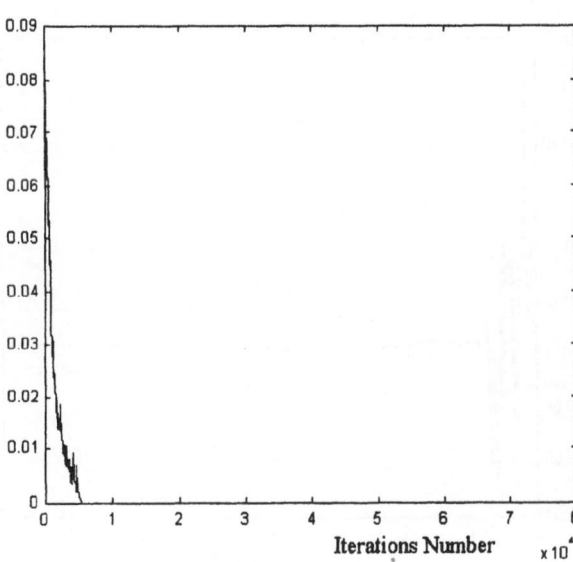

Figure 8.81: Evolution of c_n^{23}.

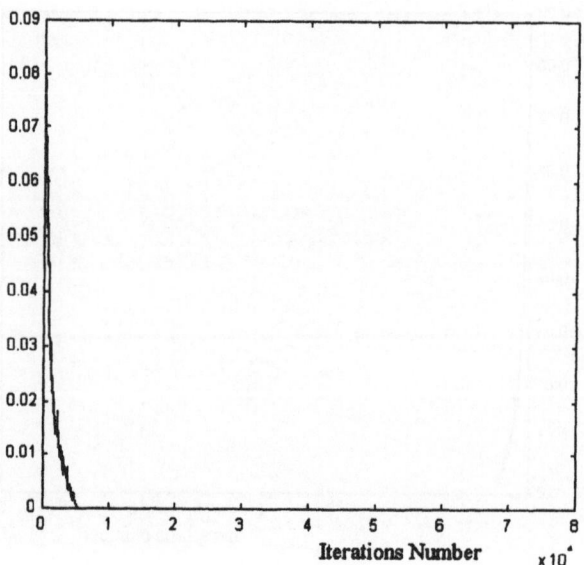

Figure 8.82: Evolution of c_n^{31}.

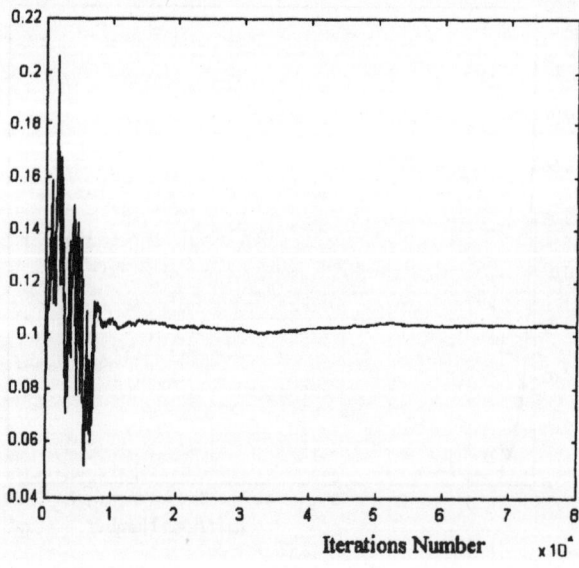

Figure 8.83: Evolution of c_n^{32}.

8.5. THE UNCONSTRAINED CASE (EXAMPLE 2)

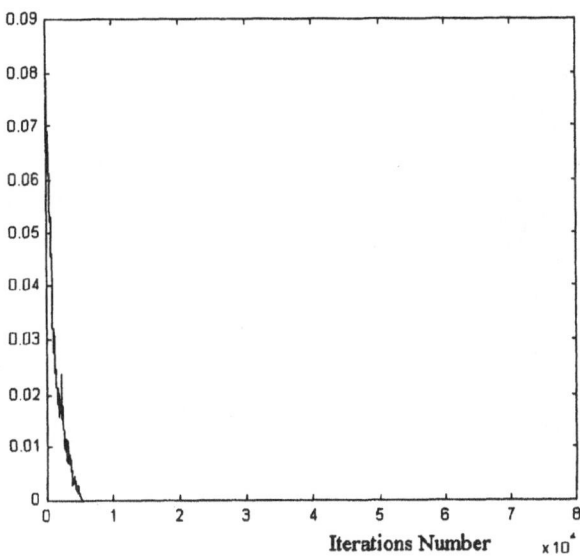

Figure 8.84: Evolution of c_n^{33}.

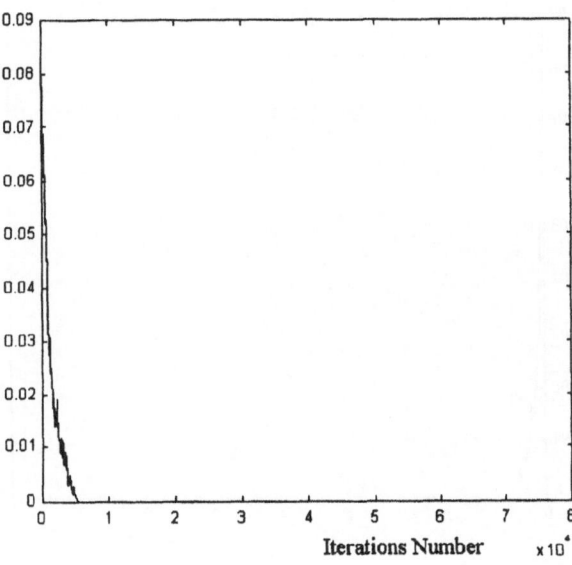

Figure 8.85: Evolution of c_n^{41}.

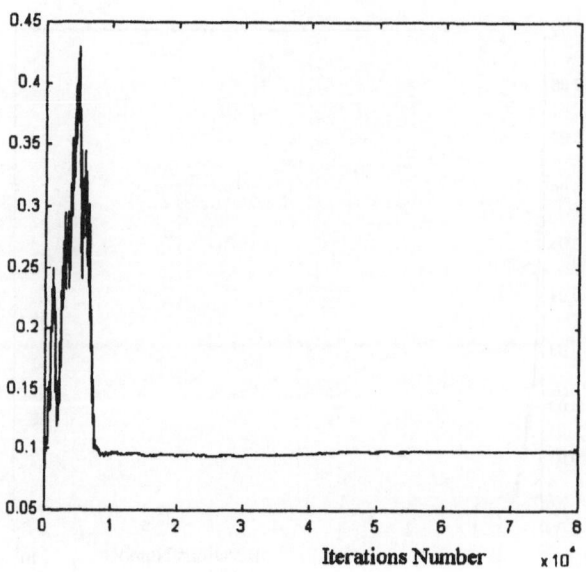

Figure 8.86: Evolution of c_n^{42}.

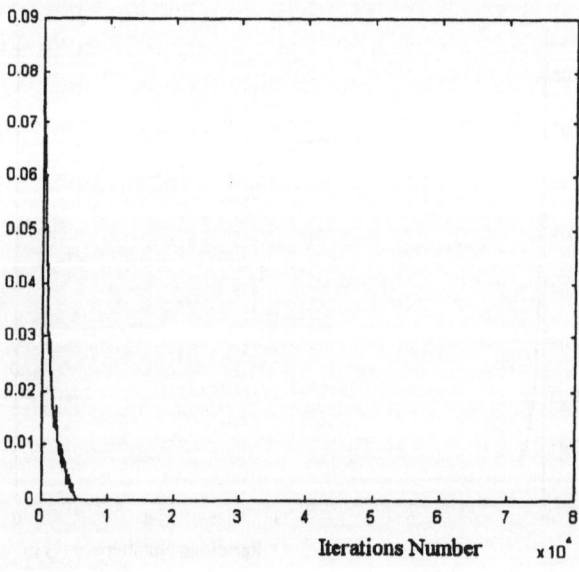

Figure 8.87: Evolution of c_n^{43}.

8.6 The constrained case (example 2)

In this section, our main emphasis will be on the analysis of the performance and effectiveness of the adaptive control algorithms dealing with the control of constrained Markov chains.

8.6.1 Lagrange multipliers approach

The matrix v^0 was selected as before. The constraint v^1 was chosen equal to:

$$v^1 = \begin{bmatrix} 10 & -10 & 10 \\ 10 & -10 & 10 \\ 10 & -10 & 10 \\ 10 & -10 & 10 \end{bmatrix}.$$

The Matlab Optimization Toolbox leads to the following results:

$$c = \begin{bmatrix} 0 & 0.1406 & 0 \\ 0.5000 & 0.1453 & 0 \\ 0 & 0.1111 & 0 \\ 0 & 0.1030 & 0 \end{bmatrix},$$

$\tilde{V}_0(c) = 1.4532$ and $\tilde{V}_1(c) = 0$.

The design parameters associated with the adaptive control algorithm based on the Lagrange multipliers were chosen as follows:

$$\delta_0 = 0.5, \lambda_0^+ = 0.3 \text{ and } \gamma_0 = 0.006.$$

The implementation of the adaptive control algorithm outlined in chapter 5 was run on a PC. This control algorithm is easy to implement and few design parameters are associated with it. The results of the experiment presented below correspond to a simulation horizon n of $81,000$ samples (iterations). The obtained results are:

$$c_L = \begin{bmatrix} 0 & 0.1173 & 0 \\ 0.4861 & 0.1610 & 0 \\ 0 & 0.1090 & 0.0002 \\ 0 & 0.1264 & 0 \end{bmatrix},$$

$\tilde{V}_{0,L}(c) = 1.7960$ and $\tilde{V}_{1,L}(c) = -0.3618$.

These results show the efficiency of the self-learning algorithm derived on the basis of the Lagrange multipliers approach. Notice that the normalization procedure plays an important role towards the characteristic of the adaptation scheme. The loss functions Φ_n^0 and Φ_n^1 are respectively drawn in figure 8.88 and figure 8.89. In these figures on the abscissa axis we plotted the iterations number (samples), and on the ordinate axis the value of the loss functions.

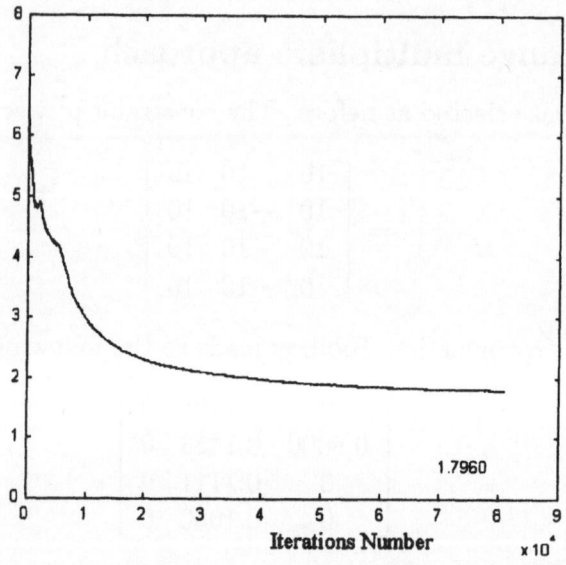

Figure 8.88: Evolution of the loss function Φ_n^0.

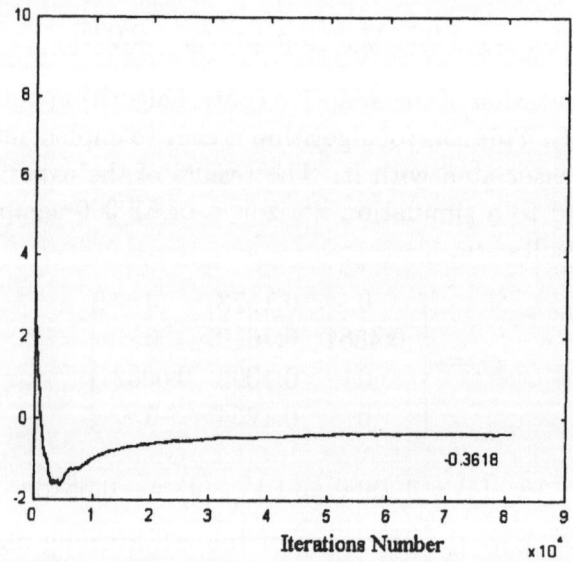

Figure 8.89: Evolution of the loss function Φ_n^1.

8.6. THE CONSTRAINED CASE (EXAMPLE 2)

The loss functions Φ_n^0 and Φ_n^1 decrease respectively to their final values: 1.7960 and -0.3618. The evolution of the components of the probability vector p_n are depicted in figure 8.90. We can observe the effect of the control actions on the controlled system. The probability $p(2)$ converges to: $p(2) = 0.6331$.

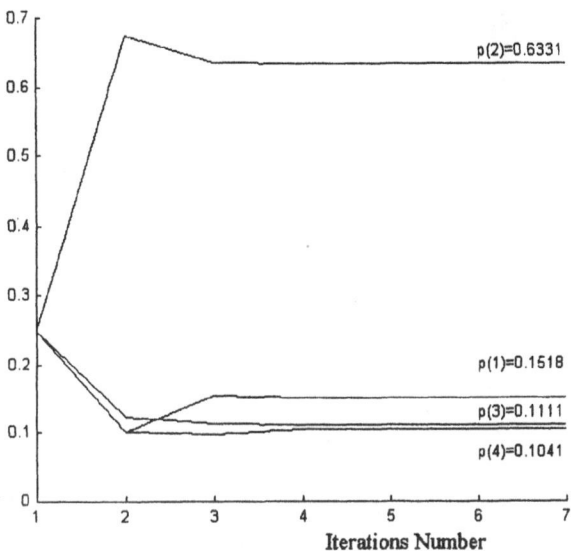

Figure 8.90: Time evolution of the state probability p_n.

These figures illustrate the performance and the efficiency of this adaptive control algorithm. The adaptive behaviour of the adaptive control algorithm is also illustrated by the evolution of the components of the matrix c_n. As before, the initial probability vector was selected as follows:

$$p_0 = [0.25\ 0.25\ 0.25\ 0.25]^T.$$

The performance of the learning algorithm does not depend on the initial value of the probability vector. Figures 8.91-8.102 plot the components of the matrix c_n versus the iterations number. These components converge respectively to:

$$(0.0, 0.12, 0.0, 0.48, 0.16, 0.0, 0.12, 0.0, 0.0, 0.13, 0.0).$$

From these figures, we can notice that the adaptive control algorithm based on the Lagrange multipliers approach achieves the desired control objective. The learning phase which gives an idea about the speed of the convergence of the algorithm is less than 10^4 samples (iterations).

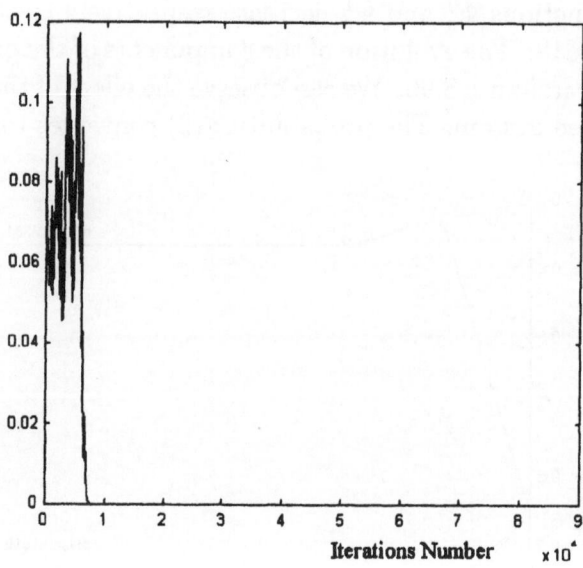

Figure 8.91: Evolution of c_n^{11}.

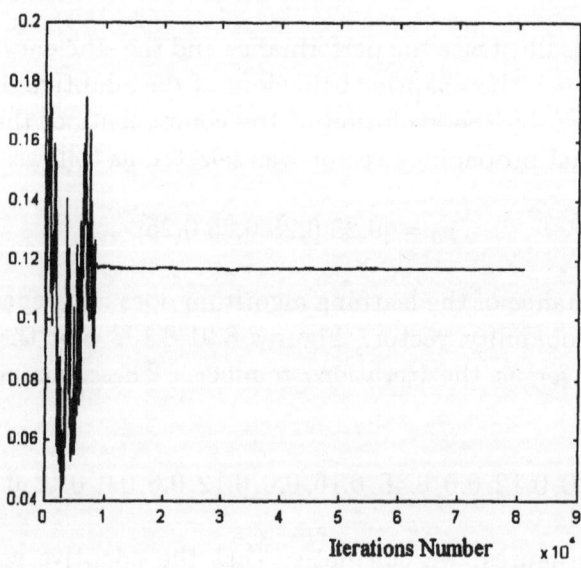

Figure 8.92: Evolution of c_n^{12}.

8.6. THE CONSTRAINED CASE (EXAMPLE 2)

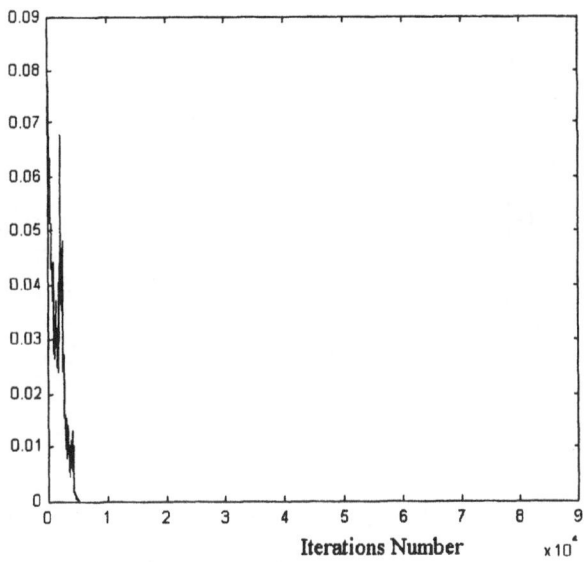

Figure 8.93: Evolution of c_n^{13}.

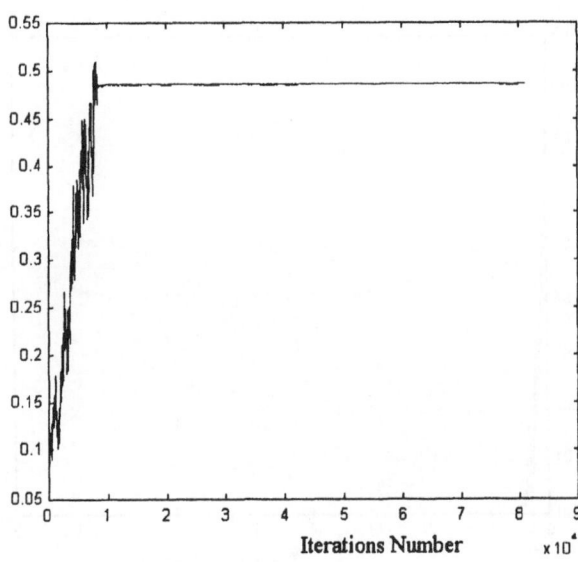

Figure 8.94: Evolution of c_n^{21}.

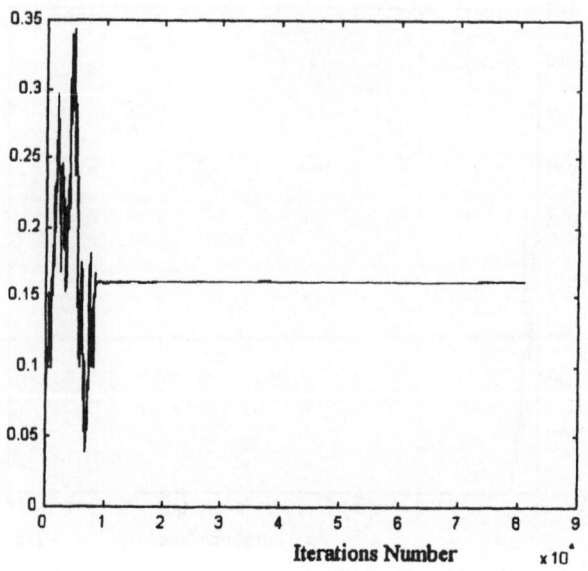

Figure 8.95: Evolution of c_n^{22}.

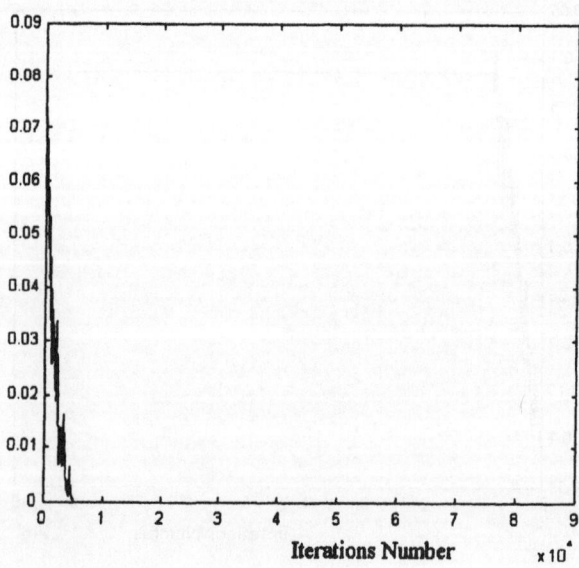

Figure 8.96: Evolution of c_n^{23}.

8.6. THE CONSTRAINED CASE (EXAMPLE 2)

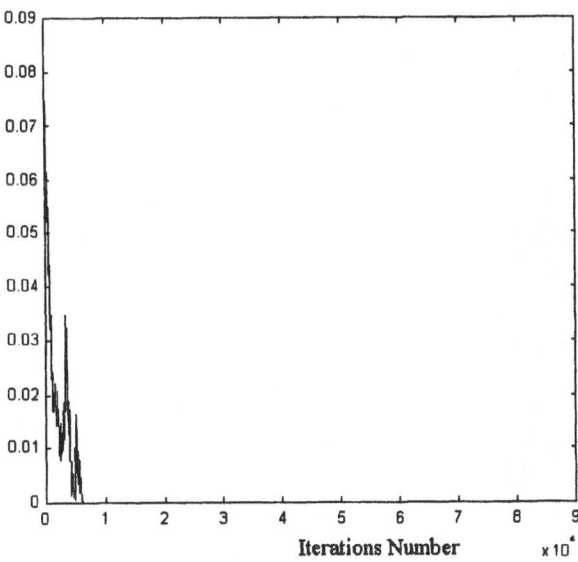

Figure 8.97: Evolution of c_n^{31}.

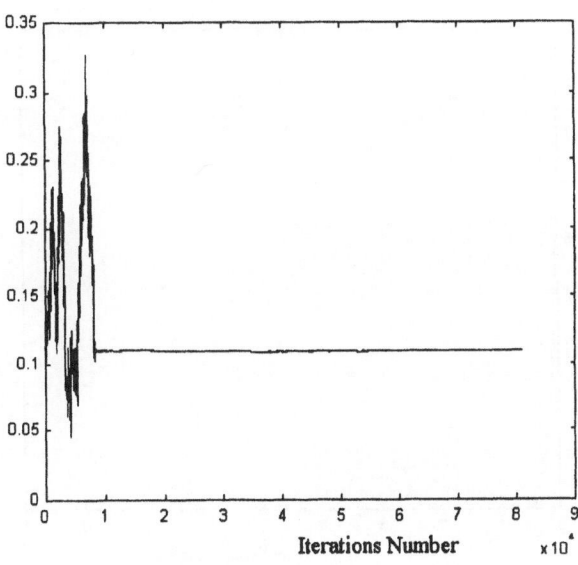

Figure 8.98: Evolution of c_n^{32}.

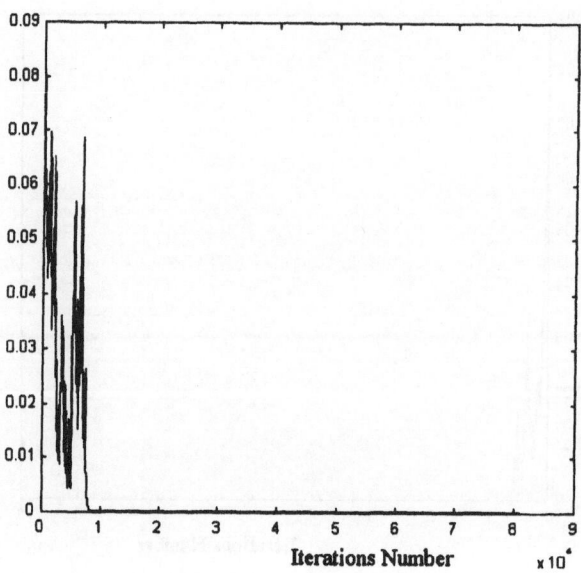

Figure 8.99: Evolution of c_n^{33}.

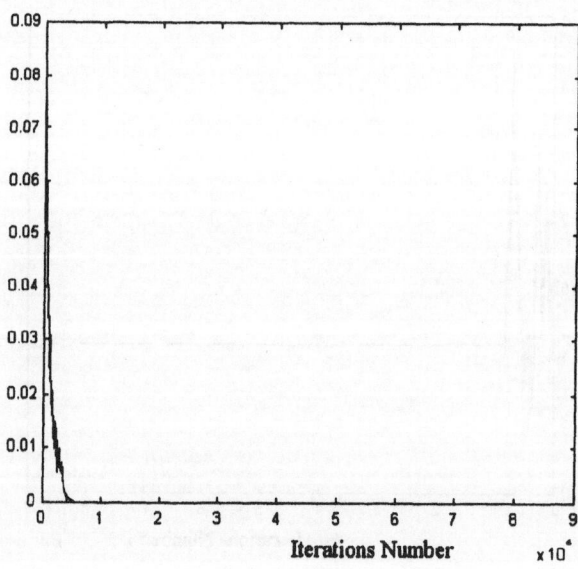

Figure 8.100: Evolution of c_n^{41}.

8.6. THE CONSTRAINED CASE (EXAMPLE 2)

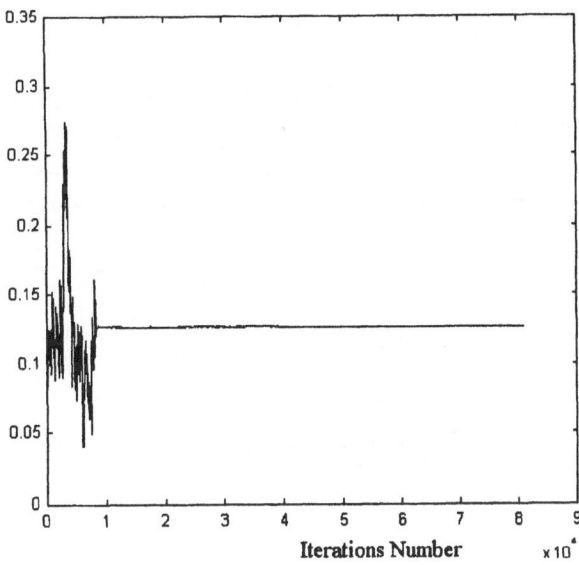

Figure 8.101: Evolution of c_n^{42}.

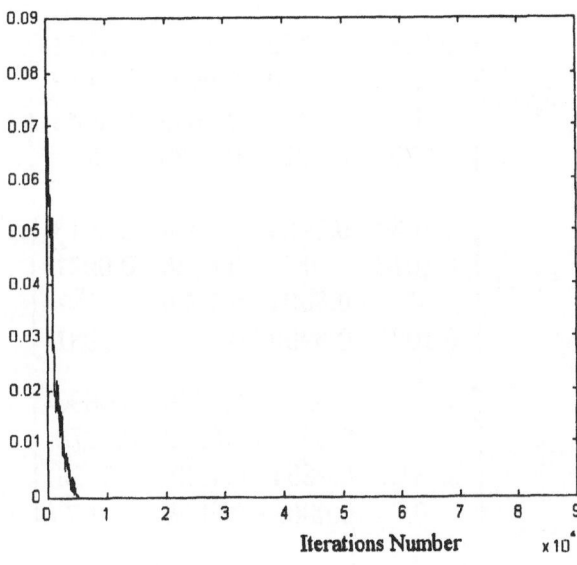

Figure 8.102: Evolution of c_n^{43}.

8.6.2 Penalty function approach

This subsection provides a comprehensive performance evaluation of the adaptive algorithm developed on the basis of penalty function approach for constrained controlled Markov chains. The simulation results reported herein were carried out over 81,000 iterations, with the following choices:

$$\delta_0 = 0.5, \mu_0 = 4 \text{ and } \gamma_0 = 0.006.$$

They lead to the following results:

$$c_P = \begin{bmatrix} 0.0227 & 0.1200 & 0.0043 \\ 0.4878 & 0.1372 & 0.0024 \\ 0.0061 & 0.1061 & 0.0024 \\ 0.0024 & 0.1059 & 0.0028 \end{bmatrix}$$

and the corresponding losses are equal to:

$$\tilde{V}_{0,P}(c) = 1.9014 \text{ and } \tilde{V}_{1,P}(c) = 0.6316.$$

These results are very close to the results obtained by the adaptive control based on the Lagrange multipliers approach.

The estimation of the probabilities and the constraints are:

$$\hat{\pi}_{ij}^1 = \begin{bmatrix} 0.6861 & 0.1160 & 0.1029 & 0.0951 \\ 0 & 0.7970 & 0.0998 & 0.1033 \\ 0 & 0 & 0.8658 & 0.1342 \\ 0.7757 & 0.1221 & 0.1023 & 0 \end{bmatrix}$$

$$\hat{\pi}_{ij}^2 = \begin{bmatrix} 0.1030 & 0.5963 & 0.0994 & 0.2013 \\ 0.8018 & 0 & 0.0986 & 0.0997 \\ 0 & 0.8004 & 0.1996 & 0 \\ 0.1029 & 0.6990 & 0.0950 & 0.1031 \end{bmatrix}$$

$$\hat{\pi}_{ij}^3 = \begin{bmatrix} 0 & 0 & 0.9110 & 0.0890 \\ 0 & 0.0870 & 0.8152 & 0.0978 \\ 0.7195 & 0.0854 & 0.1128 & 0.0823 \\ 0 & 0.0962 & 0.9038 & 0 \end{bmatrix}$$

and

$$\hat{v}^1 = \begin{bmatrix} 10.01 & -9.998 & 9.9993 \\ 9.9883 & -9.999 & 9.9978 \\ 9.990 & -10.099 & 9.998 \\ 10.00 & -10.001 & 9.9967 \end{bmatrix}.$$

8.6. THE CONSTRAINED CASE (EXAMPLE 2)

Figure 8.103 plots the evolution of the loss function Φ_n^0. This function decreases as an exponential function and tends to a limit which is equal to 1.9014. The exponential decay of Φ_n^0 is clearly revealed in figure 8.103. Figure 8.104 indicates, how Φ_n^1 evolves with the iterations number. The limiting value is equal to 0.6316.

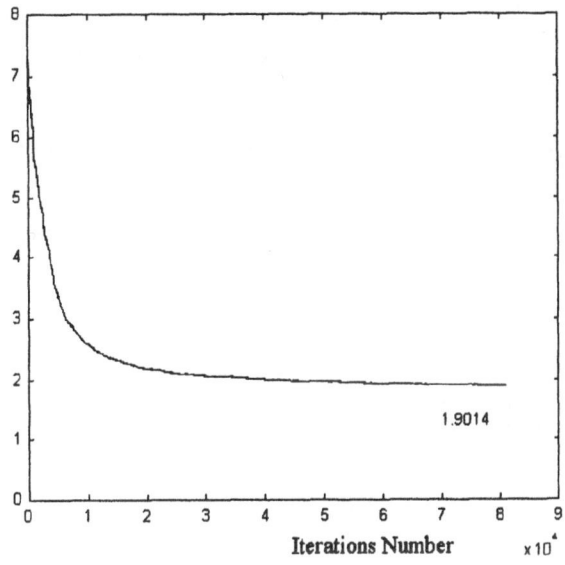

Figure 8.103: Evolution of the loss function Φ_n^0.

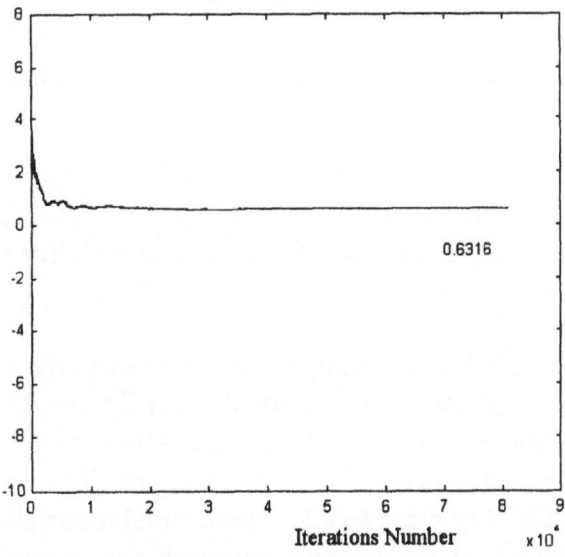

Figure 8.104: Evolution of the loss function Φ_n^1.

The nature of convergence of the algorithm is clearly reflected by figures 8.103 and 8.104. We also observe that the transient behaviour (learning period) of the algorithm is less than 10^4 iterations.

The components of the probability vector p_n are depicted in Figure 8.105. This vector tends to

$$p = [0.1511\ 0.6246\ 0.1239\ 0.1004]^T.$$

Figure 8.104 gives an image of the effect of the control actions on the controlled system.

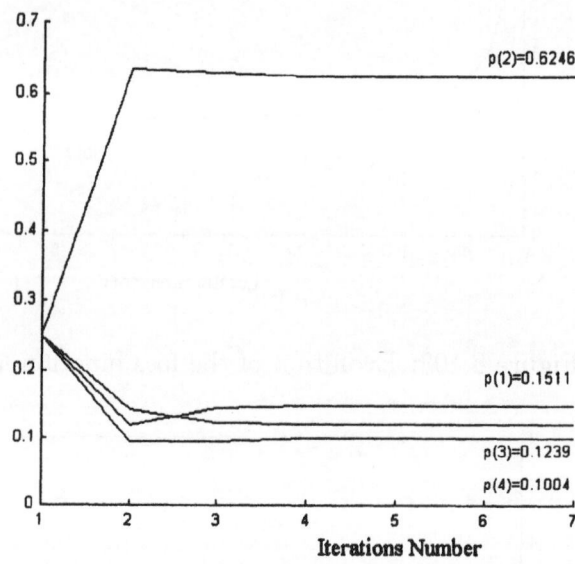

Figure 8.105: Evolution of the state probability vector p_n.

In figures 8.106-8.117 the components of the matrix c_n against the iterations number are plotted. It is evident from figures 8.106-8.117 that the time for convergence is very short. By inspection of these figures, we see that the learning period is less than 500 iterations. The component c_n^{21} tends to a value which is very close to 0.5. These simulations show that the learning algorithm based on the penalty approach can be successfully used for the adaptive control of constrained Markov chains. This algorithm which requires few tuning (design) parameters has been implemented on a PC.

8.6. THE CONSTRAINED CASE (EXAMPLE 2)

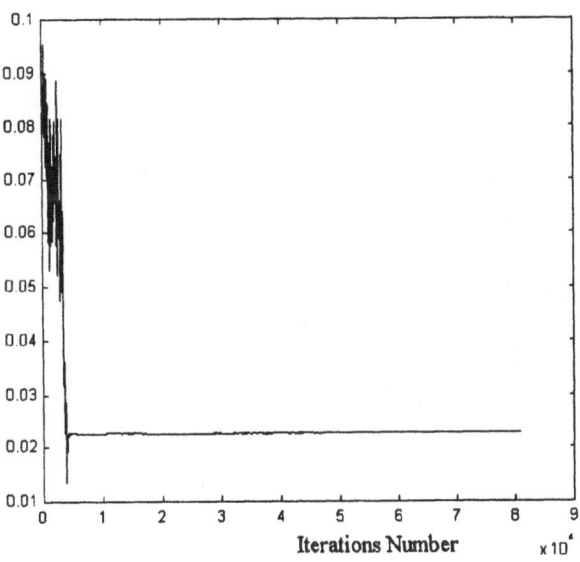

Figure 8.106: Evolution of c_n^{11}.

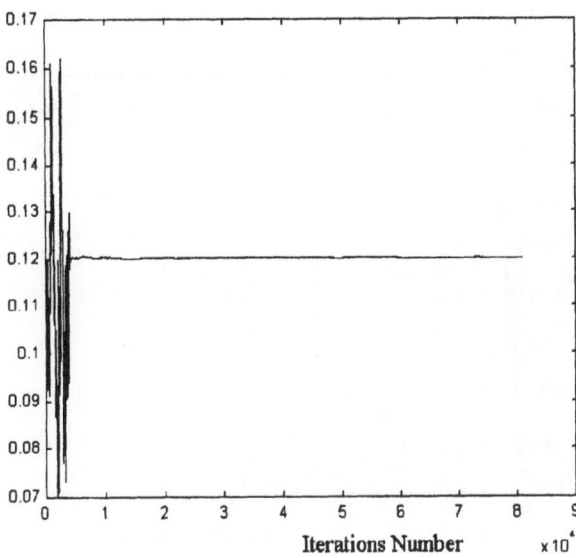

Figure 8.107: Evolution of c_n^{12}.

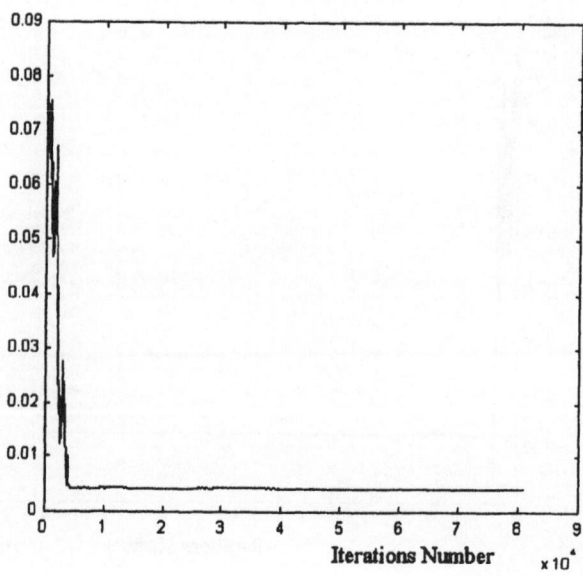

Figure 8.108: Evolution of c_n^{13}.

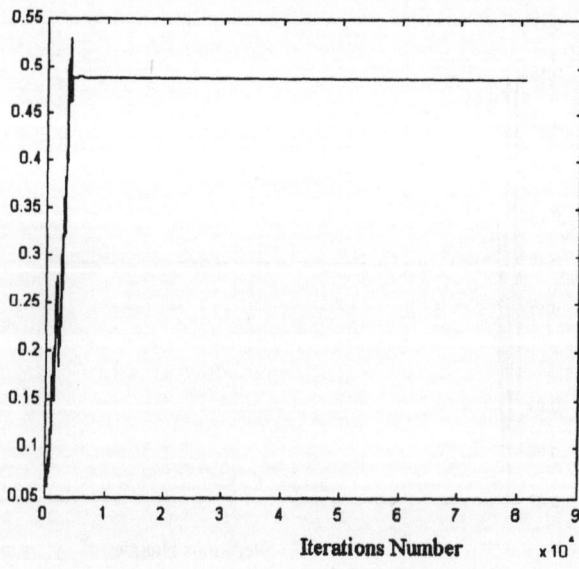

Figure 8.109: Evolution of c_n^{21}.

8.6. THE CONSTRAINED CASE (EXAMPLE 2)

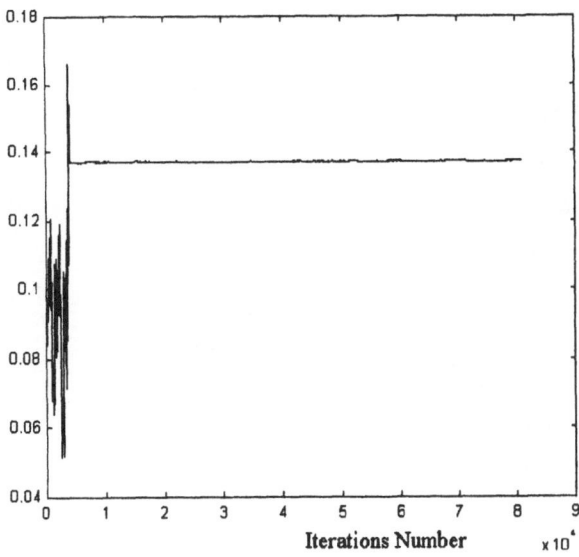

Figure 8.110: Evolution of c_n^{22}.

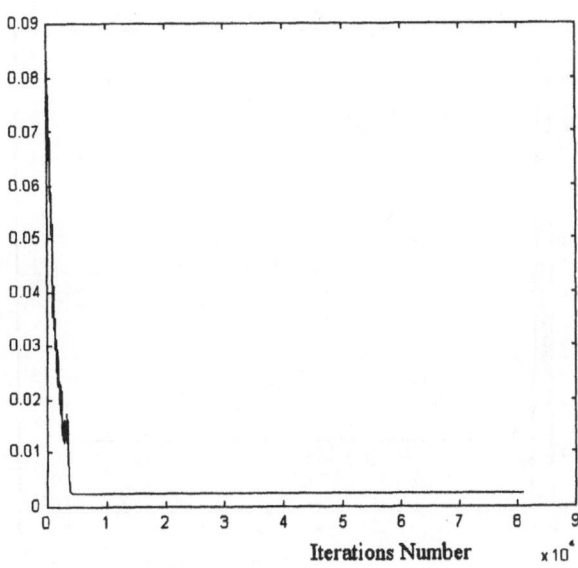

Figure 8.111: Evolution of c_n^{23}.

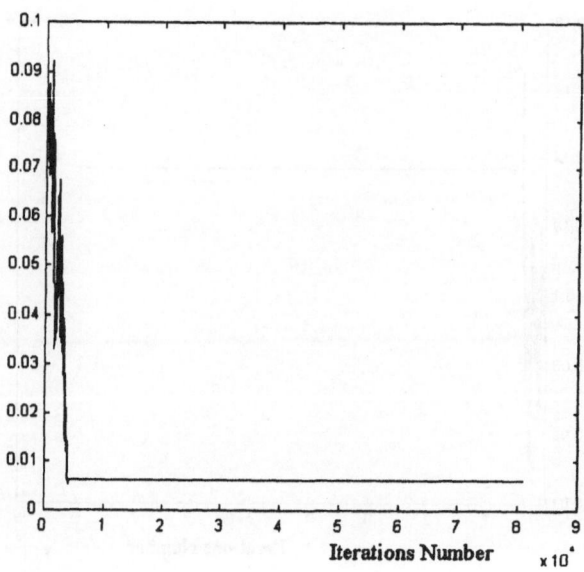

Figure 8.112: Evolution of c_n^{31}.

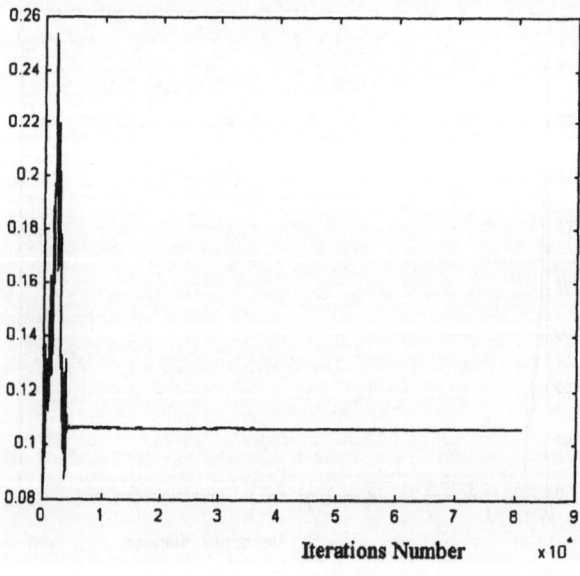

Figure 8.113: Evolution of c_n^{32}.

8.6. THE CONSTRAINED CASE (EXAMPLE 2)

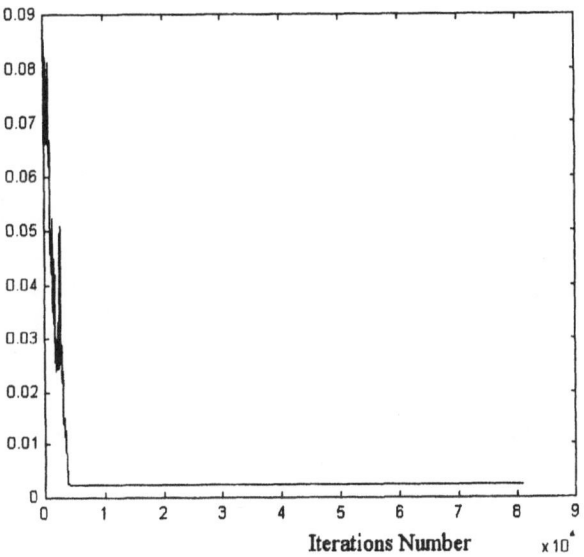

Figure 8.114: Evolution of c_n^{33}.

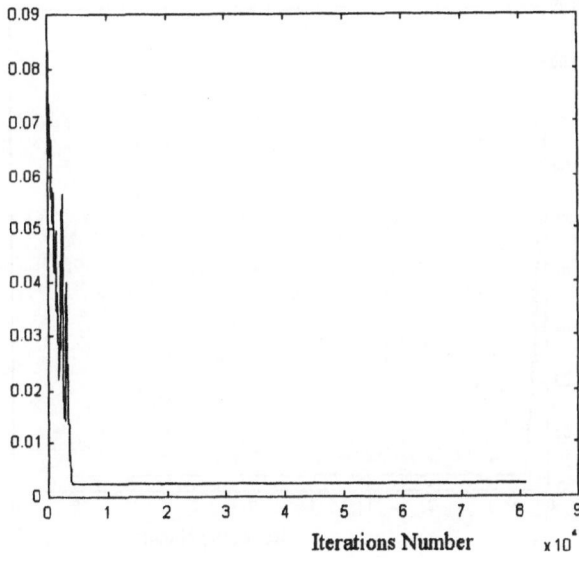

Figure 8.115: Evolution of c_n^{41}.

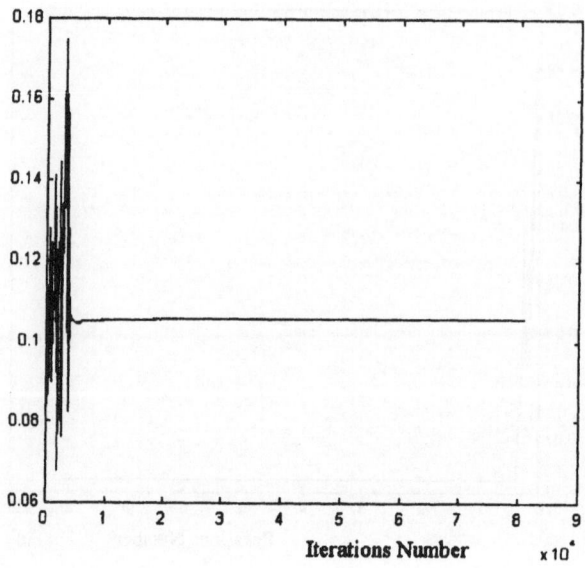

Figure 8.116: Evolution of c_n^{42}.

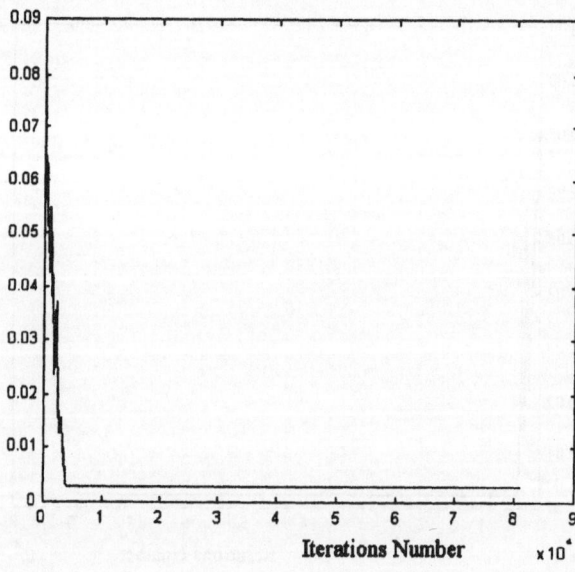

Figure 8.117: Evolution of c_n^{43}.

The nature of convergence of the algorithm is clearly reflected by figures 8.103-8.117. These figures show the feasibility and the effectiveness of the adaptive control algorithm based on the penalty function approach.

The simulation results show that the adaptive scheme achieved the desired control objective. This behaviour was expected from the theoretical results stated in the previous chapters. The above mentioned results can be explained by the adaptive structure of the algorithm. We must also note that this level of performance was achieved even when the recursive control procedure was initiated with the initial probability vector set at arbitrary values.

The convergence of an adaptive scheme is important but the convergence speed is also essential. It depends on the number of operations performed by the algorithm during an iteration as well as by the number of iterations needed for convergence. We can observe that the components of the matrix c_n converge after 10^4 iterations. The behaviour of the adaptive control algorithm based on the penalty function was expected from the theoretical results stated in the previous chapters. Lack of knowledge is overcome by learning. Note that compared to the previous approach (Lagrange multipliers) the computational time associated with this algorithm is greater than the computational time associated with the Lagrange multipliers approach in which the transition matrices do not have to be estimated.

8.7 Conclusions

The Lagrange multipliers and the penalty function approaches can be successfully used to solve the adaptive control problem related to both unconstrained and constrained finite Markov chains. The learning process is relatively fast. The numerical examples presented in this chapter show that the transient phase vary between $5,000$ and $9,000$ samples (iterations). The simulations verified that the properties stated analytically also hold in practice. The Lagrange multipliers approach exhibits more attractive features than the penalty function approach. In fact, the convergence speed which is essential, depending on the number of operations performed by the algorithm during an iteration as well as on the number of iterations needed for convergence. The transition matrices have to be estimated in the adaptive control algorithm based on the penalty function approach. It should also be mentioned that the Lagrange multipliers approach is less sensitive to the selection of the design parameters than the penalty function approach. We conclude that lack of knowledge is overcome by learning.

Appendix A

On Stochastic Processes

In the first part of this appendix we shall review the important definitions and some properties concerning stochastic processes. The theorems and lemmas used in the theoretical developments presented in this book are stated and proved in the second part of this appendix.

A *stochastic process*, $\{x_n, n \in N\}$ is a collection (family) of random variables indexed by a real parameter n and defined on a probability space $(\Omega, \mathcal{F}, \mathbf{P})$ where Ω is the space of elementary events ω, $\mathcal{F}$ the basic σ-algebra and $\mathbf{P}$ the probability measure. A *σ-algebra* $\mathcal{F}$ is a set of subsets of Ω (collection of subsets). $\mathcal{F}(x_n)$ denotes the σ-algebra generated by the set of random variables x_n. The σ-algebra represents the knowledge about the process at time n. A family

$$\mathcal{F} = \{\mathcal{F}_n, n \geq 0\}$$

of σ-algebras satisfy the standard conditions

$$\mathcal{F}_s \leq \mathcal{F}_n \leq \mathcal{F}$$

for $s \leq n$, $\mathcal{F}_0$ is suggested by sets of measure zero of $\mathcal{F}$, and

$$\mathcal{F}_n = \bigcap_{s \geq n} \mathcal{F}_s.$$

A random variable is a real function defined over the a probability space, assuming real values. Hence a sequence of random variables is represented by a collection of real number sequences.

Let $\{x_n\}$ be a sequence of random variables with distribution function $\{F_n\}$ we say that:

Definition 1 $\{x_n\}$ *converges in distribution (law)* to a random variable with distribution function F if the sequence $\{F_n\}$ converges to F. This is written

$$x_n \xrightarrow{law} x.$$

Definition 2 $\{x_n\}$ *converges in probability* to a random variable x if any $\varepsilon, \delta > 0$ there exists $n_0(\varepsilon, \delta)$ such that

$$\forall n > n_0: \ P(|x_n - x| > \varepsilon) < \delta.$$

This is written

$$x_n \xrightarrow{prob} x.$$

Definition 3 $\{x_n\}$ *converges almost surely (with probability 1)* to a random variable x if for any $\varepsilon, \delta > 0$ there exists $n_0(\varepsilon, \delta)$ such that

$$\forall n > n_0 :. \ P(|x_n - x| < \varepsilon) > 1 - \delta$$

or, in other form,

$$\lim_{n \to \infty} P(|x_n - x| \geq \varepsilon) = 1.$$

This is written

$$x_n \xrightarrow{a.s.} x.$$

Definition 4 $\{x_n\}$ *converges in quadratic mean* to a random variable x if

$$\lim_{n \to \infty} \mathbf{E}\left[(x_n - x)^T (x_n - x)\right] = 0.$$

This is written

$$x_n \xrightarrow{q.m.} x.$$

The relationships between these convergence concepts are summarized in the following:
1) convergence in probability implies convergence in law;
2) convergence in quadratic mean implies convergence in probability;
3) convergence almost surely implies convergence in probability.

In general, the converse of these statements is false.

Stochastic processes as martingales have extensive applications in stochastic problems. They arise naturally whenever one needs to consider mathematical expectations with respect to increasing information patterns. They will be used to state several theoretical results concerning the convergence and the convergence rate of learning systems.

Definition 5 *A sequence of random variables $\{x_n\}$ is said to be adapted to a the sequence of increasing σ-algebras $\{\mathcal{F}_n\}$ if x_n is $\mathcal{F}_n$ measurable for every n.*

Definition 6 *A stochastic process $\{x_n\}$ is **a martingale** if*

$$E\{|\,x_n\,|\} \stackrel{a.s.}{<} \infty$$

and for any $n = 1, \ldots$

$$E\{x_{n+1} \mid \mathcal{F}_n\} \stackrel{a.s.}{=} x_n.$$

Definition 7 *A stochastic process $\{x_n\}$ is **a supermartingale** if*

$$E\{x_{n+1} \mid \mathcal{F}_n\} \stackrel{a.s.}{\leq} x_n.$$

Definition 8 *A stochastic process $\{x_n\}$ is **a submartingale** if*

$$E\{x_{n+1} \mid \mathcal{F}_n\} \stackrel{a.s.}{\geq} x_n$$

The following theorems are useful for convergence analysis.

Theorem 1 *(Doob, 1953). Let $\{x_n, \mathcal{F}_n\}$ be a nonnegative supermartingale*

$$x_n \stackrel{a.s.}{\geq} 0, \quad n = 1, \ldots$$

such that

$$\sup_n E\{x_n\} < \infty.$$

Then there exists a nonnegative random variable x (defined on the same probability space) such that

$$E\{x\} < \infty, \qquad x_n \underset{n \to \infty}{\to} x \ (a.s.).$$

Proof. For any nonnegative real variables a and b ($a < b$) let us consider

1) the random sequence $\{\beta_n\}$ where β_n represents the number of times where the process $\{\xi_n\}$ drops below a or rises above b during the time n;

2) the random sequence of times $\{\tau_n\}$ which corresponds to the "first time" when $\{\xi_n\}$ leaves the interval $[a, b]$:

$$\tau_1 := \min\left\{n \mid \xi_n < a, \xi_t \geq a \ \forall t = \overline{1, n-1}\right\},$$

$$\tau_{2k} := \min\left\{n \mid \tau_{2k-1}, \xi_n > b, \xi_t \leq b \ \forall t = \overline{\tau_{2k-1}, n-1}\right\},$$

$$\tau_{2k+1} := \min\left\{n \mid \tau_{2k}, \xi_n < a, \xi_n \geq a \ \forall t = \overline{\tau_{2k}, n-1}\right\}.$$

Consider also the characteristic function

$$\chi_n := \begin{cases} 1, & \tau_{2k-1} < n \leq \tau_{2k}, \\ 0, & \tau_{2k} < n \leq \tau_{2k+1}. \end{cases}$$

then we have

$$\sum_{t=1}^{n-1} \chi_t \left(\xi_{t+1} - \xi_t\right) \geq \begin{cases} (b-a)\beta_n(a,b), & \tau_{2k} < n \leq \tau_{2k+1}, \\ (b-a)\beta_n(a,b) + \xi_n - \xi_{\tau_{2k-1}}, & \tau_{2k-1} < n \leq \tau_{2k}. \end{cases}$$

Notice that

$$\xi_{2k-1} < a$$

we get:

$$(b-a)\beta_n(a,b) \leq \sum_{t=1}^{n} \chi_t \left(\xi_{t+1} - \xi_t\right) + \max\{0, a - \xi_n\}.$$

Taking into account that the random variable χ_t is $\mathcal{F}_t$-measurable, we derive:

$$E\{\chi_t(\xi_{t+1} - \xi_t)\} = E\{E\{\chi_t(\xi_{t+1} - \xi_t) \mid \mathcal{F}_t\}\}$$
$$= E\{\chi_t E\{(\xi_{t+1} - \xi_t) \mid \mathcal{F}_t\}\} \leq 0$$

and

$$E\{\beta_n(a,b)\} \leq (b-a)^{-1} E\{\max\{0, a - \xi_n\}\}.$$

Based on this estimate, this theorem will be proved by contradiction. Let us assume that the limit x does not exist, i.e.,

$$P\left\{\varliminf_{n\to\infty} \xi_n < \varlimsup_{n\to\infty} \xi_n\right\} > 0.$$

Hence, we conclude that there exist numbers a and b such that

$$P\left\{\varliminf_{n\to\infty} \xi_n \leq a < b < \varlimsup_{n\to\infty} \xi_n\right\} > 0$$

and, as a result,

$$P\left\{\lim_{n\to\infty} \beta_n(a,b) = \infty\right\} > 0.$$

In view of the Fatou lemma (Shiryaev) we obtain the contradiction:

$$\infty > (b-a)^{-1} \sup_n E\{\max\{0, a - \xi_n\}\} \geq \varlimsup_{n\to\infty} E\{\beta_n(a,b)\}$$

$$\geq E\left\{\varlimsup_{n\to\infty} \beta_n(a,b)\right\} = \infty.$$

The theorem is proved. ∎

Theorem 2 *(Robbins and Siegmund, 1971). Let $\{\mathcal{F}_n\}$ be a sequence of σ-algebras and x_n, α_n, β_n and ξ_n are $\mathcal{F}_n$-measurable nonnegative random variables such that for all $n = 1, 2, \ldots$ there exists $E\{x_{n+1}/\mathcal{F}_n\}$ and the following inequality is verified*

$$E\{x_{n+1} \mid \mathcal{F}_n\} \leq x_n(1 + \alpha_n) + \beta_n - \xi_n$$

with probability one. Then, for all $\omega \in \Omega_0$ where

$$\Omega_0 := \left\{\omega \in \Omega \mid \sum_{n=1}^{\infty} \alpha_n < \infty, \ \sum_{n=1}^{\infty} \beta_n < \infty\right\}$$

the limit

$$\lim_{n \to \infty} x_n = x^*(\omega)$$

exists, and the sum

$$\sum_{n=1}^{\infty} \xi_n < \infty$$

converges.

Proof. Let us consider the following sequences:

$$\tilde{x}_n := x_n \prod_{t=1}^{n-1} (1 + \alpha_t)^{-1}, \ \tilde{\beta}_n := \beta_n \prod_{t=1}^{n-1} (1 + \alpha_t)^{-1}$$

and

$$\tilde{\xi}_n := \xi_n \prod_{t=1}^{n-1} (1 + \alpha_t)^{-1}.$$

Then, based on these definitions, we derive

$$E\{\tilde{x}_{n+1} \mid \mathcal{F}_n\} = E\{x_{n+1} \mid \mathcal{F}_n\} \prod_{t=1}^{n-1} (1 + \alpha_t)^{-1} \leq \tilde{x}_n + \tilde{\beta}_n - \tilde{\xi}_n.$$

Let us introduce also the following random variables

$$u_n := \tilde{x}_n - \sum_{t=1}^{n-1} \left(\tilde{\beta}_t - \tilde{\xi}_t\right)$$

which satisfy

$$E\{\tilde{u}_{n+1} \mid \mathcal{F}_n\} \overset{a.s.}{\leq} u_n,$$

and the random time τ, defined by

$$\tau = \tau(a) := \inf\left\{n \mid \sum_{t=1}^{n-1} \tilde{\beta}_t \geq a\right\}, \ a = const > 0.$$

Taking into account the following inequalities

$$u_{\tau \wedge n} \geq \sum_{t=1}^{(\tau \wedge n)-1} \tilde{\beta}_t \geq -a$$

where

$$\tau \wedge n = \tau \chi(\tau < n) + u\chi(\tau \geq n),$$

the stopping process

$$u_{\tau \wedge n} := u_\tau \chi(\tau < n) + u_n \chi(\tau \geq n)$$

has a lower bound.

$$E\{u_{\tau \wedge (n+1)} \mid \mathcal{F}_n\} = u_\tau \chi(\tau \leq n) + \chi(\tau < n) E\{u_{n+1} \mid \mathcal{F}_n\}$$

$$\stackrel{a.s.}{\leq} u_\tau \chi(\tau \leq n) + u_n \chi(\tau > n) = u_{\tau \wedge n},$$

we conclude that $(u_{\tau \wedge n}, \mathcal{F}_n)$ is a supermantingale which has a lower bound. Hence, in view of Doob's theorem, we conclude that

$$\lim_{n \to \infty} u_{\tau \wedge n} \stackrel{a.s.}{=} u^*(\omega).$$

Taking into account that $\beta_n \geq 0$, it follows that for any $\omega \in \Omega_0$ there exists a real value a such that

$$\tau = \tau(a) = \infty.$$

As a result, we obtain

$$u_{\tau \wedge n} = u_{\infty \wedge u} = u_n,$$

and for any $\omega \in \Omega_0$

$$u_n \xrightarrow[n \to \infty]{} u^*(\omega).$$

In other words, we get

$$\tilde{x}_n + \sum_{t=1}^{n-1} \tilde{\xi}_t \xrightarrow[n \to \infty]{} u^*(\omega) + \sum_{t=1}^{\infty} \tilde{\beta}_t.$$

The sequence $\{S_n\}$ defined by

$$S_n := \sum_{t=1}^{n} \tilde{\xi}_t,$$

is monotonic. We conclude that for almost all $\omega \in \Omega_0$ this sequence has a limit $S_\infty(\omega)$. Hence, the sequence $\{\tilde{x}_n\}$ has also a limit $\tilde{x}_\infty$ and, the following relations are valid

$$x_n \xrightarrow[n\to\infty]{} \tilde{x}_\infty(\omega) \prod_{t=1}^\infty (1+\alpha_t) := x^*(\omega)$$

and

$$\sum_{t=1}^{n-1} \xi_t = \sum_{t=1}^{n-1} \tilde{\xi}_t \prod_{m=1}^\infty (1+\alpha_m) \leq S_\infty \prod_{t=1}^\infty (1+\alpha_t) < \infty.$$

The theorem is proved. ■

Lemma 1 *Let $\{\mathcal{F}_n\}$ be a sequence of σ-algebras and η_n, and θ_n, are $\mathcal{F}_n$-measurable nonnegative random variables such that*

1.

$$\sum_{t=1}^\infty \mathbf{E}(\theta_t) < \infty,$$

2.

$$\mathbf{E}(\eta_{n+1} \mid \mathcal{F}_n) \stackrel{a.s.}{<} \eta_n + \theta_n, n = 1, 2, \ldots$$

Then

$$\lim_{n\to\infty} \eta_n \stackrel{a.s.}{=} \eta.$$

with probability 1.

Proof. In view of assumption 1 and the Fatou lemma (see Shiryaev), it follows

$$\sum_{n=1}^\infty \theta_n < \infty.$$

Applying the Robbins-Siegmund theorem for

$$\alpha_n = 0, \ \beta_n = \theta_n, \ \xi_n = 0$$

we derive the assertion of this lemma. The lemma is proved. ■

Lemma 2 *Let $\{\mathcal{F}_n\}$ be a sequence of σ-algebras and η_n, θ_n, λ_n, and ν_n are $\mathcal{F}_n$-measurable nonnegative random variables such that*

1.
$$\mathbf{E}(\eta_n) < \infty,$$

2.
$$\sum_{n=1}^{\infty} \mathbf{E}(\theta_n) < \infty,$$

3.
$$\sum_{n=1}^{\infty} \lambda_n \stackrel{a.s.}{=} \infty, \quad \sum_{n=1}^{\infty} \nu_n \stackrel{a.s.}{<} \infty,$$

4.
$$\mathbf{E}(\eta_{n+1} \mid \mathcal{F}_n) \stackrel{a.s.}{\leq} (1 - \lambda_{n+1} + \nu_{n+1})\eta_n + \theta_n.$$

Then,
$$\lim_{n \to \infty} \eta_n \stackrel{a.s.}{=} 0.$$

Proof. In view of the assumptions of the previous theorem and the Robbins-Siegmund theorem (Robbins and Siegmund, 1971), it follows that

$$\eta_n \stackrel{a.s.}{\underset{n\to\infty}{\to}} \eta^*$$

and

$$\sum_{n=1}^{\infty} \lambda_{n+1}\eta_n \stackrel{a.s.}{<} \infty.$$

As

$$\sum_{n=1}^{\infty} \lambda_n \stackrel{a.s.}{=} \infty,$$

we conclude that there exists a subsequence η_{n_k} which tends to zero with probability 1. Hence $\eta^* \stackrel{a.s.}{=} 0$. The lemma is proved. ∎

Lemma 3 *Let $\{v_n\}$ be a sequence of random variables adapted to $\{\mathcal{F}_n\}$ of σ-algebras $\mathcal{F}_n$,*

$$\{\mathcal{F}_n\} \subseteq \mathcal{F}_{n+1} \, (n = 1, 2, ...),$$

such that the random variables $\mathbf{E}(v_n \mid \mathcal{F}_{n-1})$ exist and, for some positive monotonically decreasing sequence $\{a_t\}$ the following series converges:

$$\sum_{t=1}^{\infty} a_t^{-2} \eta_t \mathbf{E}\left\{(v_t - \mathbf{E}(v_t \mid \mathcal{F}_{t-1}))^2 \mid \mathcal{F}_{t-1}\right\} \stackrel{a.s.}{<} \infty. \tag{1.1}$$

Then

$$\lim_{n \to \infty} \left(\frac{1}{a_n} \sum_{t=1}^n v_t - \frac{1}{a_n} \sum_{t=1}^n \mathbf{E}(v_t \mid \mathcal{F}_{t-1})\right) = 0$$

with probability 1.

Proof. In view of the Kronecker lemma we conclude that the random sequence with elements are given by

$$\frac{1}{a_n}\sum_{t=1}^{n}(v_t - \mathbf{E}\{v_t \mid \mathcal{F}_{t-1}\})$$

tends to zero for the random events $\omega \in \Omega$ for which the following random sequence

$$S_n(\omega) := \sum_{t=1}^{n} a_t^{-1}\{v_t - \mathbf{E}(v_t \mid \mathcal{F}_{t-1})\}$$

converges. But in view of the Robbins-Siegmund Theorem and the assumptions of this lemma we conclude that this limit exists for almost all $\omega \in \Omega$. Indeed,

$$\mathbf{E}\{S_n^2 \mid \mathcal{F}_{n-1}\} \stackrel{a.s.}{=} S_{n-1}^2 + a_n^{-2}\mathbf{E}\{v_n - \mathbf{E}\{(v_n \mid \mathcal{F}_{n-1})\}^2 \mid \mathcal{F}_{n-1}\}$$

and applying the Robbins-Siegmund theorem, we obtain the result. The lemma is proved. ∎

Corollary 1 *For*

$$v_n = \chi_n = \{0; 1\} \quad \text{and} \quad a_n = n \quad (n = 1, 2, ...)$$

and without considering assumption (1.1), it follows:

$$\lim_{n \to \infty} \left(\frac{1}{n}\sum_{t=1}^{n}\chi_t - \frac{1}{n}\sum_{t=1}^{n}\mathbf{P}\{\chi_t = 1 \mid \mathcal{F}_{t-1}\}\right) = 0.$$

with probability 1.

Lemma 4 *Let $\{v_n\}$ be a sequence of random variables adapted to $\{\mathcal{F}_n\}$ of σ-algebras $\mathcal{F}_n$,*

$$\{\mathcal{F}_n\} \subseteq \mathcal{F}_{n+1} \quad (n = 1, 2, ...),$$

such that the random variables $\mathbf{E}(v_n \mid \mathcal{F}_{n-1})$ exist and, the following series

$$\sum_{t=1}^{\infty} t^{-2}\eta_t \mathbf{E}\{(v_t - \mathbf{E}(v_t \mid \mathcal{F}_{t-1}))^2 \mid \mathcal{F}_{t-1}\} \stackrel{a.s.}{<} \infty$$

converges with probability 1, where the deterministic sequence $\{\eta_t\}$ satisfies

$$\lim_{n \to \infty} \left(\frac{\eta_n}{\eta_{n-1}} - 1\right) n := \lambda < 2.$$

Then

$$\eta_n S_n^2 \stackrel{a.s.}{\underset{n\to\infty}{\to}} 0$$

where

$$S_n := \frac{1}{n} \sum_{t=1}^{n} \{v_t - \mathbf{E}(v_t \mid \mathcal{F}_{t-1})\},$$

or, in other words,

$$S_n \stackrel{a.s.}{=} o(\eta_n^{-\frac{1}{2}}).$$

Proof. For large enough n, we have

$$\mathbf{E}\left\{S_n^2 \mid \mathcal{F}_{n-1}\right\} \stackrel{a.s.}{=} S_{n-1}^2 \left(1 - \frac{1}{n}\right)^2$$

$$+n^{-2}\mathbf{E}\left\{(v_n - \mathbf{E}(v_n/\mathcal{F}_{n-1}))^2 \mid \mathcal{F}_{n-1}\right\} = S_{n-1}^2 \left(1 - \frac{2 + o(1)}{n}\right)^2$$

$$+n^{-2}\mathbf{E}\left\{(v_n - \mathbf{E}(v_n \mid \mathcal{F}_{n-1}))^2 \mid \mathcal{F}_{n-1}\right\}.$$

For $W_n := \eta_n S_n^2$, the previous relation leads to

$$\mathbf{E}\{W_n \mid \mathcal{F}_{n-1}\} \stackrel{a.s.}{=} W_{n-1}\left(1 - \frac{2 - \lambda + o(1)}{n}\right)$$

$$+\eta_n n^{-2}\mathbf{E}\left\{(v_n - \mathbf{E}(v_n \mid \mathcal{F}_{n-1}))^2 \mid \mathcal{F}_{n-1}\right\}.$$

In view of the Robbins-Siegmund theorem and the assumptions of this lemma, it follows:

$$W_n \stackrel{a.s.}{\underset{n\to\infty}{\to}} 0.$$

The lemma is proved. ∎

Corollary 2 *If in the assumptions of this lemma*

$$\eta_n = O^*(n^\varkappa),$$

then we have

$$\lambda = \varkappa,$$

and for $\varkappa < 2$ we obtain:

$$S_n \stackrel{a.s.}{=} o(n^{-\frac{\varkappa}{2}})$$

Lemma 5 *Let $\{u_n\}$ be a sequence of nonnegative random variables u_n measurable with respect to the σ-algebra $\mathcal{F}_n$, for all $n = 1, 2...,$. If the random variables*

$$\mathbf{E}(u_n \mid \mathcal{F}_n) \ \forall \ n = 1, \ 2, ...$$

exist and the following inequality

$$\mathbf{E}(u_{n+1} \mid \mathcal{F}_n) \leq u_n(1 - \alpha_n) + \beta_n$$

holds with probability 1, where $\{\alpha_n\}$ and $\{\beta_n\}$ are sequences of deterministic variables such that

$$\alpha_n \in (0, 1], \quad \beta_n \geq 0, \sum_{n=1}^{\infty} \alpha_n = \infty,$$

and for some nonnegative sequence

$$\{\nu_n\} \ (\nu_n > 0, \ n = 1, 2, ...)$$

the following series converges

$$\sum_{n=1}^{\infty} \beta_n \nu_n < \infty$$

and, the limit

$$\lim_{n \to \infty} \frac{\nu_{n+1} - \nu_n}{\alpha_n \nu_n} := \mu < 1$$

exists, then

$$u_n = o_\omega(\frac{1}{\nu_n}) \overset{a.s.}{\to} 0$$

when $\nu_n \to \infty$.

Proof. Let $\tilde{u}_n$ be the sequence defined as

$$\tilde{u}_n := u_n \nu_n.$$

Then, based on the assumptions of this lemma, we derive

$$\mathbf{E}(\tilde{u}_{n+1} \mid \mathcal{F}_n) \overset{a.s.}{\leq} \tilde{u}_n(1 - \alpha_n)(\frac{\nu_{n+1}}{\nu_n}) + \nu_{n+1}\beta_n$$

$$= \tilde{u}_n(1 - \alpha_n)(\frac{\nu_{n+1} - \nu_n}{\nu_n} - 1) + \nu_{n+1}\beta_n$$

and

$$\mathbf{E}(\tilde{u}_{n+1} \mid \mathcal{F}_n) \overset{a.s.}{\leq} \tilde{u}_n \left[1 - \alpha_n(1 - \mu + o(1))\right] + \nu_{n+1}\beta_n.$$

Then, from this inequality and Robbins-Siegmund theorem (Robbins and Siegmund, 1971), we obtain

$$\nu_n \tilde{u}_n \overset{a.s.}{\to} 0,$$

which is equivalent to

$$u_n = o_\omega\left(\frac{1}{\nu_n}\right) \overset{a.s.}{\to} 0.$$

The lemma is proved. ∎

The following lemmas turn out to be useful for deterministic numerical sequences analysis.

Lemma 6 *(the matrix version of the Abel's identity). For any matrices*

$$A_t \in R^{m \times k}, \quad B_t \in R^{k \times l}$$

and any integer numbers n_0 and n ($n_0 \leq n$) the following identity

$$\sum_{t=n_0}^{n} A_t B_t = A_n \sum_{t=n_0}^{n} B_t - \sum_{t=n_0}^{n} [A_t - A_{t-1}] \sum_{s=n_0}^{t-1} B_s$$

holds.

Proof. For $n = n_0$ we obtain:

$$A_{n_0} B_{n_0} = A_{n_0} B_{n_0} - [A_{n_0} - A_{n_0-1}] \sum_{s=n_0}^{n_0-1} B_s = A_{n_0} B_{n_0}.$$

The sum

$$\sum_{s=n_0}^{n_0-1} B_s$$

in the previous equality is zero by virtue of the fact that the upper limit of this sum is less than the lower limit.

We use induction. We note that the identity (Abel's identity) is true for n_0. We assume that it is true for n and prove that it is true for $n+1$:

$$\sum_{t=n_0}^{n+1} A_t B_t = \sum_{t=n_0}^{n} A_t B_t + A_{n+1} B_{n+1}$$

$$= A_n \sum_{t=n_0}^{n} B_t - \sum_{t=n_0}^{n} [A_t - A_{t-1}] \sum_{s=n_0}^{t-1} B_s + A_{n+1} B_{n+1}$$

$$= \left(A_{n+1} \sum_{t=n_0}^{n} B_t + A_{n+1} B_{n+1} \right)$$

$$- \left((A_{n+1} - A_n) \sum_{t=n_0}^{n} B_t + \sum_{t=n_0}^{n} [A_t - A_{t-1}] \sum_{s=n_0}^{t-1} B_s \right)$$

$$= A_{n+1} \sum_{t=n_0}^{n+1} B_t - \sum_{t=n_0}^{n+1} [A_t - A_{t-1}] \sum_{s=n_0}^{t-1} B_s$$

The identity (Abel's identity) is proved. ∎

Lemma 7 *(Nazin-Poznyak, 1986).* Let $\{\gamma_n\}$ and $\{g_n\}$ be deterministic sequences satisfying
$$0 < \gamma_{n+1} \leq \gamma_n \quad (n = 1, 2, ...)$$

and
$$G := \sup_n \left| \sum_{t=1}^{n} g_t \right| < \infty.$$

Then, for any $n = 1, 2, ...$ the following inequality
$$\left| \sum_{t=1}^{n} g_t \gamma_t^{-1} \right| \leq 2G \gamma_n^{-1}$$

holds and, as a result, for nonnegative sequence $\{h_n\}$ satisfying
$$\lim_{n \to \infty} h_n \gamma_n = \infty,$$

we have
$$\lim_{n \to \infty} \frac{1}{h_n} \sum_{t=1}^{n} \frac{g_t}{\gamma_t} = 0.$$

Proof. Using Abel identity

$$\sum_{t=1}^{n} \alpha_t \beta_t = \alpha_n \sum_{t=1}^{n} \beta_t - \sum_{t=1}^{n} (\alpha_t - \alpha_{t-1}) \sum_{\tau=1}^{t-1} \beta_\tau, \quad \alpha_0 = 0$$

for
$$\alpha_t := \frac{1}{\gamma_t} \text{ and } \beta_t := g_t \quad (t = 1, 2, ...)$$

we derive:
$$\left|\sum_{t=1}^{n}\frac{g_t}{\gamma_t}\right| = \left|\frac{1}{\gamma_t}\sum_{t=1}^{n}g_t - \sum_{t=1}^{n}\left(\frac{1}{\gamma_t}-\frac{1}{\gamma_{t-1}}\right)\sum_{\tau=1}^{t-1}g_\tau\right|$$
$$\leq \frac{1}{\gamma_t}\sup_{n}\left|\sum_{t=1}^{n}g_t\right| + \sum_{t=1}^{n}\left(\frac{1}{\gamma_t}-\frac{1}{\gamma_{t-1}}\right)\sup_{m}\left|\sum_{\tau=1}^{m}g_\tau\right|$$
$$= \frac{2}{\gamma_n}\left(\sup_{n}\left|\sum_{t=1}^{n}g_t\right|\right),$$

for any $n = 1, 2, \ldots$. The lemma is proved. ∎

Lemma 8 *(Toeplitz) Let $\{a_n\}$ be a sequence of nonnegative variables such that for all $n \geq n_0$*

$$0 < b_n := \sum_{t=1}^{n}a_t \underset{n\to\infty}{\to} \infty,$$

and $\{x_n\}$ is the sequence which converges to x^*, i.e.,

$$x_n \underset{n\to\infty}{\to} x^*.$$

Then,

$$b_n^{-1}\sum_{t=1}^{n}a_t x_t \underset{n\to\infty}{\to} x^*.$$

Proof. Let us select $\varepsilon > 0$ and n_0 such that for all $n \geq n_0$

$$|x_n - x^*| \leq \varepsilon,$$

then, we conclude

$$\left|b_n^{-1}\sum_{t=1}^{n}a_t x_t - x^*\right| \leq b_n^{-1}\sum_{t=1}^{n}a_t|x_t - x^*|$$
$$\leq b_n^{-1}\sum_{t=1}^{n_0-1}a_t|x_t - x^*| + b_n^{-1}\sum_{t=n_0}^{n}a_t\varepsilon$$
$$\leq b_n^{-1}n_0\max_{t=\overline{1,n_0}}a_t|x_t - x^*| + \varepsilon.$$

The last inequality is valid for any $\varepsilon > 0$ and in view of the property

$$b_n \underset{n\to\infty}{\to} \infty,$$

we get the proof of this lemma. The lemma is proved. ∎

Corollary 3 *If*

$$x_n \underset{n\to\infty}{\to} x^*$$

then

$$n^{-1}\sum_{t=1}^{n}x_t \underset{n\to\infty}{\to} x^*.$$

Proof. This result directly follows from this lemma for
$$a_n = 1 \quad (n = 1, 2, ...).$$

The corollary is proved. ∎

Lemma 9 *(Kronecker) Let $\{b_n\}$ be a sequence of non-negative variables such that*
$$b_n \underset{n\to\infty}{\to} \infty, \quad b_{n+1} \geq b_n \ \forall n \geq 1,$$
and $\{x_n\}$ be a sequence of variables such that the sum
$$\sum_{n=1}^{\infty} x_n$$
converges. Then
$$b_n^{-1} \sum_{t=1}^{n} b_t x_t \underset{n\to\infty}{\to} 0.$$

Proof. In view of Abel's identity, we derive:
$$b_b^{-1} \sum_{t=1}^{n} b_t x_t = S_t + b_n^{-1} \sum_{t=0}^{n-1} (b_t - b_{t+1}) S_t = S_t - b_n^{-1} \sum_{t=1}^{n-1} a_t S_t,$$
where
$$S_t := \sum_{m=1}^{t} b x_m, \ S_0 := b_0 = 0 \text{ and } a_t := b_{t+1} - b_t \geq 0.$$
Observe that
$$S_t \underset{n\to\infty}{\to} \sum_{t=1}^{\infty} x_t := S,$$
we have
$$b_n^{-1} \sum_{t=1}^{n-1} a_t S_t \underset{n\to\infty}{\to} S,$$
and
$$S_n - b_n^{-1} \sum_{t=1}^{n-1} a_t S_t \underset{n\to\infty}{\to} 0.$$

The lemma is proved. ∎

The Matlab^{TM} programs are presented in the following appendix.

Appendix B

Matlab Programs

This appendix contains a set of Matlab programs ready for use. It is not necessary to use any Toolbox except the Optimization Tool box for the implementation of the lp command (Linear Programming). The first program concerns the mechanization of the adaptive control algorithms based on the Lagrange multipliers approach. The second program correspond to the Matlab coding of the self-learning control algorithms developed on the basis of the penalty function approach. Both unconstrained and constrained controlled Markov chains are dealt with. The third program is dedicated to the linear programming problem. The source codes include comments and declarations. In the coding flexibility and readability have been emphasized rather than compactness and speed.

Lagrange multipliers approach

% Program LAGR.M

% Adaptive control algorithm based on the Lagrange

% multipliers approach

clear all

% Load probabilities and constraints

load pis17; load rest33;

% K number of states and N number of actions

```
[K, N] = size(v0);
% M number of constraints
M = 1;
% Iterations number
it = 10000;
% Number of iterations to be displayed
disp = 200;
% Initial values and optimal parameters
%==========================
n0 = it;
l0p = 0.3;
d0 = 0.5;
gl = 0.1;
e0 = 1/600;
% Optimal parameters
ep = 1/6;
de = ep;
la 0;
ga = 2/3;
%==========================
Initial conditions for variables
%==========================
% Evolution of each component of the matrix cil
for i = 1:K
    for j = 1:N
```

```
            eval(['cil' num2str(i) num2str(j) ' = zeros(1,it);'])
        end
    end
action = zeros(1,it);
cil = 1/(N*K)*ones(K,N);
lj=0.1*(rand(K,1)-0.5);
var = 1;
if M > 0
    lam = - rand(1,M);
else
    lam = 0;
    eta = 0;
end
for i = 1:M+1
    eval(['pi' num2str(i-1) ' = 0;'])
    eval(['phi' num2str(i-1) ' = zeros(1,it);'])
end
% Evaluation of Sigma
Sig0 = 0.5 + max(max(abs(v0)));
for i = 1:M;
    eval(['sig(' num2str(i) ') = 0.5+max(max(abs(v' num2str(i) ')));'])
end
if M == 0, sig = 0; end
gc = g0;
lnp = l0p;
```

```
dn = d0;

en = e0;

nf = it;

% The algorithm

%==========================

tic

n = 1;

while n < it

% Use corollaries 1 & 2 for the selection of the parameters

if n > n0

    gc = g0/((n - n0)^ga);

    lnp =lop*(1 + ((n - n0)^la)*log(n-n0));

    dn = d0/((n - n0)^de);

    en = e0/(1 + ((n - n0)^ep)*log(n-n0));

end

pn = sum(cil')';

% Estimation of alpha

w = rand;

[alfa] = coninp(pn,w);

dal = 1/pn(alfa)*cil(alfa,:);

% Estimation of beta

w = rand;

[beta] = coninp(dal',w);

pij = eval(['pij' num2str(beta)]);

action(n)=beta;
```

MATLAB PROGRAMS

```
% Estimation of gamma
w = rand;
[gama] = coninp(pij(alfa),:);
% Estimation of eta
eta0 = (rand - 0.5)*var + v0(alfa,beta);
for i = 1:M
    eval(['eta' num2str(i) ' ') = (rand - 0.5)*var +
    v' num2str(i) '(alfa,beta);'])
end
xi = eta0 - (lj(alfa) - lj(gama) + lam*eta' + dn*cil(alfa,beta);
an = (2*(sig0 + (2 + sum(sig))*lnp)/en +N*K/(N*K - 1)*dn)^(- 1);
bn = (sig0 + (2 + sum(sig))*lnp)*an;
zeta = (an*xi + bn)/cil(alfa,beta);
x = zeros(K,1);
x(gama) = 1;
psi0 = pn -dn*lj - x;
psim = eta - dn*lam;
lnp = l0p*(1 + ((n + 1)^la)*log(n + 1));
lj = (lj +gl*psi0);
for i = 1:K
    if (lj(i) > lnp)
        lj(i) = lnp;
    end
    if (lj(i) < - lnp)
        lj(i) = - lnp;
```

```
        end
    end
    lam = lam + gl*psim;
    for i = 1:M
        if (lam(i) > lnp)
            lam(i) = lnp;
        end
        if (lam(i) < 0)
            lam(i) = 0;
        end
    end
end
% Compute the matrix C
for i = 1:K
    for j = 1:N
        if (i == alfa) & (j == beta)
            cil(alfa,beta) = cil(alfa,beta) +gc*(1 - cil(alfa,beta) - zeta);
            eval(['cil' num2str(i) num2str(j) '(n) = cil(alfa,beta);' ])
        else
            cil(i,j) = cil(i,j) +gc*(cil(i,j) - zeta/((N*K) - 1));
            eval(['cil' num2str(i) num2str(j) '(n) = cil(i,j);' ])
        end
    end
end
% Evaluation of the loss function
pi0 = (1 - 1/n)*pi0 + 1/n*eta0;
```

```
        phi0(n) = pi0;
        for i = 1:M
            eval(['pi' num2str(i) '= (1 - 1/n)*pi' num2str(i) '+ 1/n*eta(i);' ])
            eval(['phi' num2str(i) '(n) = pi' num2str(i) ';' ])
        end
        n = n + 1;
    end
toc
```

Penalty function approach

```
% Program PENAL.M
% Adaptive control algorithm based on the Penalty
% Function approach
clear all
% Load probabilities and constraints
load pis17; load rest34;
% K number of states and N number of actions
[K, N] = size(v0);
% M number of constraints
M = 1;
% Iterations number
it = 10000;
var = 1;
% Optimal parameters
de = 1/6;
```

```
mu = 1/6;

ga = 2/3;

% Initial values

disp = 200;

n0 = it;

d0 = 0.5;

mu0 = 4;

gc0 = 0.006;

gc = gc0;

dn = d0;

mun = mu0;

nf = it;

% Evaluation of sigma

sig0 = 0.5 + max(max(abs(v0)));

for i = 1:M

    eval(['sig' num2str(i) ' ) = 0.5 + max(max(abs(v' num2str(i) ')));' ])

end

if M == 0, sig = 0; eta = 0; end

% Initial conditions

%==========================

for i = 1:K

    for j = 1:N

        eval(['cil' num2str(i) num2str(j) ' = zeros(1,it);' ])

    end

end
```

MATLAB PROGRAMS

```
action = zeros(1,it);
cil = 1/(N*K)*ones(K,N); % For cil
for i = 1:M + 1
    eval(['pi' num2str(i - 1) ' = 0;' ])
    eval(['phi' num2str(i - 1) ' = zeros(1,it);' ])
end
% Initial conditions for pij
for l = 1:N
    for i = 1:K
        d = rand(1,K);
        d = d/sum(d);
        eval(['pie' num2str(l) '(i,:) = d;' ]);
    end
    eval(['S' num2str(l) ' = zeros(K,K);' ]);
end
for m = 1:M
    eval(['teta' num2str(m) ' = zeros(K,N);' ]);
end
%==========================
tic
n = 1;
while n < it
% Use corollaries 1 & 2 for the selection of the parameters
if n > n0
    gc = gc0/((n - n0)^ga);
```

```
        mun = mu0/(1 + (n - n0)^mu*log(n - n0));
        en = e0/((n - n0)^ep);
        dn = d0/(1 + ((n - n0)^de)*log(n - n0));
end
pn = sum(cil')';
% Estimation of alfa
w = rand;
[alfa] =coninp(pn,w);
dal = cil(alfa,:)/pn(alfa);
% Estimation of beta
w = rand;
[beta] =coninp(dal,w);
pij = eval(['pij' num2str(beta)]);
action(n) = beta;
% Estimation of gamma
w = rand;
[gama] =coninp(pij(alfa,:)',w);
% Estimation of eta
eta0 = (rand - 0.5)*var + v0(alfa,beta);
for i = 1:M
     eval(['eta(i) = (rand - 0.5)*var + v' num2str(i) '(alfa,beta);' ])
end
ss = zeros(1,K);
for j = 1:K
     for i = 1:K
```

```
            for l = 1:N
                ss(j) = ss(j) + eval(['pie' num2str(l) '(i,j)*cil(i,l)']);
            end
        end
    end
    xx = zeros(1,K);
    eval(['xx = - pie' num2str(beta) '(alfa,:);']);
    xx(alfa) = 1 + xx(alfa);
    if M > 0
        sss = zeros(1,M);
        for m = 1:M
            for i = 1:K
                for l = 1:N
                    eval(['sss(m) = sss(m) + teta' num2str(m) '(i,l)*cil(i,l);']);
                end
            end
            if sss(m) < 0
                sss(m) = 0;
            end
        end
    else
        sss = 0;
    end
    xi = mun*eta0 + (pn' - ss)*xx' + sss*eta' + dn*cil(alfa,beta);
    an = (2*(mun*sig0 + 4 + sum(sig.^2))/en + N*K/(N*K - 1)*dn)^(-1);
```

```
bn = an *(mun*sig0 + 4 + sum(sig.^2));
zeta = (an*xi + bn)/cil(alfa,beta);
eval(['S' num2str(beta) '(alfa,gama) = S' num2str(beta) '(alfa,gama) +
    1;']);
ssinv = eval(['sum(S' num2str(beta) '(alfa,:))';]);
% Compute the matrix C
for j = 1:N
  for i = 1:K
    if( i == alfa) & (j == beta)
      cil(alfa,beta) = cil(alfa,beta) + gc*(1 - cil(alfa,beta) - zeta);
      eval(['cil' num2str(i) num2str(j) '(n) = cil(alfa,beta);' ])
    else
      cil(i,j) = cil(i,j) - gc*(cil(i,j) - zeta/((N*K) - 1));
      eval(['cil' num2str(i) num2str(j) '(n) = cil(i,j);' ])
    end
  end
end
% Estimation of pij
xx = zeros(1,K);
xx(gama) = 1;
eval(['pie' num2str(beta) '(alfa,:) = pie' num2str(beta) '(alfa,:)
- (pie' num2str(beta) '(alfa,:) - xx)/ssinv;']);
% Estimation of teta
for m = 1:M
  eval(['teta' num2str(m) '(alfa,beta) = teta' num2str(m)
```

```
'(alfa,beta) (teta' num2str(m) '(alfa,beta) - eta(m))/ssinv;']);
end
pi0 = (1 - 1/n)*pi0 + 1/n*eta0;
phi0(n) = pi0;
for m = 1:M
    eval(['pi' num2str(i) ' = (1 - 1/n)*pi' num2str(i) ' + 1/n *eta(i);' ])
    eval(['phi' num2str(i) '(n) = pi' num2str(i) ';' ])
end
end
toc
```

Linear Programming Problem

```
% LPMC.M
% Linear Programming Problem
clear al
% K number of states and N number of actions
[K, N] = size(v0);
% M number of constraints
M =1;
% Load probabilities and constraints
load pis17; load rest 34
[K, N] = size(v0);
% Transform the problem into vector form
%========================
% Transform the matrix v0 into a vector form
```

```
f = zeros(K*N,1);
for i = 1:K*N
   iy = floor((i - 1)/N + 1;
   ix = rem(i - 1,N) + 1;
   f(i) = v0(iy,ix);
end
b = zeros(1 + K + M + K*N,1);
b(1) = 1;
% Construct the matrix A using the probabilities
% and the constraints
%===========================
A = zeros(1 + K + M + K*N);
% First row of the matrix A
A(1,:) = ones(1,K*N);
% The Probabilities (starting from the second row)
for i = 2:2+K
   for j = 1:K*N
      ix = rem(j - 1,N) + 1;
      pn = eval(['pij' num2str(ix)]);
      iy = floor(j - 1)/N) + 1;
      if iy == i - 1
         A(i,j) = pn(iy,i - 1) -1;
      else
         A(i,j) = pn(iy,i - 1);
      end
```

        end
end

% The constraints (starting from the $(K + 2)^{st}$ row)

for i = 1 + K +1:1 + K + M

  vv =eval(['v' num2str(i - K - 1)']);

  for j = 1:K*N

    ix = floor((j - 1)/N) + 1;

    iy = rem(j - 1,N) + 1;

    A(i,j) = vv(ix,iy);

  end

end

% The constraint $C \geq 0$

A = [A; - 1 eye(K*N,K*N);

% Define the lower and upper bounds

vlb = 0;

vub = 1;

% Initialization of C

c0 = 0.1/(K*N)*ones(K*N,1);

% Solve the Linear Programming Problem

cilv = lp(f,A,b,vlb,vub,c0,1 + K);

for i = 1:K*N

  ix = floor((i - 1)/N) + 1;

  iy = rem(i - 1,N) + 1;

  cil(ix,iy) = cilv(i);

end

cil

% Probability condition ($\sum_{i=1}^{K} \sum_{l=1}^{N} c^{il} = 1$)

sumcil = sum(sum(cil));

% Compute V0(c)

V0c =f*cilv;

%Compute V1(c) and V2(c)

for i = 1:M

    eval(['vc' num2str(i) '= sum(sum(v' num2str(i) '.*cil))']);

end

Index

A
absorbing, 7
accessible state, 6
action, 18, 47, 55, 74, 126, 172
adaptation rate, 181, 186
adaptive, 6, 76, 90, 98, 129, 134, 142, 154, 160, 179, 182
adaptive control, 77
algebra, 2–5, 19, 96
aperiodic, 11, 24, 53, 70, 94, 123, 145
average, 69, 100, 117, 141, 163, 168

B
Borel, 2
Borel-Cantelli lemma, 79, 155, 157, 185
Bush-Mosteller, 48, 56, 75, 118, 128, 141, 153

C
coefficient of ergodicity, 12–13
communicating, 7, 10, 18, 21–22, 184
consistency, 102, 141
convergence rate, 63, 83, 107, 136, 162

E
ergodic, 10–11, 22, 24, 49, 52, 57, 70, 77, 129, 167–168, 172, 179, 182–183
expectation, 3–4, 58, 61, 78, 81, 101, 131, 156

F
frozen, 89–90, 94, 168

G
general type, 25, 182
gradient, 87, 89, 156, 168, 173, 183

H
homogeneous, 9, 13, 18, 21
homomorphism, 109

I
inequality problem, 182
inequality type problem, 167
irreducible, 10, 21

L
Lagrange, 48, 53–54, 58, 62, 118, 123–124, 130, 135
law of large numbers, 50, 79, 105
learning, 48, 52, 54, 57, 62, 74, 82, 118, 126, 134, 142, 150, 160

Lebesgue, 4
Lipshitz, 53, 71, 148
loss function, 47–48, 51, 69, 98, 100, 118, 121, 129, 142–143, 174
Lyapunov, 57, 77, 129, 154

M

Markov, 5–6, 48, 57, 69, 88, 111, 142, 168, 172, 181–182
mean squares, 58, 62, 77, 82, 130, 134–135, 155, 160

N

non-return, 6, 10, 22, 183
non-singular, 19, 23, 52, 169, 183
normalization, 54, 73–74, 125, 149

O

optimization, 27–28, 53, 57, 62, 70–71, 76, 82, 87, 118, 122, 124, 135, 143, 147–148, 154, 167, 182, 185

P

penalty function, 69–70, 78, 82, 141, 147, 155
periodicity, 10
probability, 2–3, 8, 77, 82, 97, 102, 134, 142, 155, 169
programming problem, 64, 70, 72, 84, 90, 111, 118, 122, 136, 144, 162
projection, 73, 87–90, 99, 168, 173, 183

R

randomized control, 19
recurrent, 7
regular, 11, 24, 26, 71, 94, 124, 171, 179
Robbins-Siegmund, 62, 81, 93, 97, 134, 155, 160
Rozanov, 13, 94, 171

S

saddle point, 53, 124
simplex, 53, 71, 124, 147
Sragovitch, 21
state, 5, 7, 10, 74, 88, 91, 94, 118, 127
stationary distribution, 12–13
stochastic matrix, 9, 74, 169, 171
strategy, 18

T

Toeplitz, 17, 105, 176–177
transient, 7
transition matrix, 8–9, 13, 20–21, 52, 87, 94, 123, 145, 171
Tsetlin, 28